Applied Geotechnics for Construction Projects 1

To my mother, Fatima Ezohra.
To Leïla-Marie, Michaël-Hassen,
To Hajer, Thérèse, Siwar, Catherine and Yannick.
To my brothers and sisters.
To all my friends
from Dieppe to Velaux,
from Tunis to Gabès.

This book is dedicated to the memory of my father Salem Dhouib, Hassen Ben El Hadj Salem Dhouib (my grandfather, the wise one), Hélène Dyerick-Urbanski (my second mother in France, the generous one), Pierre Bertin, the Oranais, my friend and brother always, and to Christiane Bertin-Guigues, friend and mother for eternity.

Series Editor
Gilles Pijaudier-Cabot

Applied Geotechnics for Construction Projects 1

Soil and Experimental Data

Ammar Dhouib

iSTE

WILEY

English edition first published 2022 in Great Britain and the United States by ISTE Ltd and John Wiley & Sons, Inc.

French edition published by Presses de l'École Nationale des Ponts et Chaussées, Paris, France
1st edition © Presses de l'École Nationale des Ponts et Chaussées 2016
2nd edition (revised and updated) © Presses de l'École Nationale des Ponts et Chaussées 2021

ISTE Ltd
27-37 St George's Road
London SW19 4EU
UK

www.iste.co.uk

John Wiley & Sons, Inc.
111 River Street
Hoboken, NJ 07030
USA

www.wiley.com

Library of Congress Control Number: 2022941482

British Library Cataloguing-in-Publication Data
A CIP record for this book is available from the British Library
ISBN 978-1-78630-775-0

Cover images: **Photo 1** (background image): VINCI Construction Terrassement (VCT) earthworks site on a railway line. Photo by Pascal Le Doaré (supplied by photographer for a fee). **Photo 2** (top center): Pont Jacques Chaban-Delmas vertical-lift bridge in Bordeaux, France. Photo by Richard Nourry (courtesy of photographer). Architect: Cabinet Lavigne (MM. Charles and Thomas Lavigne), Groupement entreprises (VINCI) – Representing Jean-François Roubinet. Project owner: Communauté Urbaine de Bordeaux (CUB). **Photo 3** (middle right): Institut Gustave Baguer, Asnières sur Seine, France. Photo by Augusto da Silva (courtesy of photographer). Entreprise: GTM BATIMENT VINCI – Representing Jean-Yves Cojean, President. **Photo 4** (bottom right): Pavilion of Michel and Camille Richaud, Velaux, France. Photo by the author. DR M. & Mme Richaud. **Photo 5** (bottom center): Shielded trench with horizontal struts. Photo by the author. VINCI Construction France site (DR VCF). **Photo 6** (bottom left): CMA Tower, Marseille, France. Photo by Govin Sorel (courtesy of photographer). Architect: Zaha Hadid.

Contents

Foreword

This book is the work of a practicing engineer who has spent more than 30 years in the field of geotechnics applied to construction works. It is therefore already the result of a whole range of experiences, made up of numerous studies and works, as well as research, university chairs and publications in this discipline.

Ammar Dhouib, who received his doctorate in soil mechanics from the University of Lille in 1987, has since acquired extensive experience from design offices and companies specializing in geotechnics (Terrasol, Fondasol, Louis-Ménard, Solen and Keller Fondations Spéciales). He is currently chief engineer and geotechnical expert at VINCI Construction France, assisting and consulting for subsidiaries of the VINCI Construction group.

In addition to all of the studies he has carried out for foundations or other geotechnical works, he is also known for having jointly published a book in 2004 with Jean-Pierre Magnan and Philippe Mestat, on the improvement of soils in situ and another book in 2005 with Francis Blondeau, on stone columns.

The present work by Ammar Dhouib has the advantage of being both comprehensive and detailed while remaining quite traditional in its presentation. It includes 12 chapters ranging from the definition of soils, their surveys and the determination of geotechnical parameters to the design of structures (foundations and retaining walls in particular, tunnels do not come under such scrutiny). It is worth noting that there is an important chapter devoted to the effects of water in soils, which we know can lead to damage and failure when not all aspects of a geotechnical structure have been properly studied. As for the chapter on soil improvement, it was only natural that it should be included, given both the author's knowledge in this field and the importance that this aspect now has in the foundations of construction works.

This book also presents very instructive elements for all those interested in geotechnics. These are the practical applications and examples given at the end of each chapter to illustrate and explain the principles, as well as the difficulty of geotechnical sizing. The author presents real cases that have been measured, which allow the sizing to be compared with reality.

This method is not new, it is the so-called "observational method", generally reserved for complex or large structures, for which it is necessary to avoid any disruption. The author also shows cases of excessive deformations, or even the beginnings of failure in structures and gives the causes.

Ammar Dhouib's book therefore appears to us to be of interest to a wide range of geotechnical scientists, from students to qualified engineers, including building owners, project managers and court experts.

Philippe GUILLERMAIN†[1]

François SCHLOSSER

1 Philippe suddenly passed away on May 7, 2016, without being able to attend the presentation ceremony of the first edition of the book on September 16, 2016, at the FNTP.

Entrepreneur’s Tribune

Geotechnics is at the Heart of Our Projects

“Geotechnics is at the heart of our projects” – is this just an idea or an irreversible “truth” of the moment? It is indeed the current “reality” because, nowadays, “good” soils are becoming rare, projects are more complex and regulations are more developed.

Good sites are rare. Between industrial wastelands and marshy grounds, the soil is often aquiferous, soft or compressible, and its significant deformations under the weight of the works often become, in this context, incompatible with the allowable displacements of the structure.

The projects are complex, ranging from towers of several dozen or even several hundred meters in height, sometimes located in the middle of sensitive existing structures, to deep excavations (70 m for the open excavation of the ODEON tower in Monaco), bridges in aquatic sites (Normandy Bridge in France, Charílaos Trikoúpis Bridge in Rion Antirion Greece) and tunnels in soft soils and sensitive urban environments (the tunnel under the city of Toulon is a perfect example).

Regulations have undergone considerable development and are becoming more and more demanding because of the new French and European standards, such as the standardized geotechnical missions and the “Eurocodes”.

Limiting ourselves to sizing the foundations within this framework, both with rigor and common sense, is no longer sufficient in the face of today’s requirements of quality, conformity and respect for deadlines and budget, on the one hand, and safety and environmental imperatives on the other hand.

The geotechnical engineer must think of alternative foundation solutions, size them and optimize them with four points in mind: to ensure the quality, solidity, stability and durability of the structure. The geotechnical engineer will seek substantial savings on the foundations for the company and the client, and will be concerned with societal and sustainable development criteria.

This book is dedicated to students, engineers in design offices and companies, legal experts in geotechnics, insurers and financiers. It exposes the fundamental bases of soil mechanics, and illustrates and develops practical examples in the fields of geotechnics, in the broadest sense. In these fields, the geotechnical engineer can propose, study and justify foundation variants by replacing, for example, piles by soil improvement, in order to make the best use of the properties of the ground, and therefore of natural resources. The man of art will propose alternative design solutions to retain structures and ensure the stability of the excavations and existing constructions. These alternative solutions are calculated, optimized and justified by various approaches of analytical calculations and numerical modeling in deformation, which are undergoing major developments. The geotechnical engineer must also understand better risk management, whether related to soil defects or to the new requirements of modern projects.

Pascal LEMOINE
Eric DURAND

Preface

In the foreword to this book, Pascal Lemoine of the *Fédération Nationale des Travaux Publics* (French National Federation of Public Works, FNTP) and Eric Durand of the *Fédération Française du Bâtiment* (French Building Federation, FFB) wrote: "Geotechnics is at the heart of our projects"!

The term "Geotechnics" is new, but geotechnics itself is not, as our ancestors have always known how to "integrate" soil and foundations together in order to build bridges to cross rivers, roads to bring cities closer together, dams to irrigate plains and buildings to house the peoples of Africa, the Americas, Asia, Europe and Oceania.

Without looking in the "*Grand Larousse*" or the "*Petit Robert*"[1], geotechnics can have many definitions and is based on many rules. We happily acknowledge some of them:

– Geotechnics is a "marriage" between soil and foundation; this marriage can only be successful when the soil is "healthy" and in good condition, and the foundation is not overly constraining, so as to ensure the quality, solidity, stability and durability of the structure.

– Like medicine, geotechnics requires targeted and adapted investigations, where the soil must be examined well and carefully identified geologically, and mechanically characterized well. The geotechnical engineer can then establish an accurate diagnosis and suggest basic rules to design and size their project, like a doctor who issues a prescription in order to cure their patient.

1 These are the two most well-respected French language dictionaries, the equivalent of the Merriam-Webster Dictionary and the Oxford English Dictionary.

– Geotechnics is directly related to the place where the geotechnical engineer is located, like the farmer who loves their land, meditates on it, ploughs it and sows it. The engineer is attached to their projects to establish them in the ground, while seeking to optimize the foundations of their project, keeping in mind the safety of the men and women during construction (the work phase), the stability and durability of the work over time and the safety of people who will live there from far or near (the service phase).

– Geotechnics is not an exact science; it resembles the art of "cooking", a very refined and elaborate cooking, rich enough to be the most beautiful of practical sciences that allow engineers to excel over time, like how soil strengthens by "aging" and how cooks become real "chefs" after trying many recipes, concocting new dishes and serving many meals.

That being said, there is no geotechnics without soil mechanics, continuous media mechanics and resistance of materials. Like human beings, soil is very complicated and complex, in other words, "vicious"[2], because:

– soil has many parameters to be identified, recognized and classified: its moisture content, plasticity, grain size and surrounding voids, compressibility, creep, cohesion and friction;

– soil has several behaviors: inelastic, viscoelastic, plastic, dilatant and contracting, swelling and shrinking and subject to creep over time under constant stress;

– it is sensitive to the water that circulates within it and to external agents (chemical aggression, frost and water circulation) that can destabilize its structure (gullying and collapse) and cause it to evolve over time (swelling, shrinking and creep);

– the soil is subject to all kinds of anomalies. Natural anomalies include pockets of dissolution (e.g. gypsum, which dissolves like sugar[3]), karsts, "collapsible"[4] soils, etc. Examples of artificial anomalies are underground quarries, marl pits in Normandy, catiches in the North of France and so on.

---

2 The term "vicious soil" or "soil defect" is particularly used by judges and lawyers to designate an anomaly, a hazard, a "new fact" detected in the soil by the project actors. The term "defect" is often an argument put forward by lawyers to defend their clients in legal expertise in order to specify that, for example, the anomaly encountered in the soil is a hazard, and therefore a "defect" inherent in the soil.

3 Gypsum glows in the sun (Figure 1.12 in Volume 1, Chapter 1) and dissolves in water within minutes like sugar, hence the term "saccaroid gypsum".

4 "Collapsible" soils are a recent discovery in the big disaster (without end) of the Grand Littoral in Marseille where, following the collapse of the foundation piles, it was noted that the soils, applied under the same loading constraint, undergo deformations by the imbibition effect.

Through studies and the elaboration of projects, university chairs, conferences and national and international congresses, we noticed that practicing engineers, students, pupil-engineers, beginners and even young experienced engineers have difficulty in understanding the meaning of geotechnical parameters. They do not master the methods which are used to determine these parameters, nor the way to use them in simple methods of calculation and dimensioning. They have not mastered how to introduce them in more complex models using, for example, modeling and calculations in finite elements (“CESAR-LCPC”, “FLAC-3D” and “PLAXIS”), in finite differences, by the boundary method, or even in software of traditional rupture calculation like “TALREN”, or in retaining structures software based on the principle of the coefficient of reaction like “RIDO”, “PAROI”, “DENEBOLA-LCPC” or “k-Réa”. Some young people, with their infatuation with computers, have good mastery of computer software and great speed in building complex geometric models, and use results that are sometimes erroneous because of a bad choice of parameters or the models are inadequate to accurately describe and follow the behavior of the soil, which can indeed be very difficult, very complex and very “vicious”.

Works dealing with geotechnics are relatively numerous in France. The four volumes of this book constitute, first and foremost, a practical and useful guide for beginners or experienced engineers, for students and student-engineers, for project managers and for insurance and justice experts specialized in Geotechnics: Section C125-SOL (soil).

Resulting from conferences in national and international congresses and symposiums, at multiple university chairs of its author, at the *Ecoles polytechniques de Lille, d’Orléans et de Paris Sorbonne* (Polytechnic Schools of Lille, Orleans and Paris Sorbonne), at the *Ecole centrale de Lille* (Central School of Lille), at the *Ecole Nationale des Ingénieurs de Tunis* (Higher National Engineering School of Tunis, ENIT) and the universities of Tunis, at the Hassania School in Casablanca, Morocco, and at the *Ecole Nationale des Ponts et Chaussées* in Paris, this book in four volumes is largely enriched by several practical applications, generally resulting from concrete projects studied by the author and his collaborators and/or his students in end-of-study works, from TERRASOL, via FONDASOL and SOLEN, to VINCI. Applications and projects are marked by the the author’s own practical experience, since 2007, of judicial expertise before judicial and administrative courts.

Developed between 2008 and 2021, and then taken up again before being further enriched by a specific chapter related to Geotechnical Expertise for its re-edition, this book in four volumes is articulated in four main themes:

– Definition of the soil, choice of the geotechnical parameters and methods of their determination, mainly from tests and investigations on site, and incidentally from laboratory tests.

– Fundamental relations and laws of soil behavior, ranging from elastoplasticity (simplistic hypothesis because the soil is not elastic) to soil creep, making it possible to understand the formulations introduced in the calculation and numerical modeling software that continues to develop in the field of "modern" geotechnics.

– Applications to foundations, retaining structures, backfill and embankment, soil improvement and underground structure projects, with a reminder of the simple physical rules and an introduction to the various standards, references and rules in place: from the DTU, via the leaflets, to the Eurocodes; references are compared, analyzed and commented on, particularly in Volume 3, Chapters 1 to 3, which is dedicated to the foundations of construction projects.

– Feedback: by means of "what not to do" where, at the end of the chapter, as well as in the new Chapter 3 in Volume 2, entitled "Geotechnical expertise", some cases of litigation and claims due to several factors are presented: bad studies, unfinished or even erroneous design, lack of follow-up and control of the execution, bad execution, not to mention in some, fortunately relatively rare, cases encountered in particular in judicial expertise, "fraudulent execution (poor workmanship and/or non-execution)".

First of all, we wanted to present all the geotechnical parameters involved in the calculation and dimensioning of structures, as well as simplify the basic principles and fundamental relations of soil mechanics, then to present and comment on an overview of practical project examples that is as complete as possible, in order to best cover the field of geotechnics applied to construction projects. This is a vast field that is both simple and complex, and is not simplified by the recent European codes, called "Eurocodes", and the new national application standards (NAN) of Eurocode 7, which are deliberately only referred to in the applications in order to make using this book as easy as possible. We consider that a student, a beginner or even an experienced engineer, as well as a project manager or a building expert focused on geotechnics, will easily understand the meaning of a safety coefficient of 3 on the point and 2 on the skin friction when sizing a pile, before encountering Eurocode errors and the NAN of Eurocode 7 where "partial" weighting factors and "partial" safety coefficients are linked together, often without any basis. This complicates the approaches, methods and traditional references related to geotechnics.

After the edition and publication of the book on September 16, 2016 by the *Presses de l'Ecole Nationale des Ponts et Chaussées*, under the aegis of the FNTP and the IREX (Institute for Applied Research and Experimentation in Civil

Engineering), the FFB, the *Compagnie des Experts près de la Cour d'Appel de Versailles* (Company of Experts at the Court of Appeal of Versailles, CECAV) and the *Conseil Pont Formation*, we have received feedback from various readers and users of the book, including observations and opinions that are, on the whole, very positive. We therefore considered it beneficial to re-edit the book, correcting a few "typos" and taking into account some very useful comments. Naturally, the new edition retains the entire framework of the book, but it includes several illustrations and enriches the appendices in almost every chapter.

Thus, at the request of some geotechnical colleagues from design offices and judicial and insurance experts, whose kind messages have delighted us, we have written and completed the book with a new chapter (Volume 2, Chapter 3) that is simple, rich and well-illustrated, entitled "Geotechnical Expertise". The purpose of this detailed chapter is to:

– Present, analyze and comment on examples of concrete projects with or without disorder in the context of geotechnical construction projects, and how to search for simple foundation solutions with an objective of "economic realism".

– Define the bases and focus of the geotechnical expertise ordered by a competent court (known as "judicial expertise") and the role of the geotechnical expert appointed by this same court, formerly called the "judicial expert" and more recently the "justice expert".

– Conclude with some general geotechnical information, lessons and rules to remember.

Volume 2, Chapter 3 is intended to enrich and give the book a truly practical aspect. It is a repository for feedback from daily geotechnical work and expertise that deserves to be exposed to the young engineers and geotechnical experts of the current and future generations.

Contrary to the first version, each volume of the new edition concludes with an index in order to make the use of this book, and the search for technical information in the 12 chapters and 1,100 pages, easier and more convenient.

Acknowledgments

Our first thanks go to VINCI, in particular VINCI Construction France, its management and staff throughout France and around the world, who have enabled us to work, from the call for tender to the "turnkey" handover, on diverse and varied, and sometimes exceptional, projects with men and women who count.

The Entrepreneur's Tribune, which introduces the book, testifies to the interest shown by the FNTP, represented by Pascal Lemoine, and the FFB, represented by Eric Durand, in the completion and publication of this work. To these two federations, pioneers of the construction industry in France, and to their representatives Pascal Lemoine and Eric Durand, we extend our sincere thanks.

The *Compagnie des Experts près la Cour d'Appel de Versailles* (Company of Experts at the Court of Appeal in Versailles, CECAV) has warmly accompanied us in the presentation of the book. We extend our deepest thanks to its president, Anne-Marie Pruvost-Paris, and to all of its members.

We would also like to thank Professor François Schlosser, our ENPC "Master" at TERRASOL since 1984, and Philippe Guillermain, our sponsor at the courts since 2007, for writing the preface for this book and enriching the text with their valuable advice.

Mr. Louis Demilecamps, former director of the *Direction des Ressources Techniques et Développement Durable* (Technical Resources and Sustainable Development Department, DRD) at VINCI Construction France, has encouraged us since 2008 and has energetically contributed to the publication of this book. We would like to send our warmest thanks.

A special thought goes out to Michel Khouri, who advised us for a long time to deepen the relations between soil and structure through the reinforcements to be introduced in the concrete of piles, footings, rafts and retaining walls.

Many people have contributed in one way or another through their end-of-studies work, internships and direct collaboration. We give our sincere thanks to all of them here: Cécilia Guibert, Sophie Jacquemin, Thomas Defoy, Olivier Payan, Grégoire Priol, Benjamin Leroi, Lina Bawji, Marilyse Dupraz, Sophie Lelièvre, Ludovic Boucaux, Anne Cotte, Isabelle Decker, Joan Mimica, Karim El Jouhari, Jihane Laboudi, Hélène Roulet, Fabienne Magnon, Laurence Oettli, Laurent Soyer, Bilge-Beryl Aksoy, Annouar Siala, Paul Lacrampe and Jese Andriamboavonjy.

ISTE's translation, production and editing teams have shown great professionalism in the work accomplished in close collaboration and synergy with the author for the preparation of the four volumes of this book. The author would like to thank them very warmly and sincerely.

There are many warm and kind testimonies, spoken and/or written, received after the publication of the 2016 book. We would like to thank (in alphabetical order): Mr. Alba Jean-Michel (SOL-ESSAIS), Mr. Bataille Arnaud (ESIRIS), Mr. Brulé Stéphane (Ménard), Mr. Caporali Pascal (SOL CONSEIL), Mr. Delhomel Alain (SNCF), Mr. Fourmentraux Hugues (President of VINCI Construction France), Mr. Gallet de Saint-Aurain Jean-Marc (SEMOFI), Mr. Gambin Michel, Mr. Guichet Richard (BEFES Fondations & Structures), Mr. Huillard Xavier (President of VINCI), Mr. Llobet Lionel (COFEX-VINCI), Mr. Schmitt Pierre (SOLETANCHE-BACHY), Mr. Jérôme Stubler (President of VINCI Construction), Mr. Tadbir Eric (GINGER CEBTP), Mr. Henry Thonier, Mr. François Vahl and Mr. Jérôme Varillon (VINCI Construction Terrassement).

Symbols and Notations

Basic sizes and notations

Type	Designation	Symbol	Dimension	Unit
Geometry	Width	B, a, b	L	m
	Length	L, l	L	m
	Thickness	B, e	L	m
	Height	H, h	L	m
	Depth	D, d, Z	L	m
	Diameter	B, D, d, Φ	L	m^2
	Area, Section	A, S	L^2	m^2
	Volume	V	L^3	m^3
Mass	-	m, m', M	M	kg
Density	-	ρ	ML^{-3}	Mg/m^3
Bulk density	-	γ	ML^{-3}	kN/m^3
Force	Vertical, axial force	N, Q, Q_v, V	M (MLT^{-2})	kN
	Horizontal, transversal force	H, Q_h, T	M (MLT^{-2})	kN
	Normal force, traction	N, T	M (MLT^{-2})	kN
	Parasite horizontal pressure	P	M (MLT^{-2})	kN
	Weight	G, W	M (MLT^{-2})	kN
Moment	Bending, overturning, tilting moment	M	ML (ML^2T^{-2})	kN.m
Stress	Normal stress, total stress	σ	ML^{-2} ($ML^{-1}T^{-2}$)	kPa
	Effective stress	σ'	ML^{-2} ($ML^{-1}T^{-2}$)	kPa
	Shear stress	τ (= τ')	ML^{-2} ($ML^{-1}T^{-2}$)	kPa

Displacement	Vertical, horizontal, lateral	u, v, w, s	L	m
Deformation	Vertical, horizontal	ε	-	None
Time	-	t	T	s
Velocity	-	v	LT^{-1}	m/s
Acceleration	Horizontal, vertical, nominal Of gravity	a g	LT^{-2} LT^{-2}	m/s^2 m/s^2
Action	Durable Variable Due to water Accidental	G Q F_w F_A	$M\ (MLT^{-2})$ $M\ (MLT^{-2})$ $M\ (MLT^{-2})$ $M\ (MLT^{-2})$	kN kN kN kN
Factor	Of partial safety Of safety on materials Of safety on actions Of investable weight	F, Γ γ_m γ_a $\gamma_G, \gamma_Q, \psi_i$	- - - -	- - - -

Capital letters

Symbol	Designation	Unit	Dimension	Chapters
A	Area (Section)	m^2	L^2	**Vol. 4:** 1, 2
A	Section of an inclusion	m^2	L^2	**Vol. 4:** 2
A, B, C	Soil category according to fascicle 62, title V	-	-	**Vol. 3**
A, B, C, D, R, F	Soil classes according to the GTR guide in France	-	-	**Vol. 1:** 1
A, B, C, D, E, S_1, S_2	Soil class according to Eurocode 8	-	-	**Vol. 3** **Vol. 4:** 2
A_a	Active earth pressure cohesion term	kPa	ML^{-2}	**Vol. 4:** 1
A_a	Reinforcement section of foundation reinforcement	m^2	L^2	**Vol. 3** **Vol. 4:** 2
A_c	Conical tip section (Cone static penetration test: CPT)	m^2	L^2	**Vol. 1:** 3
A_c	Solid section of a column	m^2	L^2	**Vol. 4:** 2
A_l, A_L	Lateral surface (of a shaft foundation)	m^2	L^2	**Vol. 3**
A_{IN}	Abrasiveness of the rock	-	-	**Vol. 1:** 1
A_p	Solid section at the peak of a foundation element	m^2	L^2	**Vol. 3**

A_p	Passive earth pressure cohesion term	kPa	ML^{-2}	**Vol. 4:** 1
A_r	Reduced section of an inclusion (due to corrosion)	m^2	L^2	**Vol. 4:** 2
A_s	Section of the soil column around an inclusion	m^2	L^2	**Vol. 4:** 2
A_u	Useful section of a reinforcement element	m^2	L^2	**Vol. 4:** 1
B	Width of footing, slurry trench pile or pile diameter	m	L	**Vol. 3**
B_0	Reference width of a foundation ($B_0 = 0.6$)	m	L	**Vol. 3**
B_q	Pore pressure coefficient	-	-	**Vol. 1:** 3
B'	Compressed width of a foundation under eccentric load	m	L	**Vol. 3**
C	Quantity of the corresponding states: c.Cotanφ	-	-	**Vol. 2:** 2 **Vol. 4:** 1
C	Covering of a tunnel	m	L	**Vol. 4:** 3
C_c	Compression index: $-\Delta e/\Delta \lg\sigma'$	-	-	**Vol. 1:** 2 **Vol. 2:** 2 **Vol. 3**
C_e	Group effect minority coefficient: $C_e = Q_l^N / Q_l^1$	-	-	**Vol. 3**
C_f	Coefficient of form and stiffness of raft foundations	-	-	**Vol. 3**
CM	Coefficient of earthquake magnitude	-	-	**Vol. 3** **Vol. 4:** 2
C_p	Back-passive pressure force	kN	M	**Vol. 4:** 1
C_s	Swelling index: $-\Delta e/\Delta \lg\sigma'$	-	-	**Vol. 1:** 2 **Vol. 2:** 2 **Vol. 3**
CPT	Static penetrometer: "Cone Penetration Test"	-	-	**Vol. 1:** 3
CPU	Static penetrometer with pore pressure measurement ("Piezocone")	-	-	**Vol. 1:** 3
C_U	Hazen uniformity coefficient: D_{60}/D_{10}	-	-	**Vol. 1:** 1
C_z	Curvature coefficient: $D^2_{30}/(D_{60}.D_{10})$	-	-	**Vol. 1:** 1
CRR (τ_l or τ_{max})	Cyclic Resistance Ratio	kPa	ML^{-2}	**Vol. 3** **Vol. 4:** 2
$CRR_{7,5}$	Cyclic Resistance Ratio for a magnitude M of 7.5	kPa	ML^{-2}	**Vol. 3** **Vol. 4:** 2

CSR (τ_m)	Cyclic Stress Ratio	kPa	ML^{-2}	**Vol. 3** **Vol. 4:** 2
C_α	Secondary consolidation index (compression): $d\varepsilon/\Delta lgt$	-	-	**Vol. 1:** 2 **Vol. 2:** 2
$C_{\alpha e}$	Creep index: $(1 + e_0).C\alpha$	-	-	**Vol. 2:** 2
CEC	Critical State Curve (Cam-Clay Model)	-	-	**Vol. 1:** 2
CEL	Limit State Curve (Cam-Clay Model)	-	-	**Vol. 1:** 2
CO_3Ca	Carbonate content	%	-	**Vol. 1:** 1
D	Diameter (of grains, piles, water test cavity)	m	L	**Vol. 1:** 1, 3 **Vol. 3**
D	Embedding of a foundation in the soil	m	L	**Vol. 3**
D, D_0	Length of a deep foundation	m	L	**Vol. 3** **Vol. 4:** 2
D	Diameter of an inclusion	m	L	**Vol. 4:** 2
D	Zone of influence of a drain	m	L	**Vol. 2:** 2
D_c	Critical depth of a foundation	m	L	**Vol. 1:** 1
D_c	Average diameter of a column	m	L	**Vol. 4:** 2
D_e	Equivalent (or effective in Eurocodes) embedment of a foundation	m	L	**Vol. 1:** 1
D_f	Diameter of a boring	m	L	**Vol. 4:** 1, 2
D_r	Relative density (Density index I_d)	%	-	**Vol. 1:** 1
D_s	Sealing diameter ($D_s=\alpha.D_f$)	m	L	**Vol. 4:** 1, 2
DU	Hardness of the rock	-	-	**Vol. 1:** 1
D_{10}, D_{15}, D_{50}	Grain diameter at 10%, 15%, 50% sieve pass	m	L	**Vol. 1:** 1
$[D]$	Stiffness matrix $\{F\}=[D].\{U\}$	-	-	**Vol. 1:** 2
E	Modulus of deformation (Young's)	kPa	ML^{-2}	**Vol. 1:** 1, 2, 3 **Vol. 3**
E_c	Spherical pressuremeter modulus (at the base of the foundation)	kPa	ML^{-2}	**Vol. 3**
E_c	Modulus of deformation of a column	kPa	ML^{-2}	**Vol. 4:** 2
E_d	Deviatoric pressuremeter modulus	kPa	ML^{-2}	**Vol. 3**
E_{dyn} (E_{vd})	Dynamic modulus (measured at the dynaplaque)	kPa	ML^{-2}	**Vol. 1:** 2, 3

E_e	Equivalent deformation modulus	kPa	ML^{-2}	**Vol. 4:** 2
E_{ij}	Instantaneous modulus of concrete (short term)	kPa	ML^{-2}	**Vol. 1:** 2 **Vol. 3**
E_M	Menard pressuremeter deformation modulus	kPa	ML^{-2}	**Vol. 1:** 2 **Vol. 3** **Vol. 4:** 1, 2
E_{oed}	Oedometer modulus: $E_{oed}=\frac{(1-\nu)}{(1+\nu)(1-2\nu)}.E$	kPa	ML^{-2}	**Vol. 1:** 2
E_{oed}	Oedometer modulus: $E_{oed}=\alpha_1.q_c$	kPa	ML^{-2}	**Vol. 3**
E_s	Conventional soil deformation modulus	kPa	ML^{-2}	**Vol. 1:** 2 **Vol. 3** **Vol. 4:** 1, 2
E_s	Secant modulus of the soil: $k_s.B$	kPa	ML^{-2}	**Vol. 3** **Vol. 4:** 2
E_{vj}	Deferred modulus of concrete (long term)	kPa	ML^{-2}	**Vol. 1:** 2 **Vol. 3**
E_u	Modulus of deformation of undrained soil	kPa	ML^{-2}	**Vol. 1:** 2
E_{V1}	Modulus of first loading with plate bearing	kPa	ML^{-2}	**Vol. 1:** 3
E_{V2}	Modulus of second loading with plate bearing	kPa	ML^{-2}	**Vol. 1:** 3
E_1, E_2, $E_{3,5}$, …	Pressuremeter modulus per ground trenches or the practical calculation of E_c and E_d	kPa	ML^{-2}	**Vol. 3**
E'	Drained soil deformation modulus	kPa	ML^{-2}	**Vol. 1:** 2
E(x), F(x), G(x), H(x)	Characteristic function of load transfer on piles under lateral pressure	kN and m	M and L	**Vol. 3** **Vol. 4:** 2
EA, EA_p, EA_u	Axial stiffness of a foundation element	kN	M	**Vol. 3** **Vol. 4:** 1, 2
EI, EI_e, EI_p	Bending stiffness of a foundation element	$kN.m^2$	ML^2	**Vol. 3**
EI_e	Bending stiffness of a non-continuous curtain wall	$kN.m^2$	ML^2	**Vol. 4:** 1
SLS	Serviceability Limit States	-	-	**Vol. 3** **Vol. 4:** 1, 2
ULS	Ultimate Limit States	-	-	**Vol. 3** **Vol. 4:** 1, 2

ES	Sand Equivalent	%	-	**Vol. 1:** 1
F (Γ)	Safety coefficient	-	-	**Vol. 2:** 2 **Vol. 4:** 2
F_A	Accidental action	kN	M	**Vol. 3**
F_c	Euler critical load	kN	M	**Vol. 4:** 1
F_{min} (Γ_{min})	Minimum safety coefficient	-	-	**Vol. 2:** 2 **Vol. 4:** 1, 2
F_n	Negative frictional force	kN	M	**Vol. 3**
F_s	Safety factor for liquefaction	-	-	**Vol. 3** **Vol. 4:** 2
F_W	Water action	kN	M	**Vol. 3**
$F_n(b)$	Negative skin friction force on an isolated foundation element	kN	M	**Vol. 3**
$F_n(\infty)$	Negative skin friction force on a foundation group	kN	M	**Vol. 3**
F(σ)	Load function: $F(\sigma) < 0$ (elastic medium), $F(\sigma) = 0$ (boundary), $F(\sigma) > 0$ (failure)	-	-	**Vol. 1:** 2
F(p,q)	Load function depending on p and q	-	-	**Vol. 1:** 2
$\vec{F}$ ($\{F\}$)	Force vector (of components f_x, f_y and f_z)	kN	M	**Vol. 1:** 2
F_R (or R_f)	Friction ratio: f_s/q_c (CPT)	%	-	**Vol. 1:** 3
FS	Sand friability	-	-	**Vol. 1:** 1
G	Durable action	kN	M	**Vol. 3**
G	Weight	kN	M	**Vol. 3**
G	Shear modulus (G_{dyn}: dynamic)	kPa	ML^{-2}	**Vol. 1:** 2
G_0	Shear modulus at the shearing curve origin	kPa	ML^{-2}	**Vol. 1:** 2
G_{max}	Maximum shear modulus	kPa	ML^{-2}	**Vol. 1:** 2
G(z)	Law of mobilization of the lateral displacement of a compressible layer: G(z) method	-	-	**Vol. 3**
G_s	Secant shear modulus	kPa	ML^{-2}	**Vol. 1:** 2
G_1, G_2, G_3, G_4, G_5	Standard geotechnical missions (NF P94-500 standard)	-	-	**Vol. 1:** 3
H	Horizontal force	kN	M	**Vol. 1:** 2
H	Thickness of the aquifer	m	L	**Vol. 2:** 1

H	Half thickness of a drainage layer	m	L	**Vol. 2:** 1
H_0	Horizontal force on a foundation element	kN	M	**Vol. 3**
H_i	Force in a reinforcement element (shore)	kN	M	**Vol. 4:** 1
I	Inertia of a foundation/reinforcement element	m^4	L^4	**Vol. 4:** 1
I_P	Plasticity Index: $I_P = w_L - w_P$)	-	-	**Vol. 1:** 1 **Vol. 3**
I_c	Consistency index: $I_c = (w_L - w)/I_P$	-	-	**Vol. 1:** 1 **Vol. 3**
IC	Rock Continuity Index	-	-	**Vol. 1:** 1
IPI	Immediate Proctor Index	-	-	**Vol. 1:** 1
K	Compressibility modulus (K_{dyn}: dynamic)	kPa	ML^{-2}	**Vol. 1:** 2
K	Proportionality coefficient: $\sigma_v = K.\sigma_h$			**Vol. 3** **Vol. 4:** 1
$K.\text{Tan}\delta$	Negative skin friction calculation factor	-	-	**Vol. 3**
K_a	Active earth pressure coefficient			**Vol. 4:** 1
K_I	Initial state pressure coefficient (= or $\neq K_0$)	-	-	**Vol. 4:** 1
K_h	Horizontal stiffness of a footing	kN/m	ML^{-1}	**Vol. 3**
K_p	Passive earth pressure coefficient	-	-	**Vol. 3** **Vol. 4:** 1, 2
K_{pc}	Passive earth pressure coefficient of the material of a column		-	**Vol. 4:** 2
K_p	Slope of the mobilization line at the point (pile)	kPa/m	ML^{-3}	**Vol. 3**
K_τ	Slope of the line of mobilization of the friction	kPa/m	ML^{-3}	**Vol. 3**
K_W	Coefficient of Westergaard reaction	-	-	**Vol. 1:** 3
K_v	Vertical stiffness of a footing (of a raft)	kN/m	ML^{-1}	**Vol. 3**
K_{HH} (K_f)	Horizontal stiffness of a pile	kN/m	ML^{-1}	**Vol. 3**
K_{HM} = (K_{MH})	Coupling stiffness of a pile	kN.m/m	ML^{-1}	**Vol. 3**
K_{MM}	Bending stiffness of a pile	kN.m/rad	ML^{-1}	**Vol. 3**
K_{VV}	Vertical stiffness of a pile	kN/m	ML^{-1}	**Vol. 3**
K_θ	Rotational stiffness of a footing	kN.m/rad	M.L	**Vol. 3**

K_ω	Torsional stiffness of a footing	kN.m/rad	M.L	**Vol. 3**
K_0	Coefficient of earth pressure at rest	-	-	**Vol. 1:** 2 **Vol. 3** **Vol. 4:** 1
K_{0c}	Coefficient of resting earth of a column material	-	-	**Vol. 4:** 2
L	Width of a foundation	m	L	**Vol. 3**
L_L	Free length of a tie-rod	m	L	**Vol. 4:** 1
L_S	Length of tie-rod sealing (of a micropile)	m	L	**Vol. 4:** 1
L_t	Total length of a tie-rod	m	L	**Vol. 4:** 1
M, M'	Mass	kN	M	**Vol. 1:** 2, 3 **Vol. 3**
M, M_0	Couple, Moment at the head of a foundation	kN.m	ML	**Vol. 1:** 3 **Vol. 3**
M	Earthquake magnitude (Richter scale)	-	-	**Vol. 3** **Vol. 4:** 2
M(x)	Moment function of x in an inclusion	kN.m	ML	**Vol. 4:** 2
M_m	Actional moment	kN.m	ML	**Vol. 2:** 2
M_{max}	Maximum moment in a foundation element	kN.m	ML	**Vol. 4:** 1, 2
M_p	Maximum moment of a foundation element	kN.m	ML	**Vol. 3**
M_p	Plasticity moment of a foundation element	kN.m	ML	**Vol. 3**
M_r	Resisting moment	kN.m	ML	**Vol. 2:** 2
M_r	Ultimate moment in an inclusion	kN.m	ML	**Vol. 4:** 2
M_0	Moment at the head of an inclusion	kN.m	ML	**Vol. 4:** 2
M(z)	Moment as a function of depth z	kN.m	ML	**Vol. 3** **Vol. 4:** 2
MO	Organic material content	%	-	**Vol. 1:** 1 **Vol. 3**
MPT	Menard Pressuremeter Test			**Vol. 1:** 3
N	Normal component of the resultant R	kN	M	**Vol. 2:** 2
N	Number of foundation elements in group	-	-	**Vol. 3**
N, N(σ)	Normal force (axial)	kN	M	**Vol. 1:** 2 **Vol. 2:** 2
N, N(x)	Traction in a "reinforced earth" strip	kN	M	**Vol. 4:** 2
N_c	Value of the characteristic design force	kN	M	**Vol. 4:** 2
N_c, $N_c(\varphi)$	Bearing capacity factor (cohesion)	-	-	**Vol. 1:** 3 **Vol. 3**

N_d	Number of blows measured with a dynamic penetrometer	-	-	**Vol. 1:** 3
N_{fl}	Resistance to soil/strip friction TA	kN	M	**Vol. 4:** 2
N_m	Number of SPT blows: $N_1 + N_2$	-	-	**Vol. 1:** 3
N_{max}	Maximum traction in a "reinforced earth" strip	kN	M	**Vol. 4:** 2
N_r	Mechanical resistance of an inclusion	kN	M	**Vol. 4:** 2
N_{rr}	Mechanical resistance of a reduced section inclusion (corrosion)	kN	M	**Vol. 4:** 2
N_s	Tensile strength of a "reinforced earth" strip	kN	M	**Vol. 4:** 2
N_{sl}	Frictional resistance of soil/inclusion	kN	M	**Vol. 4:** 2
N_{SPT}	Number of SPT blows	-	-	**Vol. 1:** 3 **Vol. 3**
$N_{1(60)}$	Number of SPT blows corrected for the effects of soil loads and energy	-	-	**Vol. 1:** 3 **Vol. 3** **Vol. 4:** 2
N_q, $N_q(\varphi)$	Bearing capacity factor (overload)	-	-	**Vol. 1:** 3 **Vol. 3**
N_γ, $N_\gamma(\varphi)$	Bearing capacity factor (heavy soil)	-	-	**Vol. 1:** 3 **Vol. 3**
N^*_c	Corrected bearing capacity factor (cohesion)	-	-	**Vol. 1:** 3 **Vol. 3**
N^*_γ	Corrected bearing capacity factor (heavy soil)	-	-	**Vol. 1:** 3 **Vol. 3**
N_0	Traction at the origin of a "reinforced earth" strip	kN	M	**Vol. 4:** 2
N_0, N_1, N_2	Number of SPT blows per 0.15 m	-	-	**Vol. 1:** 3 **Vol. 3**
N_θ	Additional normal force due to expansion in a metal shore	kN	M	**Vol. 4:** 1
OCR	Over Consolidation Ratio	-	-	**Vol. 2:** 2
P	Perimeter of a foundation (piles)	m	L	**Vol. 3**
P, P(z), $P_h(z)$	Resultant of lateral pressure on a foundation element	kN	M	**Vol. 3**
P_a	Active earth pressure resultant on a retaining wall	kN	M	**Vol. 4:** 1
P_f	Horizontal creep reaction on a foundation element	kN	M	**Vol. 3**
P_{fron}	Frontal reaction on a foundation element	kN	M	**Vol. 3**
P_l	Horizontal limit reaction on a foundation element	kN	M	**Vol. 3**

P_{lat}	Lateral reaction on a foundation element	kN	M	**Vol. 3**
P_p	Passive earth pressure resultant on a retaining wall	kN	M	**Vol. 4:** 1
P_{tot}	Total reaction on a foundation element: $P_{lat} + P_{front}$	kN	M	**Vol. 3**
PF	Fire Loss	-	-	**Vol. 1:** 1
Q, Q_0	Vertical load (on a foundation)	kN	M	**Vol. 2:** 2 **Vol. 3** **Vol. 4:** 2
Q	Point force (load)	kN	M	**Vol. 1:** 3 **Vol. 3**
Q	Operating overload	kN	M	**Vol. 1:** 3
Q	Variable action	kN	M	**Vol. 3**
Q	Drawdown or dewatering flow rate	m^3/s	L^3T^{-1}	**Vol. 2:** 1 **Vol. 4:** 1
Q_a	Flow rate in the Lefranc test	m^3/s	L^3T^{-1}	**Vol. 2:** 1
Q_c	Apparent point force of the static penetrometer	kN	M	**Vol. 1:** 3
Q_c	Creep load of the soil	kN	M	**Vol. 3**
Q_{cr}	Critical buckling load of a foundation element	kN	M	**Vol. 4:** 2
Q_d	Point load of a dynamic penetrometer	kN	M	**Vol. 1:** 3
Q_{SLS}	Load at Serviceability Limit States	kN	M	**Vol. 3** **Vol. 4:** 1
Q_{ULS}	Load at Ultimate Limit States	kN	M	**Vol. 3** **Vol. 4:** 1
Q_g	Uplifting force due to soil swelling	kN	M	**Vol. 3**
Q_f	Creep load	kN	M	**Vol. 3**
Q_h	Horizontal load on a foundation	kN	M	**Vol. 3**
Q_l	Limit load of soil under a foundation	kN	M	**Vol. 3** **Vol. 4:** 2
Q_{max}	Maximum load on a foundation (compression)	kN	M	**Vol. 3**
Q_{min}	Maximum load on a foundation (traction)	kN	M	**Vol. 3**
Q_p	Point load of static penetrometer	kN	M	**Vol. 1:** 3
Q_p	Point load of a shaft	kN	M	**Vol. 3**
Q_{pu}	Ultimate load at the point of a shaft	kN	M	**Vol. 3** **Vol. 4:** 1

Q_r	Failure load of reinforcement steel	kN	M	**Vol. 4:** 2
Q_{re}	Force due to soil shrinkage on a deep foundation element	kN	M	**Vol. 3**
Q_s	Frictional force with static penetrometer	kN	M	**Vol. 1:** 3
Q_s	Soil/shaft skin friction force	kN	M	**Vol. 3**
Q_{cl}	Grout/inclusion frictional resistance	kN	M	**Vol. 4:** 2
Q_{sl}	Soil/inclusion frictional resistance	kN	M	**Vol. 4:** 2
Q_{su}	Ultimate soil/shaft frictional force	kN	M	**Vol. 1:** 3 **Vol. 3** **Vol. 4:** 1
Q_t	Total static penetrometer force	kN	M	**Vol. 1:** 3
Q_U	Water action	kN	M	**Vol. 1:** 3
Q_u	Ultimate soil load (under a foundation)	kN	M	**Vol. 3**
Q_v	Vertical load on a foundation	kN	M	**Vol. 3**
Q_l^1	Ultimate load on an isolated foundation element	kN	M	**Vol. 3** **Vol. 4:** 2
Q_l^N	Limit load on N foundation elements	kN	M	**Vol. 3** **Vol. 4:** 2
Q_θ	Action due to temperature effect	kN	M	**Vol. 3** **Vol. 4:** 1
Q_1, Q_2, … Q_5	Lateral friction curves f_{soil} (according to Eurocodes)	-	-	**Vol. 3**
Q_1, Q_2, … Q_7	Unitary lateral friction curves limit q_s	-	-	**Vol. 3** **Vol. 4:** 1
R	Resultant (of forces)	kN	M	**Vol. 2:** 2 **Vol. 4:** 2
R_a	Radius of influence of drawdown by pumping	m	L	**Vol. 2:** 1
Rb (Q_{pu})	Point soil resistance under a pile	kN	M	**Vol. 3** **Vol. 4:** 1
Rb,k	Characteristic value of point resistance of soil under a pile	kN	M	**Vol. 3** **Vol. 4:** 1
R_c	Unconfined compressive strength	kPa	ML^{-2}	**Vol. 1:** 1, 2 **Vol. 4:** 2
Rc,d	ULS load under a pile	kPa	M	**Vol. 3** **Vol. 4:** 1
Rc,cr,d	SLS load under a pile	kPa	M	**Vol. 3** **Vol. 4:** 1
Rc,cr,k	Characteristic value of the creep load under a pile	kPa	ML	**Vol. 3** **Vol. 4:** 1

R_I	Linear shrinkage factor	-	-	**Vol. 2:** 2
R_f (F_R)	Friction ratio (= f_s/q_c)	-	-	**Vol. 1:** 3 **Vol. 3** **Vol. 4:** 2
R_g	Swelling ratio	-	-	**Vol. 2:** 2
RQD	Rock Quality Designation	%	-	**Vol. 1:** 1
Rs (Q_{su})	Resistance to soil/shaft skin friction	kN	M	**Vol. 3** **Vol. 4:** 1
Rs,k	Characteristic value for soil/pile lateral friction resistance	kN	M	**Vol. 3** **Vol. 4:** 1
R_t	Resistance to traction	kPa	ML^{-2}	**Vol. 1:** 2
Rv	Soil resistance under footing	kN	M	**Vol. 3**
Rv,k	Characteristic value of soil resistance under footing	kN	M	**Vol. 3**
Rv,d	Ultimate State Load (ULS and SLS) of the soil under footing	kN	M	**Vol. 3**
S	Section of a soil sample	m^2	L^2	**Vol. 1:** 1, 2
S	Storage coefficient	-	-	**Vol. 1:** 3
S	Cross-section of m x n foundation elements	m^2	L^2	**Vol. 3**
S	Wet section of an excavation	m^2	L^2	**Vol. 4:** 1
S	Soil coefficient increasing the seismic acceleration	-	-	**Vol. 3**
S'	Fictitious section of the influence of m x n foundation elements	m^2	L^2	**Vol. 3**
S_r	Degrees of saturation	%	-	**Vol. 1:** 1
S_t	Sensitivity coefficient: c_u/c_u^r (scissometer)	%	-	**Vol. 1:** 1
S(t)	Force as a function of time	kN	M	**Vol. 3**
T, T(τ)	Tangential (shear) force	kN	M	**Vol. 2:** 2
T	Tangential component of the resultant R	kN	M	**Vol. 2:** 2
T	Traction force (Tension)	kN	M	**Vol. 1:** 2
T	Period (= $2\pi / w$)	s	T	**Vol. 3**
T	Hydraulic transmissivity	m^2/s	L^2T^{-1}	**Vol. 2:** 1
T	Shear force in a stiff inclusion	kN	M	**Vol. 4:** 2
T, T(z)	Shearing force on a foundation element	kN	M	**Vol. 3** **Vol. 4:** 2

T	Tension in a tie-rod: $T = T_0 + \Delta T$	kN	M	**Vol. 4:** 1
"TA"	Reinforced earth	-	-	**Vol. 4:** 2
T_c	Shear force in a stiff inclusion	kN	M	**Vol. 4:** 2
T_A, T_B, T_C, T_D	Characteristic periods of the (elastic) earthquake response spectrum	s	T	**Vol. 3**
T_{l1}	Characteristic plasticizing resistance of the soil	kN	M	**Vol. 4:** 2
T_{l2}	Characteristic plasticizing resistance of a stiff strip	kN	M	**Vol. 4:** 2
T_{max}	Maximum force in a foundation element	kN	M	**Vol. 4:** 1
T_p	Point force of a foundation	kN	M	**Vol. 3**
T_r	Time factor (radial consolidation)	-	-	**Vol. 2:** 2
T_R	Tensile strength of the steel of a tie-rod	kN	M	**Vol. 4:** 2
T_v	Time factor (vertical consolidation)	-	-	**Vol. 2:** 2 **Vol. 3**
T_0	Force at the head of a foundation	kN	M	**Vol. 3**
T_0, T(0)	Force at the head of an inclusion	kN	M	**Vol. 4:** 2
T_0	Prestressing force in a tie-rod	kN	M	**Vol. 4:** 1
U	Force due to water	kN	M	**Vol. 2:** 2
U	Degree of consolidation	%	-	**Vol. 2:** 2 **Vol. 4:** 2
U_v	Degree of vertical consolidation	%	-	**Vol. 2:** 2
U_r	Degree of radial consolidation	%	-	**Vol. 2:** 2
U_L	Lugeon unit: $\approx 10^{-7}$ m/s	m/s	LT^{-1}	**Vol. 2:** 1
$\vec{U}$ ($\{U\}$)	Displacement vector (components u_1, u_2, u_3 or u_x, u_y, u_z)	m	L	**Vol. 1:** 2
V	Volume of a solid	m^3	L^3	**Vol. 1:** 2
V	Volume of the pressuremeter probe	m^3	L^3	**Vol. 1:** 3
V	Total volume of a soil sample	m^3	L^3	**Vol. 1:** 1
V	Volume of expansion of a cylindrical cavity	m^3	L^3	**Vol. 4:** 2
V	Excavated volume (tunnel volume)	m^3	L^3	**Vol. 4:** 3
V	Vertical force	kN	M	**Vol. 2:** 2
V	Wind action	kN	M	**Vol. 3**

V_a	Volume of air in a soil sample	m^3	L^3	**Vol. 1:** 1
V_s	Volume of grains in a soil sample	m^3	L^3	**Vol. 1:** 1
V_s	Volume of bowl due to tunnel excavation	m^3	L^3	**Vol. 4:** 3
V_w	Volume of water in a soil sample	m^3	L^3	**Vol. 1:** 1
V_0	Vertical force on a foundation element	kN	M	**Vol. 3**
V_0	Initial volume of a cylindrical cavity (boring)	m^3	L^3	**Vol. 4:** 2
VBS	Methylene blue value	-	-	**Vol. 1:** 1
W	Weight (total of a soil sample)	kN	M	**Vol. 1:** 1
W	Weight of a soil column	kN	M	**Vol. 2:** 2
W_n, W_t	Normal and tangential components of W	kN	M	**Vol. 2:** 2
W_s	Weight of grains in a soil sample	kN	M	**Vol. 1:** 1
W_w	Weight of water in a soil sample	kN	M	**Vol. 1:** 1
XA1	Aggression class of concrete (… XA2, XA3)	-	-	**Vol. 3**
Z	Depth of the surface at the axis of the tunnel	m	M	**Vol. 4:** 3

Lowercase letters

Symbol	Designation	Unit	Dimension	Chapters
a	Incorporation ratio with column in the soil	%	-	**Vol. 4:** 2
a	Expansion coefficient of the pressuremeter probe: $\Delta V/\Delta p$	cm^3/MPa	$M^{-1}L^5$	**Vol. 1:** 3
a, b	Characteristic sizes for foundations	m	L	**Vol. 3**
a, b	Foundation dimensions	m	L	**Vol. 3**
a, b	Parameter of the interaction law of the soil/shaft foundation	kN et m	M and L	**Vol. 4:** 2
a_g	Earthquake calculation acceleration	m/s^2	LT^{-2}	**Vol. 3** **Vol. 4:** 2
a_{gr}	Maximum acceleration of earthquake reference	m/s^2	LT^{-2}	**Vol. 3**
a_h	Horizontal acceleration due to earthquake	m/s^2	LT^{-2}	**Vol. 3**

a_{max}	Maximum surface acceleration of earthquake	m/s^2	LT^{-2}	**Vol. 3** **Vol. 4:** 1, 2
a_v	Coefficient of compressibility: $d\varepsilon/d\sigma' = 1/E_{oed}$	kPa^{-1}	$M^{-1}L^2$	**Vol. 2:** 2
a_v, a_{vg}	Vertical acceleration due to an earthquake	m/s^2	LT^{-2}	**Vol. 3**
b	Fictitious distance between foundation elements in a group	m	L	**Vol. 3**
b	Width of an inclusion (band)	m	L	**Vol. 4:** 2
b'	Reduced width of a band (corrosion)	m	L	**Vol. 4:** 2
c	Soil cohesion (general term)	kPa	ML^{-2}	**Vol. 2:** 2 **Vol. 3** **Vol. 4:** 1
c_i	Cohesion measured by phicometer	kPa	ML^{-2}	**Vol. 1:** 3 **Vol. 4:** 1
c_p	Back passive earth pressure	kPa	ML^{-2}	**Vol. 4:** 1
c_u	Undrained soil cohesion	kPa	ML^{-2}	**Vol. 2:** 2 **Vol. 3** **Vol. 4:** 1, 2
c_{uu}	Undrained unconsolidated soil cohesion	kPa	ML^{-2}	**Vol. 2:** 2
c_{u0}	Initial undrained soil cohesion	kPa	ML^{-2}	**Vol. 1:** 2, 3 **Vol. 2:** 2
$c_u^{\ c}$	Undrained cohesion corrected according to IP: $c_u^{\ c} = \mu_r.c_u$	kPa	ML^{-2}	**Vol. 1:** 3
$c_u^{\ r}$	Undrained cohesion of remolded soil	kPa	ML^{-2}	**Vol. 1:** 3
c_r	Radial consolidation coefficient	m^2/s	L^2T^{-1}	**Vol. 1:** 3
c_v	Vertical consolidation coefficient: $k/(a_v\gamma_w)$	m^2/s	L^2T^{-1}	**Vol. 1:** 3
c_1, c_2, c_3, c_4	Constant coefficients applied on the bearing capacity factors k_p and k_c	-	-	**Vol. 3**
c'	Effective soil cohesion	kPa	ML^{-2}	**Vol. 1:** 2, 3 **Vol. 2:** 2 **Vol. 3** **Vol. 4:** 1, 2

c'_i	Effective cohesion deduced from c_i (phicometer)	kPa	ML^{-2}	**Vol. 1:** 3
c'_R	Residual cohesion (of remolded soil)	kPa	ML^{-2}	**Vol. 2:** 2
c^*	Cohesion affected by the safety coefficient Γ_c	kPa	ML^{-2}	**Vol. 4:** 2
d	Dilatancy (of sand grains in a dense state)	-	-	**Vol. 1:** 2
d	Distance from the crest of the slope	m	L	**Vol. 3**
d	Distance between foundation elements	m	L	**Vol. 3**
dh	Variation in the head of water	m	L	**Vol. 2:** 1
dv	Elementary displacement of a foundation	m	L	**Vol. 4:** 2
e	Void ratio (between soil grains)	-	-	**Vol. 1:** 1 **Vol. 2:** 2
e	Eccentricity of the load (Q) on a foundation	m	L	**Vol. 3**
e_d	Driving of the tip of the dynamic penetrometer	m	L	**Vol. 1:** 3 **Vol. 3**
e_h	Horizontal spacing between reinforcements	m	L	**Vol. 4:** 1, 2
e_{max}	Maximum void ratio (loosest state)	-	-	**Vol. 1:** 1 **Vol. 2:** 2
e_{min}	Minimum void ratio (densest state)	-	-	**Vol. 1:** 1 **Vol. 2:** 2
e_p	Thickness of a retaining wall	m	L	**Vol. 4:** 1
e_0, e_i	Initial void index	-	-	**Vol. 1:** 1 **Vol. 2:** 2 **Vol. 3**
e_1	Elastic shortening of a pile	m	L	**Vol. 3**
f	Friction coefficient: $f = \tan\psi$	-	-	**Vol. 2:** 2
f_c	Compressive stress in concrete	kPa	ML^{-2}	**Vol. 1:** 2 **Vol. 3**
f_{cj}	Compressive stress in concrete at d-days	kPa	ML^{-2}	**Vol. 1:** 2 **Vol. 3**
f_{cmax}	Maximum capped compressive stress in concrete	kPa	ML^{-2}	**Vol. 3**
f_{c28}	Compressive stress in concrete at 28 days	kPa	ML^{-2}	**Vol. 3**
f_g	Force of gravity	kN	M	**Vol. 2:** 1

f_n	Negative unit friction on a foundation element	kPa	ML^{-2}	**Vol. 3**
f_{tj}	Tensile stress in concrete at 28 days	kPa	ML^{-2}	**Vol. 1:** 2 **Vol. 3**
f_s	Lateral friction measured with a static penetrometer (CPT)	kPa	ML^{-2}	**Vol. 1:** 3 **Vol. 3** **Vol. 4:** 1, 2
f_{soil}	Soil/deep foundation element friction	kPa	ML^{-2}	**Vol. 3** **Vol. 4:** 1, 2
f_e (f_{yk})	Elastic limit of the steel	kPa	ML^{-2}	**Vol. 3** **Vol. 4:** 2
f_0	Resonance frequency	s^{-1}	T^{-1}	**Vol. 3**
f_0	Driving of wall at depth in the soil	m	L	**Vol. 4:** 1
f^*_d	Dynamic soil/band friction “TA”	-	-	**Vol. 4:** 2
f^*_s	Apparent static soil/band friction “TA”	-	-	**Vol. 4:** 2
g	Acceleration of gravity: 9.81 m/s^2	m/s^2	LT^{-2}	**Vol. 1:** 3 **Vol. 3**
g_{max}	Maximum displacement of compressible soil	m	L	**Vol. 3**
$g_{max}(t)$	Maximum displacement of compressible soil as a function of time t	m	L	**Vol. 3**
g(z)	Horizontal displacement of soil with pile at depth z	m	L	**Vol. 3**
g(z, t)	Horizontal displacement of compressible soil as a function of time t	m	L	**Vol. 3**
h	Height of a sample	m	L	**Vol. 1:** 2
h	Thickness of a soil layer	m	L	**Vol. 1:** 2
h	Height of foundation anchor in the bearing layer	m	L	**Vol. 3**
h	Height of pile driving	m	L	**Vol. 1:** 3 **Vol. 3**
h, h(t)	Head (or potential) of water	m	L	**Vol. 2:** 1 **Vol. 4:** 1
h_a	Height on which the resultant of active earth pressure P_a is exerted	m	L	**Vol. 4:** 1
h_c	Critical foundation anchor in the bearing layer	m	L	**Vol. 4:** 1
h_c	Critical height of an excavation	m	L	**Vol. 4:** 1

h_c, h_N	Critical height, position of the neutral point N	m	L	**Vol. 3**
h_i	Thickness of the layer i	m	L	**Vol. 1:** 2
h_I	Thickness of a granular soil layer on soft soil	m	L	**Vol. 3**
h_M	Head (or potential) of water at point M	m	L	**Vol. 2:** 1
h_p	Height over which the resultant of passive earth pressure P_a is applied	m	L	**Vol. 4:** 1
h_r	Height of an earth fill	m	L	**Vol. 2:** 2 **Vol. 3** **Vol. 4:** 2
h_s	Height, length of soil/foundation friction	m	L	**Vol. 3** **Vol. 4:** 1, 2
h_{si}	Sealing height of a reinforcement in a layer i	m	L	**Vol. 4:** 2
h_w	Water height	m	L	**Vol. 1:** 2
i	Angle of inclination of a reinforcement	rad	Radian	**Vol. 4:** 1, 2
$\vec{i}$	Hydraulic gradient	-	-	**Vol. 2:** 1
i_c	Critical hydraulic gradient: γ'/γ_w	-	-	**Vol. 2:** 1
i_c i_q, i_γ	Minorizing coefficients of the bearing capacity	-	-	**Vol. 3**
I_β	Minorizing coefficient as a function of the slope angle β	-	-	**Vol. 3**
I_δ	Minorizing coefficient of the inclined load at δ	-	-	**Vol. 3**
k	Soil permeability coefficient	m/s	MT^{-1}	**Vol. 1:** 2 **Vol. 2:** 1 **Vol. 4:** 1
k (2k)	Tresca criterion	kPa	ML^{-2}	**Vol. 1:** 2, 3 **Vol. 4:** 2
k_L	Lefranc permeability coefficient	m/s	MT^{-1}	**Vol. 1:** 2 **Vol. 2:** 1
k_c	Compaction coefficient: E_{v2}/E_{v1}	-	-	**Vol. 1:** 3
k_c	Penetrometer bearing capacity coefficient	-	-	**Vol. 3** **Vol. 4:** 2
k_{cmax}	Maximum penetrometer bearing capacity coefficient	-	-	**Vol. 3**

k_{ef}	Efficiency coefficient of n micropiles in network or in group	-	-	**Vol. 4:** 2
k_p	Coefficient of pressuremeter bearing capacity	-	-	**Vol. 3** **Vol. 4:** 1, 2
k_{pmax}	Maximum coefficient of pressuremeter bearing capacity	-	-	**Vol. 3**
k_r	Axial stiffness of a foundation element	kN/m	ML^{-1}	**Vol. 4:** 2
k_s	Soil reaction coefficient	kPa/m	ML^{-3}	**Vol. 3** **Vol. 4:** 1, 2
k_{sh}	Horizontal soil reaction coefficient	kPa/m	ML^{-3}	**Vol. 3** **Vol. 4:** 1, 2
k_{sv}	Vertical soil reaction coefficient	kPa/m	ML^{-3}	**Vol. 3** **Vol. 4:** 1, 2
k_v	Stiffness of a spring in the relation: $F = -k_v.x$	kN/m	ML^{-1}	**Vol. 3** **Vol. 4:** 1
k_h, k_h, k_v	Permeability component (homogeneous anisotropic soil)	m/s	MT^{-1}	**Vol. 1:** 2 **Vol. 2:** 1
k_x, k_y, k_z	Permeability component in a three-dimensional reference mark: (x, y, z)	m/s	MT^{-1}	**Vol. 1:** 2 **Vol. 2:** 1
k_1	Minorizing coefficient depending on the type of foundation element	-	-	**Vol. 3**
k_2	Minorizing coefficient depending on the geometry of the foundation element	-	-	**Vol. 3**
k_3	Coefficient reducing the calculation value of the compressive strength of concrete (= 1 to 1.2)	-	-	**Vol. 3**
l	Length	m	L	**Vol. 1:** 2
l	Length of an inclusion	m	L	**Vol. 4:** 2
l_a	Length of soil/inclusion adhesion	m	L	**Vol. 4:** 2
l_0	Length of transfer	m	L	**Vol. 3** **Vol. 4:** 1, 2
m	Shape coefficient (of test cavity)	-	-	**Vol. 2:** 1

m	Mass of the driving hammer (dynamic penetrometer)	kN	M	**Vol. 1:** 3
m	Number of foundation elements in group	-	-	**Vol. 3**
m’	Mass of rods and accessories (dynamic penetrometer)	kN	M	**Vol. 1:** 3
$m(\lambda)$, $m(\lambda,b)$	Calculation coefficient of negative friction	-	-	**Vol. 3**
n	Number of foundation elements in group	-	-	**Vol. 3** **Vol. 4:** 2
n	Porosity	%	-	**Vol. 1:** 1
n	Stress concentration ratio	-	-	**Vol. 4:** 2
n_a	Number of “TA” strips per ml	-	-	**Vol. 4:** 2
n_0	Soil improvement factor per column	-	-	**Vol. 4:** 2
p	Pressure at the base of a foundation	kPa	ML^{-2}	**Vol. 1:** 2
p	Average pressure: $(p_1 + p_2 + p_3)/3$	kPa	ML^{-2}	**Vol. 1:** 2
p	Pressure of the probe on the boring walls	kPa	ML^{-2}	**Vol. 1:** 2
p	Soil/inclusion pressure	kPa	ML^{-2}	**Vol. 4:** 2
p, p(z)	Lateral pressure on a foundation element	kPa	ML^{-2}	**Vol. 3** **Vol. 4:** 2
p_0	Initial pressure in the ground before drilling	kPa	ML^{-2}	**Vol. 1:** 2
p_h	Water pressure around the probe	kPa	ML^{-2}	**Vol. 1:** 3
p_f	Creep pressure	kPa	ML^{-2}	**Vol. 1:** 3
p_i	Injection pressure of grout, mortar	kPa	ML^{-2}	**Vol. 4:** 2
p^*_l	Net creep pressure: $p_f - \sigma_{h0}$	kPa	ML^{-2}	**Vol. 1:** 2 **Vol. 2:** 2 **Vol. 3** **Vol. 4:** 1, 2
p_l	Pressuremeter limit pressure	kPa	ML^{-2}	**Vol. 1:** 2, 3 **Vol. 2:** 2 **Vol. 3**
p^*_l	Net pressuremeter limit pressure: $p_l - \sigma_{h0}$	kPa	ML^{-2}	**Vol. 1:** 2 **Vol. 2:** 2 **Vol. 3** **Vol. 4:** 2
p^*_{le}	Equivalent net pressuremeter limit pressure	kPa	ML^{-2}	**Vol. 3** **Vol. 4:** 1, 2

$p^*_l(z)$	Net pressuremeter limit pressure at depth z	kPa	ML^{-2}	**Vol. 3**
p^*_{li}	Limit pressure by the "inverse method"	kPa	ML^{-2}	**Vol. 1:** 2 **Vol. 2:** 2 **Vol. 3** **Vol. 4:** 2
p^*_{lh}	Limit pressure by hyperbolic extrapolation	kPa	ML^{-2}	**Vol. 1:** 2 **Vol. 2:** 2 **Vol. 3** **Vol. 4:** 2
p_r	Pressure around the probe (at a distance r)	kPa	ML^{-2}	**Vol. 1:** 3
p_r	Rupture stress under a foundation	kPa	ML^{-2}	**Vol. 1:** 3
p_u	Ultimate lateral pressure on a foundation element	kPa	ML^{-2}	**Vol. 3**
$p, p_1, \dots p_3$	Pressures acting on a retaining wall	kPa	ML^{-2}	**Vol. 3**
q	Overload	kPa	ML^{-2}	**Vol. 3** **Vol. 4:** 1
q	Deviatoric stress: $\sigma_1 - \sigma_2$ (triaxial test)	kPa	ML^{-2}	**Vol. 3**
q	Linear flow of water in the soil	$m^3/s/m$	$L^3T^{-1}L^{-1}$	**Vol. 2:** 1
q_c	Point resistance measured with a Cone static penetrometer (CPT)	kPa	ML^{-2}	**Vol. 1:** 3 **Vol. 4:** 2
q_{ce}	Equivalent point resistance (CPT)	kPa	ML^{-2}	**Vol. 1:** 3 **Vol. 3** **Vol. 4:** 2
q_{cc}	Point resistance peaked to 1.3 q_{cm} (CPT)	kPa	ML^{-2}	**Vol. 1:** 3 **Vol. 3**
q_{cm}	Average point resistance (CPT)	kPa	ML^{-2}	**Vol. 1:** 3 **Vol. 3**
q_c (z)	Point resistance measured with the Cone static penetrometer (CPT) at depth z	kPa	ML^{-2}	**Vol. 3**
q_{c1N}	Point resistance measured with the normalized static penetrometer (CPT)	kPa	ML^{-2}	**Vol. 3** **Vol. 4:** 2
q_d	Point resistance measured with a dynamic penetrometer	kPa	ML^{-2}	**Vol. 1:** 3 **Vol. 3**
q_{SLS}	Total stress at serviceability limit states	kPa	ML^{-2}	**Vol. 3** **Vol. 4:** 1
q_{ULS}	Total stress at ultimate limit states	kPa	ML^{-2}	**Vol. 3**
q'_{SLS}	Effective stress at serviceability limit states	kPa	ML^{-2}	**Vol. 3**

q'_{ULS}	Effective stress at ultimate limit states	kPa	ML^{-2}	**Vol. 3**
q^*_{SLS}	Net stress at serviceability limit states: $q_{SLS} - q_0$	kPa	ML^{-2}	**Vol. 3**
q_p	Point resistance of soil under foundation	kPa	ML^{-2}	**Vol. 3**
q_{pu}	Ultimate resistance of soil under foundation	kPa	ML^{-2}	**Vol. 3**
q_r	Failure stress under a foundation	kPa	ML^{-2}	**Vol. 4:** 2
q_{ref}	Reference stress under a foundation	kPa	ML^{-2}	**Vol. 3**
q_s	Lateral friction (limit)	kPa	ML^{-2}	**Vol. 1:** 3 **Vol. 3** **Vol. 4:** 1, 2
q_{si}	Lateral friction in layer i	kPa	ML^{-2}	**Vol. 4:** 2
q_s (z)	Lateral friction at depth z	kPa	ML^{-2}	**Vol. 3**
q_{smax}	Maximum lateral friction	kPa	ML^{-2}	**Vol. 3**
q_0	Vertical stress (total) in the soil at the base of the foundation (due to the weight of the soil)	kPa	ML^{-2}	**Vol. 3**
q'_0	Vertical stress (effective) in the soil at the base of the foundation (weight of soil removed)	kPa	ML^{-2}	**Vol. 3**
q^*_s	Lateral friction unit limit with a safety coefficient Γ_{qs}	kPa	ML^{-2}	**Vol. 4:** 2
q_t	Total resistance measured with a static penetrometer (CPT)	kPa	ML^{-2}	**Vol. 1:** 3
q_u	(Total) Failure stress under foundation	kPa	ML^{-2}	**Vol. 3** **Vol. 4:** 2
q'_u	(Effective) Failure stress under foundation	kPa	ML^{-2}	**Vol. 3**
r	Radius (of a foundation)	m	L	**Vol. 1:** 2, 3 **Vol. 3**
r_d	Earthquake stress correction factor	-	-	**Vol. 4:** 2
r_p	Radius of plasticization around the probe	m	L	**Vol. 1:** 2, 3
r_u	Pore pressure increase ratio	-	-	**Vol. 3** **Vol. 4:** 2

s	Section of a sample under water flow	m^2	L^2	**Vol. 2:** 1
s	Drawdown of water by pumping	m	L	**Vol. 2:** 1
s	Total settlement of a foundation	m	L	**Vol. 2:** 2
s_A	Settlement of a foundation element A	m	L	**Vol. 2:** 2
s_B	Settlement of a foundation element B	m	L	**Vol. 2:** 2
s_c	Primary consolidation settlement under the structure foundation	m	L	**Vol. 2:** 2
s_c	Spherical settlement (at the base of the foundation)	m	L	**Vol. 3**
s_d	Deviatoric settlement (at depth)	m	L	**Vol. 3**
s_i	Immediate settlement of a foundation	m	L	**Vol. 2:** 2
s_i	Settlement without soil improvement	m	L	**Vol. 4:** 2
s_f	Settlement with soil improvement	m	L	**Vol. 4:** 2
s_l	Settlement due to lateral deformation of the soil under the foundation	m	L	**Vol. 2:** 2
s_{max}	Maximum settlement of the settlement bowl due to tunneling	m	L	**Vol. 4:** 3
s_s	Secondary compression settlement under the foundation	m	L	**Vol. 2:** 2
s_s	Proper settlement of the soil	m	L	**Vol. 3**
s_p	Settlement of a foundation element point	m	L	**Vol. 3**
s_{pg}	Settlement of a group of foundation elements	m	L	**Vol. 3**
s_r	Reference settlement of a pile	m	L	**Vol. 3**
s_t, s_0	Settlement at the head of a shaft	m	L	**Vol. 3**
s_0	Proper settlement of the soil	m	L	**Vol. 3**
s_u	Shear stress (= c_u)	kPa	ML^{-2}	**Vol. 1:** 3
$s_x(x)$	Settlement as a function of x of the tunnel settlement bowl on the Gaussian curve	m	L	**Vol. 4:** 3
$s_z(x)$	Horizontal displacement on the tunnel settlement bowl on the Gaussian curve	m	L	**Vol. 4:** 3

s_{90}	Settlement under the foundation at 90% (U = 0.90) consolidation	m	L	**Vol. 2:** 2
s_{100}	Settlement under the foundation at 100% (U = 1.00) consolidation	m	L	**Vol. 2:** 2
t	Time	s	T	**Vol. 1:** 2 **Vol. 2:** 2
t	Thickness of a "reinforced earth" strip	m	L	**Vol. 4:** 2
t_i	Initial time	s	T	**Vol. 2:** 2
t_f	Final time	s	T	**Vol. 2:** 2
t_{50}	Time for 50% consolidation (U = 0.50)	s	T	**Vol. 2:** 2
t_{90} ($\sqrt{t_{90}}$)	Time for 90% (U = 0.90) consolidation in the Taylor representation	s	T	**Vol. 2:** 2
t_{100}	Time for 100% (U = 1.00) consolidation	s	T	**Vol. 2:** 2
t'	Reduced thickness of a "reinforced earth" strip	m	L	**Vol. 4:** 2
u	Pore pressure of water (at time t)	kPa	ML^{-2}	**Vol. 1:** 2 **Vol. 2:** 2 **Vol. 4:** 1
u, u(z)	Displacement of a foundation element	m	L	**Vol. 3**
u_h	Pore pressure at the piezocone drilling stop	kPa	ML^{-2}	**Vol. 1:** 2
u_G	Displacement of the center of gravity of a foundation element	m	L	**Vol. 3**
u_0	Pore pressure of water at the initial time	kPa	ML^{-2}	**Vol. 1:** 2
u, v, w u_1, u_2, u_3	Component displacements of the vector $\vec{U}$	m	L	**Vol. 1:** 2
u, v, w u_1, u_2, u_3	Displacements in the (x, y, z) or (1, 2, 3) reference mark	m	L	**Vol. 1:** 2
u_{max}	Threshold on displacement tolerance	m	L	**Vol. 1:** 2
v	Apparent speed of water flow in the component soil (v_x, v_y, v_z)	m/s	LT^{-1}	**Vol. 2:** 1
v	Half-width of a foundation	m	L	**Vol. 4:** 1
v_l	Vertical displacement of a foundation	m	L	**Vol. 4:** 2

v_0	Vertical displacement at the head of a foundation	m	L	**Vol. 4:** 2
v'	Real average water flow speed	m/s	LT^{-1}	**Vol. 2:** 1
w, w_n	Water content	%	-	**Vol. 1:** 1 **Vol. 2:** 2 **Vol. 3** **Vol. 4:** 2
w	Foundation settlement	m	L	**Vol. 3**
w_{opn}	Water content at Proctor Optimum	%	-	**Vol. 1:** 1
w_L	Liquidity limit	%	-	**Vol. 1:** 1 **Vol. 3**
w_P	Plasticity limit	%	-	**Vol. 1:** 1
w_{RE}	Shrinkage limit	%	-	**Vol. 1:** 1 **Vol. 2:** 2
x	Horizontal distance	m	-	**Vol. 1:** 2 **Vol. 4:** 2
x, y	Two-dimensional coordinate system	m	L	**Vol. 1:** 2 **Vol. 3** **Vol. 4:** 2
x, y, z	Three-dimensional coordinate system	m	L	**Vol. 1:** 2
x*, z*	Change of variables in an x, z system	M	L	**Vol. 3**
y, y(z)	Proper horizontal displacement of a foundation element	m	L	**Vol. 3**
z, Z	Depth, vertical distance	m	L	**Vol. 1:** 2 **Vol. 3** **Vol. 4:** 1, 2

Capital Greek letters

Symbol	Designation	Unit	Dimension	Chapters
Δc_u	Variation in undrained soil cohesion	kPa	ML^{-2}	**Vol. 2:** 2 **Vol. 3**
Δg_{max} (t)	Horizontal displacement of compressible soil due to settlement calculated at time t	m	L	**Vol. 3**
ΔH	Vertical spacing between two inclusions	m	L	**Vol. 4:** 2
Δh	Variation in height (vertical displacement)	m	L	**Vol. 1:** 2
Δh	Variation of the thickness of a soil layer	m	L	**Vol. 1:** 2

Δh	Variation of the head of water	m	L	**Vol. 2:** 1
Δl	Variation of length	m	L	**Vol. 1:** 2
Δl	Shortening of a foundation element	m	L	**Vol. 4:** 1
ΔN_d	Dynamic traction increment (reinforced earth strip) under earthquake acceleration $m = a_h/g$	kN	M	**Vol. 4:** 2
Δq	Excess load on the soil	kN	M	**Vol. 3**
ΔQ	Elementary load on a foundation	kN	M	**Vol. 4:** 2
Δr	Radius variation (radial displacement)	m	L	**Vol. 1:** 2
Δu	Variation of the pore pressure of water	kPa	ML^{-2}	**Vol. 1:** 2, 3 **Vol. 2:** 2
Δu_t	Variation of the pore pressure at time t	kPa	ML^{-2}	**Vol. 2:** 2
Δu_{ti}	Variation of the pore pressure at initial time t_i	kPa	ML^{-2}	**Vol. 2:** 2
Γ_{cu}	Safety coefficient on the undrained cohesion c_u	-	-	**Vol. 4:** 2
$\Gamma_{c'}$	Safety coefficient on the effective cohesion c'	-	-	**Vol. 4:** 2
Γ_{EI}	Safety coefficient on bending stiffness EI	-	-	**Vol. 4:** 1
Γ_{f^*s}	Safety coefficient on friction f^*_s	-	-	**Vol. 4:** 2
Γ_{qs}	Safety coefficient on friction q_s	-	-	**Vol. 4:** 2
Γ_{pl}	Safety coefficient for the pressuremeter limit pressure p_l	-	-	**Vol. 4:** 2
$\Gamma_{\varphi'}$	Safety coefficient on the effective friction angle φ'	-	-	**Vol. 4:** 2
Γ_s	Seismic safety coefficient (liquefaction)	-	-	**Vol. 4:** 2
$\Gamma_{\sigma e}$	Safety coefficient on the elastic limit σ_e	-	-	**Vol. 4:** 2
ΔV	Variation in volume of a solid	m^3	L^3	**Vol. 1:** 2
ΔV	Variation in volume of a soil sample	m^3	L^3	**Vol. 2:** 2
ΔV	Variation in the volume of water in the pressuremeter probe	cm^3	L^3	**Vol. 1:** 3

Δy, $\Delta y(z)$	Difference in pile/soil displacement: $y(z) - g(z)$	m	L	**Vol. 3**
Δs	Differential settlement between points A and B	m	L	**Vol. 3**
$\Phi_1(\delta)$, $\Phi_2(\delta)$	Minorizing potentials of the load function of δ	-	-	**Vol. 3**
θ_a	Angular deformation	rad	Radian	**Vol. 1:** 2

Lowercase Greek letters

Symbol	Designation	Unit	Dimension	Chapters
α	Menard rheological coefficient: E_M/E_s (E)	-	-	**Vol. 1:** 2 **Vol. 4:** 2
α_s	Minorizing factor of the drilling diameter	-	-	**Vol. 4:** 1, 2
α_{si}	Minorizing factor of the drilling diameter of the layer i	-	-	**Vol. 4:** 1, 2
α_0	Coefficient of thermal expansion: $N_\theta = \alpha_0.\theta$	-	-	**Vol. 4:** 1
α_1	Minorizing coefficient: $E_{oed} = \alpha_1.q_c$	-	-	**Vol. 3**
α_1, α_2	Coefficient of foundation shape	-	-	**Vol. 3**
β	Percentage of damping	%	-	**Vol. 1:** 2
β	Factor: $q_s = \beta.c_u$	-	-	**Vol. 1:** 3
β	Minorizing factor: $q_s = Min(\frac{q_c(z)}{\beta}, q_{s\max})$	-	-	**Vol. 3**
β	Angle of backfill with respect to the horizontal	rad	Radian	**Vol. 1:** 3 **Vol. 2:** 2 **Vol. 3**
β	Angle of backfill behind the retaining wall	rad	Radian	**Vol. 4:** 1
β	Settlement reduction factor	-	-	**Vol. 4:** 2
β'	Fictitious angle of the slope/horizontal	rad	Radian	**Vol. 3**
λ	Minorizing coefficient: $N_c = 7\lambda$	-	-	**Vol. 3**
λ	Coefficient of soil/shaft latching	-	-	**Vol. 3**
λ	Angle of inclination with respect to the vertical of a retaining wall (batter)	rad	Radian	**Vol. 4:** 1

λ	Deconfinement rate which represents the relative position of the tunnel section/ face	-	-	**Vol. 4:** 3
λ, μ	Lamé parameters	kPa	ML^{-2}	**Vol. 1:** 2
λ_a	Conventional inclination angle of active earth pressure	rad	Radian	**Vol. 4:** 1
λ_c	Spherical shape coefficient of a foundation	-	-	**Vol. 3**
λ_d	Deviatoric shape coefficient of a foundation	-	-	**Vol. 3**
λ_p	Conventional angle of inclination of passive earth pressure	rad	Radian	**Vol. 4:** 1
λ_{cu}	Increase factor of undrained cohesion c_u	rad	Radian	**Vol. 1:** 3 **Vol. 2:** 2
δ	Angle of inclination of the load/vertical	rad	Radian	**Vol. 3**
δ	Minorizing coefficient of undrained cohesion: $\delta.c_u$	-	-	**Vol. 3**
δ	Friction angle of the soil/foundation contact	rad	Radian	**Vol. 3**
$\varepsilon(z)$	Deformation function of z	%	-	**Vol. 4:** 2
$\bar{\varepsilon}$ (ε_{ij})	Deformation tensor: I = 1–3, j = 1–3	%	-	**Vol. 1:** 2
$\varepsilon_1, \varepsilon_2, \varepsilon_3$	Deformations in the 3 main directions	%	-	**Vol. 1:** 2
ε_v	Volumetric deformation: $\varepsilon_v = \varepsilon_1 + \varepsilon_2 + \varepsilon_3$	%	-	**Vol. 1:** 2
ε_e	Deformation of elastic origin (reversible)	%	-	**Vol. 1:** 2
ε_p	Deformation of plastic origin (irreversible)	%	-	**Vol. 1:** 2
$\dot{\varepsilon}_{pc}$	Shear deformation rate	s^{-1}	T^{-1}	**Vol. 1:** 2
$\dot{\varepsilon}_{pv}$	Volumetric deformation rate	s^{-1}	T^{-1}	**Vol. 1:** 2
ε_r	Radial deformation in a plane (r, θ)	%	-	**Vol. 1:** 3
ε_θ	Tangential deformation in a plane (r, θ)	%	-	**Vol. 1:** 3
$\dot{\varepsilon}_{vs}$	Viscous deformation rate	s^{-1}	T^{-1}	**Vol. 1:** 2
φ	Friction angle in the ground (general term)	rad	Radian	**Vol. 1:** 2 **Vol. 2:** 2 **Vol. 3** **Vol. 4:** 1
φ^*	Friction angle affected by a safety coefficient: Tanφ/F ou Tanφ/Γφ	rad	Radian	**Vol. 2:** 2 **Vol. 4:** 2

φ_{cu}	Friction angle (consolidated undrained soil)	rad	Radian	**Vol. 1:** 2 **Vol. 2:** 2
φ_{uu}	Apparent angle of friction (unconsolidated undrained soil)	rad	Radian	**Vol. 2:** 2 **Vol. 4:** 1
φ_i	Friction angle measured with a phicometer	rad	Radian	**Vol. 1:** 3 **Vol. 2:** 2 **Vol. 4:** 1
φ_I	Friction angle of a layer of granular soil on soft soil	rad	Radian	**Vol. 3**
φ'	Effective friction angle (between grains)	rad	Radian	**Vol. 1:** 2 **Vol. 2:** 2 **Vol. 3** **Vol. 4:** 1, 2
φ'_c	Friction angle of the (stone) column	rad	Radian	**Vol. 4:** 2
φ'_i	Friction angle deduced from φ_i (phicometer)	rad	Radian	**Vol. 1:** 2 **Vol. 2:** 2
φ'_R	Residual friction angle	rad	Radian	**Vol. 2:** 2
γ	Volume weight of the soil	kN/m^3	ML^{-3}	**Vol. 1:** 1 **Vol. 3** **Vol. 4:** 1, 2
γ_b	Partial safety coefficient on the force of the ULS of a pile tip	-	-	**Vol. 3** **Vol. 4:** 1
γ_{cr}	Partial safety coefficient on the load of the SLS of a shaft	-	-	**Vol. 3** **Vol. 4:** 1
γ_d	Volume weight of the dry soil	kN/m^3	ML^{-3}	**Vol. 1:** 1
γ_i	Volume weight of the layer i	kN/m^3	ML^{-3}	**Vol. 1:** 1
γ_r	Volume weight of backfill	kN/m^3	ML^{-3}	**Vol. 2:** 2
$\gamma_{r,\,d1}$	Partial safety coefficient on the resistance value	-	-	**Vol. 3** **Vol. 4:** 1
$\gamma_{r,\,d2}$	Partial safety coefficient on the resistance value	-	-	**Vol. 3** **Vol. 4:** 1
γ_{Rdv}	Partial safety coefficient on the characteristic resistance value under surface foundation	-	-	**Vol. 3**
γ_{Rv}	Partial safety coefficient at the SLS and ULS under surface foundation	-	-	**Vol. 3** **Vol. 4:** 1
γ_s	Volume weight of the solid grains	kN/m^3	ML^{-3}	**Vol. 1:** 1

γ_s	Partial safety coefficient on the lateral friction force of ULS of the pile in compression	-	-	**Vol. 3** **Vol. 4:** 1
$\gamma_{s,\,cr}$	Partial safety coefficient on the lateral friction force of the ULS of the pile in tension	-	-	**Vol. 3** **Vol. 4:** 1
$\gamma_{s,\,t}$	Partial safety coefficient on the lateral friction force of the ULS of the pile in tension	-	-	**Vol. 3** **Vol. 4:** 1
γ_{sat}	Volume weight of saturated soil	kN/m^3	ML^{-3}	**Vol. 1:** 1
γ_w	Volume weight of water	kN/m^3	ML^{-3}	**Vol. 1:** 1
γ'	Soil buoyant unit weight	kN/m^3	ML^{-3}	**Vol. 1:** 1 **Vol. 3**
γ'_i	Soil buoyant unit weight in layer i	kN/m^3	ML^{-3}	**Vol. 1:** 1
γ_c	Cyclic deformation (distortion)	%	-	**Vol. 1:** 2
γ_r	Reference cyclic deformation	%	-	**Vol. 1:** 2
$\gamma_{r\theta}$	Shear deformation in a plane (r, θ)	%	-	**Vol. 1:** 3
γ_m	Partial safety coefficient	-	-	**Vol. 1:** 2 **Vol. 2:** 2 **Vol. 4:** 2
γ_{s3}	Coefficient of method	-	-	**Vol. 1:** 2 **Vol. 2:** 2 **Vol. 4:** 2
γ, γ_θ	Distortion	radian	rad	**Vol. 1:** 2
γ_0	Reference distortion	radian	rad	**Vol. 1:** 2
γ_{oct}	Octahedral shear deformation defined by: $\gamma_{oct} = \frac{1}{3}\sqrt{(\varepsilon_1-\varepsilon_2)^2+(\varepsilon_2-\varepsilon_3)^2+(\varepsilon_3-\varepsilon_1)^2}$	%	-	**Vol. 1:** 2
$\vec{\gamma}$	Acceleration vector (components γ_1, γ_2, γ_3)	m/s^2	LT^{-2}	**Vol. 1:** 2
$\dot{\gamma}$	Distortion rate	s^{-1}	T^{-1}	**Vol. 1:** 2
ν	Poisson's ratio	-	-	**Vol. 1:** 2 **Vol. 3**
ν_u	Poisson's ratio of undrained soil	-	-	**Vol. 1:** 2
ν'	Poisson's ratio of drained soil	-	-	**Vol. 1:** 2
ψ	Angle of friction between grains: f = tanψ	rad	Radian	**Vol. 1:** 2
ψ	Angle of dilatancy	rad	Radian	**Vol. 1:** 2

ρ	Bulk density of the soil	t/m^3	ML^{-3}	**Vol. 1:** 1
ρ_d	Dry density of the soil	t/m^3	ML^{-3}	**Vol. 1:** 1
ρ_s	Density of soil particles	t/m^3	ML^{-3}	**Vol. 1:** 1
ρ_w	Density of water	t/m^3	ML^{-3}	**Vol. 1:** 1
ρ_p	Reducing coefficient of metal pile tip	-	-	**Vol. 3**
ρ_s	Reducing coefficient of friction of metal pile	-	-	**Vol. 3**
μ^*_0	Friction coefficient at the head of a "TA" wall	-	-	**Vol. 4:** 2
μ^*_1	Friction coefficient from 6 m in a "TA" wall	-	-	**Vol. 4:** 2
η	Viscosity coefficient (Maxwell damper)	kPa/s	ML^{-1}T^{-2}	**Vol. 1:** 2
σ	Normal stress	kPa	ML^{-2}	**Vol. 1:** 2 **Vol. 2:** 2 **Vol. 4:** 2
σ	Total stress in the soil	kPa	ML^{-2}	**Vol. 1:** 2 **Vol. 2:** 2
σ	Meyerhof stress under foundation	kPa	ML^{-2}	**Vol. 3**
σ_c, σ'_c	Stress at head of column	kPa	ML^{-2}	**Vol. 4:** 2
σ_{cl}, σ'_{clim}	Stress limit at head of column	kPa	ML^{-2}	**Vol. 4:** 2
σ_e	Elastic stress (limit)	kPa	ML^{-2}	**Vol. 1:** 2 **Vol. 4:** 1, 2
σ_g	Swelling stress (of a clay)	kPa	ML^{-2}	**Vol. 2:** 2
σ_h	Horizontal stress (or confined pressure round an inclusion)	kPa	ML^{-2}	**Vol. 2:** 2 **Vol. 4:** 2
σ_h	Horizontal stress acting on a retaining wall: $\sigma_h = K.\gamma z$	kPa	ML^{-2}	**Vol. 4:** 1
σ_{hl}, σ'_{hlim}	Lateral stress of the soil around a column	kPa	ML^{-2}	**Vol. 4:** 2
σ_{ha}	Horizontal active earth pressure: $\sigma_{ha} = K_a.\gamma z$	kPa	ML^{-2}	**Vol. 4:** 1
σ_{hp}	Horizontal passive earth pressure: $\sigma_{hp} = K_p.\gamma z$	kPa	ML^{-2}	**Vol. 4:** 1
σ_{h0}	Horizontal stress of the earth at rest: $\sigma_{h0} = K_0.\gamma z$	kPa	ML^{-2}	**Vol. 4:** 1
$\sigma_h(z)$	Horizontal stress as a function of z	kPa	ML^{-2}	**Vol. 3**

σ_{max}, σ_M	Maximum stress under foundation	kPa	ML^{-2}	**Vol. 3** **Vol. 4:** 2
σ_{min},σ_m	Minimum stress under foundation	kPa	ML^{-2}	**Vol. 3** **Vol. 4:** 2
σ_M	Stress at the peak of the block around the tunnel	kPa	ML^{-2}	**Vol. 4:** 3
σ_R	Stress mobilized in a spring (mechanical analogy)	kPa	ML^{-2}	**Vol. 2:** 2
σ_{oct}	Normal octahedral stress: $(\sigma_1 + \sigma_2 + \sigma_3)/3$	kPa	ML^{-2}	**Vol. 1:** 2
σ_r	Stress due to a backfill	kPa	ML^{-2}	**Vol. 1:** 2 **Vol. 2:** 2 **Vol. 3**
σ_r, σ_{rr}	Radial stress	kPa	ML^{-2}	**Vol. 1:** 2
σ_s	Residual stress on the soil	kPa	ML^{-2}	**Vol. 4:** 2
σ_v	Vertical stress in the soil: $\sigma_v = \gamma.z$	kPa	ML^{-2}	**Vol. 1:** 2 **Vol. 2:** 2 **Vol. 4:** 2
σ_{v0}	Total initial vertical stress in the soil	kPa	ML^{-2}	**Vol. 1:** 2 **Vol. 2:** 2, 3
σ_z	Vertical stress in the soil (depth z)	kPa	ML^{-2}	**Vol. 1:** 2 **Vol. 2:** 2
$\sigma_v(x)$	Vertical stress as a function of x	kPa	ML^{-2}	**Vol. 3**
$\sigma_v(z)$	Vertical stress as a function of z	kPa	ML^{-2}	**Vol. 4:** 2
σ_{va}	Vertical stress generating the active earth pressure	kPa	ML^{-2}	**Vol. 4:** 1
σ_{vp}	Vertical stress generating the passive earth pressure	kPa	ML^{-2}	**Vol. 4:** 1
σ_{v0}	Vertical stress in the soil at rest	kPa	ML^{-2}	**Vol. 4:** 1 **Vol. 4:** 2
σ_θ, $\sigma_{\theta\theta}$	Tangential stress	kPa	ML^{-2}	**Vol. 1:** 2
σ_0	Stress applied to the ground by the structure	kPa	ML^{-2}	**Vol. 4:** 2
σ_1, σ_2, σ_3	Principal normal stresses	kPa	ML^{-2}	**Vol. 1:** 2 **Vol. 2:** 2
$\bar{\sigma}$ (σ_{ij})	Stress tensor: i = 1–3, j = 1–3	kPa	ML^{-2}	**Vol. 1:** 2
σ’	Effective stress (in the soil grains)	kPa	ML^{-2}	**Vol. 1:** 2 **Vol. 2:** 2
$\sigma'_h(z)$	Effective horizontal stress around a shaft	kPa	ML^{-2}	**Vol. 3**

σ'_{ha}	Effective horizontal active earth pressure: $\sigma'_{ha} = K_a.\gamma' z$	kPa	ML^{-2}	**Vol. 4:** 1
σ'_{hp}	Effective horizontal passive earth pressure: $\sigma'_{hp} = K_p.\gamma' z$	kPa	ML^{-2}	**Vol. 4:** 1
σ'_{h0}	Effective horizontal stress of the earth at rest: $\sigma'_{h0} = K_0.\gamma' z$	kPa	ML^{-2}	**Vol. 4:** 1
σ'_p	Pre-consolidation stress	kPa	ML^{-2}	**Vol. 3**
$\sigma'_v(z)$	Vertical stress around a shaft	kPa	ML^{-2}	**Vol. 3**
$\sigma'_{v0}(z,r)$	Vertical stress at distance r from the shaft	kPa	ML^{-2}	**Vol. 1:** 2 **Vol. 2:** 2, 3
σ'_{va}	Effective vertical stress in active earth pressure	kPa	ML^{-2}	**Vol. 4:** 1
σ'_{vp}	Effective vertical stress in passive earth pressure	kPa	ML^{-2}	**Vol. 4:** 1
σ'_{v0}	Effective vertical stress of the earth at rest	kPa	ML^{-2}	**Vol. 4:** 1, 2
σ_v'	Effective vertical stress	kPa	ML^{-2}	**Vol. 1:** 2 **Vol. 2:** 2
σ_{v0}'	Initial effective vertical stress	kPa	ML^{-2}	**Vol. 1:** 2 **Vol. 2:** 2
τ, τ'	Shear stress ($\tau=\tau'$)	kPa	ML^{-2}	**Vol. 1:** 2
$\tau, \tau(z)$	Law of mobilization of lateral friction	kPa	ML^{-2}	**Vol. 3** **Vol. 4:** 2
τ	Shear stress at the soil/wall interface	kPa	ML^{-2}	**Vol. 4:** 1
τ_h	Earthquake-induced shear stress	kPa	ML^{-2}	**Vol. 4:** 2
τ_l	Shear stress limit at the soil/reinforcement interface	kPa	ML^{-2}	**Vol. 4:** 2
τ_l, τ_{lim}	Shear stress limit	kPa	ML^{-2}	**Vol. 2:** 2 **Vol. 4:** 2
τ_l ou τ_{max} (CRR)	Shear resistance ("Cyclic Resistance Ratio")	kPa	ML^{-2}	**Vol. 3** **Vol. 4:** 2
τ_m (CSR)	Mobilized shear stress ("Cyclic Stress Ratio")	kPa	ML^{-2}	**Vol. 3** **Vol. 4:** 2
τ_{max}	Maximum shear stress	kPa	ML^{-2}	**Vol. 1:** 2 **Vol. 2:** 2 **Vol. 4:** 1, 2
τ_R	Residual shear stress	kPa	ML^{-2}	**Vol. 2:** 2

τ_{oct}	Octahedral shear stress defined by: $\tau_{oct}=\frac{1}{3}\sqrt{(\sigma_1-\sigma_2)^2+(\sigma_2-\sigma_3)^2+(\sigma_3-\sigma_1)^2}$	kPa	ML^{-2}	**Vol. 1:** 2
τ_θ	Tangential shear stress	kPa	ML^{-2}	**Vol. 1:** 2
τ_0	Initial value of lateral friction τ, $\tau(z)$	kPa	ML^{-2}	**Vol. 1:** 2
$\tau*$	Equivalent shear stress	kPa	ML^{-2}	**Vol. 4:** 2
ω	Rotation of the current meter (scissometer)	rad	Radian	**Vol. 1:** 3
ω	Length reduction factor	-	-	**Vol. 4:** 2
ω_0	Reference rotation	rad	Radian	**Vol. 1:** 3

Introduction

From the Pharaohs of Egypt, through the Greek antiquity of Plato, the creator of the lines to measure land and Pythagoras, tracing triangles without a square[1], to Leonardo da Vinci[2] up until the present day, humans have not stopped innovating and building.

From the Pyramids of Egypt bulit more than 3,000 years ago, the Pont du Gard built more than 2,000 years ago, the Tower of Pisa (built in 1173) and the Mosque of Kairouan (erected in 1170), to the bridges of Millau in France (put into service in 2004) and Charílaos Trikoúpis Bridge in Rion Antirion in Greece[3], the towers of Paris-la-Défense in France, Manhattan in the United States and the Burj Khalifa in the United Arab Emirates[4], humans have not ceased to build heavy structures and buildings that the ground carries and supports for eternity!

These mega-projects have required, and continue to require, increasingly sophisticated means of handling. Our ancestors had no shortage of ideas for lifting or pulling heavy loads. The Pharaohs stacked bricks on top of each other to create work platforms while they were erecting the columns of the pyramid structure, and these bricks were removed once the building was completed. In ancient times, the Greeks used machines in the form of wooden assemblies: the first, which the Greeks called an "Acrobaticon", was a machine composed of stakes driven into the ground, two blocks (top and bottom) and several cables, some of which allowed the machine

1 Pythagoras' idea was to take three rulers: of 3, 4 and 5 feet which, when all joined at the ends, form a right triangle, a mathematically exact triangle given that: $5^2 = 3^2 + 4^2$.

2 The man with multiple passions: painter, architect, scientist, technician, etc. (Born in 1452 in Florence, Italy, and died in 1519 in Amboise, France)

3 A 2,500 m bridge on the sea in Greece, inaugurated in summer 2004 for the Olympic Games of Athens.

4 The highest tower in the world (850 m).

to be lifted, and others that could be operated by a crowd of men to lift heavy loads, like our cranes today. The second machine, which was used in the temples and in public works, like the construction machines of our time that are used to pull heavy objects, was called a “Banauson”.

In all projects, past or present, small or large, humans have sought harmony between soil and foundation. This involves assessing the criteria of the project, determining the resistance of the soil, verifying the presence or absence of water in the soil and establishing the foundation, taking into account the adjoining elements: this is “Geotechnics”.

In France, the first foundations of soil mechanics go back to Coulomb (1736–1806), by means of the fracture of soil, and the relationship between the stresses that prevail there and the soil’s own shear characteristics. In Dijon, H. Darcy (1803–1858) expressed the average flow velocity of water in soil as a function of the hydraulic gradient and the permeability of the soil. J. Boussinesq (1842–1929), the academic, developed elasticity to calculate the stresses and settlements under structures. A. Caquot (1948–1965), J. Kérisel (1969–1973) and E. Absi (1990) developed the active earth pressure-passive tables for the calculation of retaining walls. Louis Menard, through the invention of the pressuremeter in 1964, figured out how to simplify the mechanics of the soil and developed simple rules, in particular for the calculation of foundations.

As for the works devoted to the practice of soil mechanics and the calculation of foundations, Cassan (1980) deepened the techniques of testing, particularly on site, and also determined the geotechnical parameters on site, the choice of adequate parameters and the rules of calculation of foundations. Schlosser (1983) simplified the basic concepts of soil mechanics and its application to foundations. Philipponnat (1997) developed the basics of soil mechanics and the methods of sizing foundations and earth structures.

As an extension of these references, the present book is a practical manual. It is the product of the author’s university professorships, proven by a large number of practical applications and current projects dealt with by the author, his collaborators and his students in design offices and in end-of-studies work in various design offices. The latter are specialized in geotechnics and companies, and are the birthplace of the author’s knowledge: TERRASOL, via FONDASOL and SOLEN to VINCI. The work is also “cross-referenced” with the judicial expertise borrowings practiced by the author since the end of 2007 (oath before the first chamber of the Court of Appeal of Versailles on November 17, 2007).

The four volumes that make up this book have 12 chapters in total.

Volume 1, Chapter 1 is devoted to the definition, identification and classification of soils, using characteristic quantities and one-dimensional state parameters. This chapter is illustrated by six practical applications ranging, on the basis of simple laboratory identification tests, from particle size analysis to the "preliminary" characterization of the plastic and swelling potential of soils, the search for organic soils and the "qualitative" assessment of the risk of liquefaction.

Volume 1, Chapter 2 exposes the elementary bases of the mechanics of continuous media by explaining the concepts of stresses and deformations, as well as the fundamental relations of "stresses – strains" in plane deformations, plane stresses and axisymmetric conditions. It presents the rheological models of soil behavior most commonly used in practice (Mohr-Coulomb, hyperbolic model and Cam-Clay model) under static loads, and recalls the creep mechanism and the dynamic behavior of soils. It includes a theoretical application for the calculation of vertical stresses in a horizontal soil mass of homogeneous aquifer sand, a simplified analytical demonstration for the case of tensioned and bent beams to "popularize" the numerical resolution of soil problems by finite elements. Three practical numerical models in finite elements by "PLAXIS – 2D" of a highway embankment on soft soils ("soft soils with consolidation"), a retaining wall (Mohr-Coulomb model) with anchoring by active tie-rods and passive buttresses, and raft foundations, end this chapter.

Volume 1, Chapter 3 is designed for the determination of the geotechnical parameters of the soils that are essential to the study of the geotechnical projects of construction. This chapter introduces the general methodology and the geotechnical missions standardized in France (NF P94-500). After the inventory of the various tests of measurements of the geotechnical parameters, it presents and comments on all of the on-site tests: scissometer building site test ("vane test"), cone static penetration test (CPT) and dynamic test with the lost point (PDA and PDB) or with drill sampling; SPT ("standard penetration test"), with the Louis Ménard pressuremeter test (MPT), shearing test with the phicometer and test of the superficial load with the plate bearing test. This chapter presents the different soil classifications according to various references (Fascicle 62, Title V, Eurocodes 7 and 8, and PS92). It interprets typical real profiles from each test studied and gives several practical correlations between various modes of investigation undertaken on the same sites.

Volume 2, Chapter 1 discusses the hydraulic properties of soils. It underlines the basic concepts of water in soil, Darcy's law, Laplace's equation and the flow forces in soil. It details the principles of "large volume measurement" of soil permeability using the pumping test, and "local measurement" using the Lefranc test and injection using the Lugeon test. As applications, three real trials related to these three tests are presented, analyzed and commented on. This chapter ends with practical and graphical data on the permeability of some typical soils and piezometry.

Volume 2, Chapter 2 deals with the elementary bases of soil mechanics applied in geotechnics: from soil consolidation and swelling-shrinkage phenomena of clay soils, to the notion of short and long term, the shear strength of soils, the practical rules of slope stability calculation, according to the traditional calculations, and the Ultimate Limit States in comparison with the rules of the Eurocodes and the standards of national application of the Eurocode 7. Three practical applications, involving the justification of an embankment on soft soil, the analysis of a disaster on a pavilion related to the swelling of clays and the study of slope stability in purely granular and purely cohesive soil, enrich this chapter. Practical correlations between compressibility and shear parameters of soils make up this chapter.

Volume 2, Chapter 3, presents, analyzes and comments on examples of concrete projects, with or without damage, in the context of geotechnical construction projects, and how to seek simple foundation solutions with the objective of "economic realism". It defines the basis and focus of the geotechnical expertise ordered by a competent court ("*judicial expertise*"), and the role of the geotechnical expert appointed by this same court ("*judicial expert*"). This chapter concludes with some information, lessons and rules to remember, as well as promote, in geotechnical engineering in general.

Volume 3, Chapters 1 to 3, is the "backbone" of this book. It is devoted to the foundations of construction works. It introduces the analogies and differences between foundations. It presents shallow foundations on footings and rafts, and deep foundations on piles and slurry trench piles. For each type of foundation, the behavior of the soil under the foundations, the bearing capacity of the foundations, the calculation of settlements and special cases are developed. Five practical applications enrich this chapter, which include a residential building on shallow footings, an industrial building on deep pile foundations, a building on a raft foundation, a road embankment on soft soil and an engineering structure on piles subjected to parasitic pressure due to an adjacent earth embankment. In each application, "traditional" rules (DTU and fascicles) and "new" rules (Eurocodes and national application standards of Eurocode 7) are deliberately used for the design and justification of foundations, in order to compare different design and calculation methods and approaches, as well as to present and compare the "old" concepts, which are simpler to design, in comparison with the recent Eurocodes rules that are more difficult to understand, especially for students and young engineers.

Volume 4, Chapter 1 introduces and develops the notion of active earth pressure and passive earth pressure, analyzes and details the behavior of retaining structures, presents and comments on the sizing approaches by the traditional rupture calculation, the method based on the reaction coefficient and finite element modeling. The specific cases of shielded excavations and the stability of subgrades are then discussed. A panorama of five practical applications is successively discussed, including a weighted retaining wall in homogeneous soil under static and dynamic actions according to the "usual" and "new" rules (Eurocodes and national application standards of Eurocode 7), a sheet pile curtain by traditional rupture calculation, an excavation supported by passive buttresses and an excavation project supported by a diaphragm wall, as well as active tie-rods, all justified according to the usual calculations (TA 95 and TA 2020 Recommendations) and by the application of the recent rules (Eurocodes and national application standards of Eurocode 7). Illustrations of a sheet pile wall, damaged by rainfall, and of a diaphragm wall, which suffered excessive displacements and led to very costly damage, conclude this chapter, along with photos that are commented on in a personal and objective way.

Volume 4, Chapter 2 examines soil reinforcement and soil improvement techniques, the pillars of our research and teaching for the past 35 years. As a foreword, this chapter recalls the historical and geographical context of these techniques, the field and the limits of their applications. In the first part, this chapter develops the principle, behavior and sizing rules of the "Reinforced earth" and "nailing" processes. It studies the behavior of micropiles under axial and horizontal loads and, with respect to buckling, the group effect and design approach for micropiles. Two applications of a "mixed" wall, reinforced by metal strips in "reinforced earth", nails and a bed of active tie-rods and a crane on a raft foundation based on a group of micropiles are presented, analyzed and commented on. In a second part, a scope of *in situ* soil improvement techniques are discussed, as well as improvements in mass (vibro compaction, dynamic compaction) and inclusions: soil-cement columns by jet ("jet-grouting"), stone columns, stiff inclusions, lime-cement columns, as well as other processes such as consolidation by pre-loading or by vacuum, soil freezing and traditional injections. Some economic data and sizing approaches are provided. Three applications of highway embankment on stone columns, an industrial building slab on vertical stiff inclusions topped by stone columns and a study of the risk of liquefaction before and after soil improvement by stone columns are presented, analyzed and commented on. This chapter is also enriched by a "not to do" section, concerning a big disaster that had diligently carried out judicial expertise on five residential buildings, founded on footings resting on stone columns.

Volume 4, Chapter 3 deals with the behavior of underground structures in a concise manner. It analyzes the notion of convergence-confinement and the evolution of stresses around a tunnel, in addition to presenting the approaches of tunneling support and studying two concrete examples of calculation of settlements on the surface ("subsidence"). This subject is often of concern to judicial experts who are faced, particularly in preventive summary proceedings, with disorders in constructions following deep excavations and trenches (tunnels, work on underground networks).

The *normes nationales Françaises* (French national standards, NF), European standards (NF EN) and international standards (NF EN ISO), as well as the various reference documents in effect, are listed at the end of the book for reference purposes. The application of the Eurocodes is translated into transcriptions by the national standards bodies of each country: for example, Eurocode 7 – Part 1 is transcribed as DIN EN 1997-1 in Germany, NBN EN 1997-1 in Belgium, NF EN 1997-1 in France and BS EN 1997-1 in the UK.

In France, by virtue of the decree of January 26, 1984, the standard is considered as "a reference". It is therefore neither a law nor a regulation. Its application becomes compulsory when one party, among other signatories of the same foundation contract, has referred to it. However, the application of a standard may be mandatory when it is related to the prevention of risks, and damage to people and property. It should be noted, however, that once the client has provided a standard geotechnical mission, even minimal G1, in terms of standard NF P94-500 (Volume 1, Chapter 3), the signatories of the foundation contract (project manager, contractor and subcontractors) *are required* to carry out all of the missions that are incumbent on them. In this context, it is however necessary to distinguish the nature of the construction contracts:

– *Construction contract* (only): the project leader and their design team are the designers of the project, the company being the executor. Missions G1, G2 and G4 are the responsibility of the project leader and should preferably be carried out by the same geotechnician[5]. Mission G3 is entrusted to the company except for "contractual provisions contrary" to the contract (based on the PRO phase: G2), mission G3 should be carried out by a geotechnician that is independent from the company.

– *Design and build contract*: the project leader owes the cooperative teams a book of surveys and tests[6], meeting all of the criteria and requirements of the project in terms of foundations, low levels, retaining walls, slope, water, permeability, flow rates and jointness, and all of this done in the form of a factual report of surveys and tests, or within the framework of a minimal G1-type mission, for example. The

5 It is advisable to retain a single geotechnician for these three phases to guarantee the homogeneity of the geotechnical rendering and avoid contradictory "orders" from several "non-experts".

6 Sometimes, the project leader does not provide anything to the future designers of their project, and authorizes the design team to carry out the surveys and tests necessary for the cooperation, in return for remuneration on the overall contract or direct remuneration after the cooperation (case where the contract is not acquired).

design team takes charge of the G2 design mission. The company, even if it is part of the design team, ensures its G3 mission. The supervision of the G4 mission is always the responsibility of the project leader[7].

Finally, the same standards and rules in force must be applied to the various components of the construction project (soil and foundations, structure).

Throughout the book, we have made a point of presenting illustrated and documented practical examples, as well as a few examples of failures at the end of some chapters and in Chapter 3 of Volume 2, in order to provide geotechnical engineers, students, future beginners or experts with objective and straightforward testimonies, so as to highlight the following essential points:

1) Geotechnics is "Mother Nature", bigger and stronger than all, soil "defects"[8] are as abundant as those in humans. The engineer must keep in mind that geotechnical hazards and risks are potentially present at each stage of the construction project: from the preliminary design study (missions G1, G2-AVP) where there are high risks, and sometimes even major ones that question the feasibility of the project, via the project study (design phase or "PRO" phase: mission G2-PRO), where there are potential risks, and its execution (project execution phase or EXE phase: mission G3), up to its supervision and "life" in time (project geotechnical monitoring and supervision phase: mission G4), where geotechnical risks and hazards still "remain", because "0" risk does not exist in geotechnics.

2) Reducing geotechnical risks and hazards, first of all, requires geotechnical studies that are well adapted to the geotechnical context of the site, and to the project criteria. The engineer must use every capacity they have to convince the project leaders, who are new and eager to save a few thousand euros for geotechnical studies. This would make it possible to carry out well-adapted and serious studies in order to reduce these geotechnical risks and hazards, by applying and respecting all of the steps of the project development, and ultimately avoid burying concrete in the ground. In France and all over the world, although we talk a lot about expensive "non-quality" (more than 3% of the GDP in France), we do not talk enough about "over-quality"!

---

7 The case law considers that the "first one concerned" about the "Quality, Solidity, Stability, and Durability of the work" is the owner of the work, so it is up to them to ensure that their "baby" (project) grows well.

8 The notion of "soil defect" is notably used by judges and lawyers to designate an anomaly, a hazard, a "new fact" detected in the soil by the project actors. The term "defect" is often an argument put forward by lawyers to defend their clients in judicial expertise, in order to specify that, for example, the anomaly encountered in the soil is a hazard, a new fact, and therefore a "defect" that is inherent in the soil on the land belonging to the project owner, just like the "defect" that prevails there.

3) The geotechnical studies and investigations must be well supervised on site, with necessary adjustments according to the results obtained and the expected objectives. They must lead to a good summary of geotechnical data and calculation hypotheses, with the necessary hindsight according to the variability and dispersion of the results. Geotechnical calculations, no matter how simple or complicated, are useless if the basic inputs are based on vague and approximate assumptions. For example, what is the use of optimizing the bearing capacity factor (k_p) under a surface foundation (by trying to increase it from 0.80 to 0.90, or even 1.00) or for a pile (by setting it at 1.60 instead of 1.45) if the pressuremeter limits the results from tests that have been stopped "early", or carried out in remolded soil, or even badly interpreted? This is unfortunately the case in several projects, where we had to redo the tests in order to optimize the structures and avoid burying the concrete of shallow and deep foundations or retaining walls!

4) And finally, there is the famous execution of geotechnical projects! Even if the geotechnical investigations are well done on site, and the data and hypotheses of the geotechnical calculation models are well studied, finely analyzed, well interpreted, matured and reflected, it is the execution and the follow-up of the execution which are of prime importance. A pile that is poorly anchored in a bearing layer does not work the soil enough and ends up settling. A stone column floating under important loads leads to excessive settling of the structure. A diaphragm wall poorly anchored in a continuous and massive bedrock allows water to pass through and increases dewatering flows. Poorly sealed tie-rods in underwater sands do not work to their expected capacity and cause harmful displacements to the retaining walls and existing structures. All of the cases that we have just mentioned have led to very costly claims in France and have been the subject of long and endless "expert fights"! Their causes differ from one case to another. However, from the sometimes meager and ill-adapted on-site investigations, via incomplete theoretical studies, to the "bad" executions (faulty workmanship and/or no workmanship), the judicial experts seek the "truth" by dissecting the parts one by one, in order to technically assess the responsibilities of the parties and allow the judge of the merits to legally attribute the accountabilities and distribute the percentages!

5) In an untidy world, depending on the country and the continent, it is time to establish the principle of the "GMP (Guaranteed Maximum Price)", both in "private" and "public" contracts, so that the actors of the project reap the benefits of their "common" work: from the client who imposes and expects results to the contractor who puts in the means to guarantee the expected results. The "obligation of results" and the "obligation of means" cannot be imposed at the same time: in other words, we can impose the "obligation of results", but not the "obligation of means" to achieve them.

1

Soil: Definition, Identification and Classification

Made up of grains, water and air, soil is like man with his skeleton, blood and oxygen. The two come together in their complexity and "defects".

Ammar DHOUIB

1.1. Definition of a soil

Soil is a mixture of solid grains in which water and air circulate. This combination makes soil a heterogeneous and very complex material. Figure 1.1 illustrates the different components of a soil.

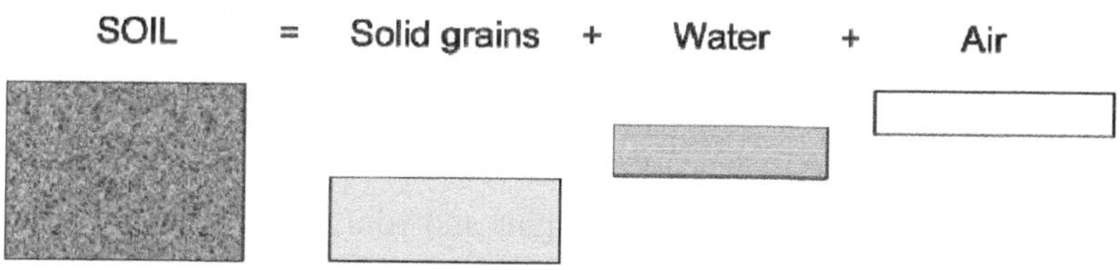

Figure 1.1. *Different components of a soil. For a color version of this figure, see www.iste.co.uk/dhouib/geotechnics1.zip*

1.1.1. *Characteristic sizes*

Figure 1.2 shows an elementary volume of soil where air occupies the volume V_a, water V_w and solid grains V_s. The solid grains and water have weights W_s and W_w, respectively. The air is non-weighted ($W_a = 0$).

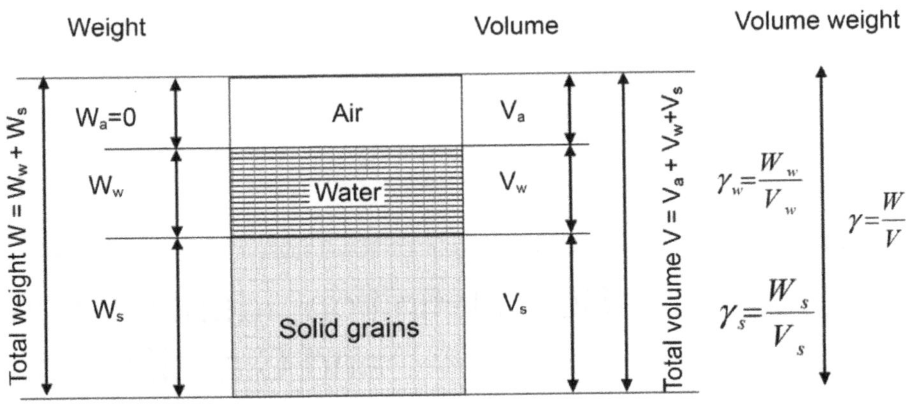

Figure 1.2. *Diagram of an elementary volume of soil*

The volume of voids (V_v) between solid grains is occupied by air and water, and is worth:

$$V_v = V_a + V_w \quad [1.1]$$

The total volume (V) is made up of three soil components:

$$V = V_a + V_w + V_s \quad [1.2]$$

The total weight of the soil sample (W) is composed of the weight of the water (W_w) and the solid grains (W_s), in other words, ($W_a = 0$):

$$W = W_w + W_s \quad [1.3]$$

1.1.1.1. *Specific weights*

Specific weights, depending on the weights and volumes made up by the soil and its components, are dimensional parameters that are as follows:

– Total specific weight of the soil (in the total volume occupied by solid grains, water and air), γ:

$$\gamma = \frac{W}{V} \quad [1.4]$$

– Specific weight of the solid grains (in the volume V_s made up by these grains), γ_s:

$$\gamma_s = \frac{W_s}{V_s} \qquad [1.5]$$

– Specific weight of dry soil (in the total volume initially made up by its three components), γ_d:

$$\gamma_d = \frac{W_s}{V} \qquad [1.6]$$

– Specific weight of the water in the soil sample, γ_w:

$$\gamma_w = \frac{W_w}{V_w} \qquad [1.7]$$

– Buoyant unit weight (of the ground subjected to the pressure of water: "Archimedes' force"), γ':

$$\gamma' = \gamma - \gamma_w \qquad [1.8]$$

1.1.1.2. *One-dimensional parameters*

The one-dimensional (or dimensionless) parameters quantify the different phases of a soil, as well as characterize its state of moisture (its water content and degree of saturation) and compactness (void index and porosity). These parameters are defined by the following:

– The water content (w) is the ratio of the weight of water (W_w) contained in the soil sample to the weight of solid grains (W_s) moistened by this water, in other words, in %:

$$w = \frac{W_w}{W_s} \times 100\% \qquad [1.9]$$

– The degree of saturation (S_r) of the soil is the ratio of the volume of water (V_w) to the volume of voids (V_v) occupied by this water in the soil sample, in %:

$$S_r = \frac{V_w}{V_v} \times 100\ \% \qquad [1.10]$$

For a saturated soil, all voids are filled with water and S_r is equal to 100%.

– The void index (e) gives an idea of the significance of the voids present in the soil sample and characterizes the state of its compactness. It is expressed as the ratio of the void volume (V_v) to the volume of solid grains (V_s) of the soil, in other words:

$$e = \frac{V_v}{V_s} \quad [1.11]$$

When the soil is in a dense state, the void index is low. On the other hand, it is high when the soil is loose, which can lead to a void volume greater than the solid grain volume and, therefore, a void index greater than 1.

– Porosity (n) is a parameter that is analogous to the void index, which refers to the ratio of void volume (V_v) to the total volume (V) of the soil sample, that is to say:

$$n = \frac{V_v}{V} \quad [1.12]$$

The porosity is always lower than the unit.

1.1.1.3. *Relations between dimensionless parameters*

Not all dimensionless parameters are independent, and the main relations between some of them are as follows:

$$n = \frac{e}{1+e} \quad [1.13]$$

$$\gamma = (1+w) \times \gamma_d \quad [1.14]$$

$$\gamma_s = \frac{1+e}{1+w} \times \gamma \quad [1.15]$$

$$\gamma_d = (1-n) \times \gamma_s \quad [1.16]$$

$$\gamma' = \frac{\gamma_s - \gamma_w}{\gamma_s} \times \gamma_d \quad [1.17]$$

1.1.1.4. *Order of magnitude of parameters*

1.1.1.4.1. Dimensional parameters: Specific weights

The specific weight depends on the compactness of the soil. Loose soil (such as sands) or compressible soil (fine soils) has a low specific gravity, compared to compacted soil[1], which has a higher specific weight. Table 1.1 shows orders of magnitude of the specific weights of some typical soils and other miscellaneous building materials.

	γ_{min} (kN/m³) (Not very compact soil)	γ_{max} (kN/m³) (Compact soil)
	"Normal" soils	
Sand	18	20
Sand and gravel	19	21
Silt	17	20
Clay	17	19
Marl	19	22
Marl-limestone	20	23
Chalk	17	20
	Organic or non-organic compressible soils	
Peat	12–14	-
Mud	14–16	-
	Other materials	
Weathered rock	20	23
Reinforced concrete	25	-
Steel	78.5	-
Aluminum	27	-
Lead	114	-
Bentonite	12	14
Sugar	9.5 (in bulk)	16 (in bags)
Water	10	

Table 1.1. *Order of magnitude of the total specific weight of soils and some other typical materials*

1 A soil that is compact or hard, as opposed to a soil that is not very resistant or load-bearing.

With the exception of soils with organic particles (peat, mud), the specific weight of solid grains (γ_s) varies in relatively small ranges where:

$$26\,kN/m^3 < \gamma_s < 28\,kN/m^3$$

1.1.1.4.2. Dimensionless parameters (one-dimensional parameters)

Table 1.2 summarizes some orders of magnitude of water content (w), void index (e) and porosity (n) of some typical soils:

Soil types	***w*** **(%)**	***e***	***n***
Silt	15–25	0.45–0.55	0.4–0.6
Clay	25–35	0.5–0.9 (1)	0.3–0.5
Sand	8–12	0.5 (2)–0.9 (3)	0.35–0.45
Peat	200–600	High: 2–9	0.8–1.0
Mud	70–120	1–2	0.7–0.8
Marl	10–30	0.4–0.6	0.35–0.45
Chalk	25–30	0.6–0.9	0.4–0.5

(1) 3.38–13.5 for Mexico clay (volcanic origin, n = 0.77–0.93); (2) Dense sand; (3) Loose sand.

Table 1.2. *Dimensionless parameters of typical soils*

COMMENTS.–

1) The dimensionless parameters are very important because they allow us to understand the state of soil compactness (dense or loose grains) in which the solid skeleton of the soil is found, and the quantities of water and air that the soil contains.

2) These are general data to give an order of magnitude of the phenomena; laboratory identification tests are necessary for soil samples taken on site.

1.2. Soil types

According to the size of the grains, we can distinguish between:

1) fine soils;

2) granular soils;

3) intermediate soils.

Figure 1.3 illustrates the two types of soils according to the increasing size of their constituent grains, namely:

– first, fine soils (soils with small particles: <0.06 mm);

– second, granular soils (grain sizes increasing from 0.06 mm to bigger than 60 mm).

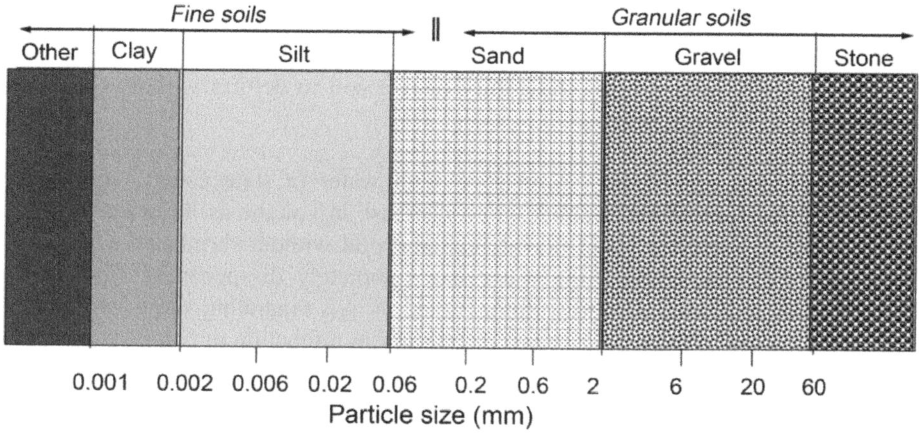

Figure 1.3. *Soils according to particle size*

1.2.1. *Granular soils*

A granular soil consists of grains with a diameter greater than 0.06 mm. Between these grains, there are voids in which water can circulate freely without being absorbed. Thus, the geotechnical properties of granular soils, such as sands, are almost identical whether these sands are dry, wet or saturated. These properties essentially depend on the size of the grains and the state of their compactness: loose (non-tight grains) or dense (tight grains).

1.2.2. *Fine soils*

Fine soils are composed of fine particles that are smaller than 0.06 mm, arranged in layers relative to one another. The compactness of fine soils mainly depends on their mineralogical composition (findings on content of water-sensitive minerals), their water content and their structure.

Depending on the water content of a fine soil, its mechanical properties evolve from those of a solid (low water content) to those of a liquid (high water content). This evolution is defined in a "conventional" way by limits as follows:

– limit of liquidity, noted as w_L, which separates the liquid state from the plastic state;

– limit of plasticity, noted as w_P, which separates the plastic state from the solid state with water.

The plasticity limit reflects the ability of fine soil to deform without breaking, like playdough between fingers.

The fine soil can be in a solid state with water (a state called "solid with shrinkage", in other words, a solid with a decrease in volume as the water escapes from the soil) or without water (a state called "solid without shrinkage", without a decrease in volume because the water has completely disappeared). Thus, it is customary to introduce a solidity limit, noted as w_S, separating these two states. Figure 1.4 illustrates the various states of fine soil as a function of increasing natural water content (w) in the soil.

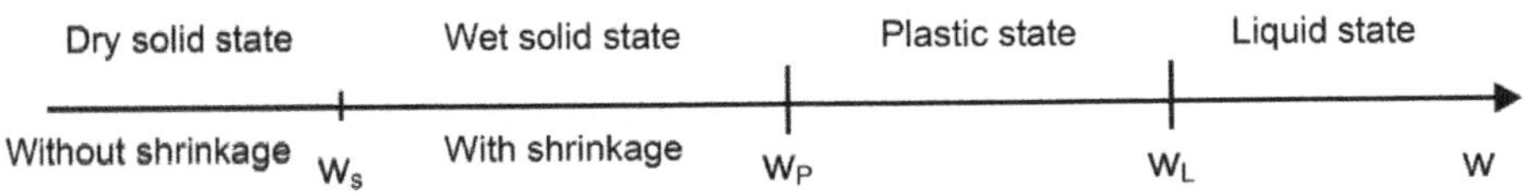

Figure 1.4. *Various fine soil states*

Fine soil is characterized by its plasticity index I_P, which is the difference between the two main limits of liquidity (w_L) and plasticity (w_P), namely:

$$I_P = w_L - w_P \qquad [1.18a]$$

This parameter is very important because it characterizes the rate of plasticity of the soil and its swelling potential (section 1.4.5, presented later in this chapter and in section 2.4 of Volume 2, Chapter 2 of this series of books). The liquidity limit (I_L) is also defined as:

$$I_L = \frac{w - w_P}{w_L - w_P} = \frac{w - w_P}{I_P} \qquad [1.18b]$$

The limits that describe the fine soil state in Figure 1.4 are known as *Atterberg limits*.

Table 1.3 lists characteristic value ranges of the Atterberg limits (w_L and w_P) and plasticity index (I_P) for some soil types.

Soils	w_L	w_P	I_P
Silt from Orly	35–40	22–28	10–14
Green clay (Sannoisian)	60–70	30–35	40–50
Plastic clay (Sparnacian)	70–85	30–40	40–60
Provins clay	100–200	50–100	50–100
Flanders clay	80–100	40–50	40–50
Boston clay	40–47	22–25	18–22
London clay	70–80	25–30	45–50
Mexico clay	400–500	350–400	100–150
Peat from Bourgoin (Isère)	66	37	29
Mud from Gran (Morbihan)	125	43	82
Mud from Martrou	83	38	45

*These are general data for the order of magnitude; laboratory identification tests (identifying the nature of the soil) are necessary for samples taken on site.

Table 1.3. *Orders of magnitude of the characteristic parameters**

1.3. Laboratory identification tests

1.3.1. *Parameters and standards*

There are many soil identification tests. The most important are as follows:

– determination of the water content w (Standards NF P94-049 and 050);

– measurement of the specific weight: γ, γ_d or γ_s (Standards NF P94-053, 054 and 059);

– grain size analysis by sieving (Standards NF P94-041 and 056);

– grain size analysis by sedimentation (Standard NF P94-057);

– measurement of the value of the VBS methylene blue (Standard NF P94-068).

The following tests are specific to granular soils:

– the equivalent of ES sand (Standards XP P18-597 and 598);

– research on the relative density D_r (function of the maximum and minimum void indexes), also called “Density index”: I_D.

And the following are specific to fine soils:

– measurement of the Atterberg limits (Standards NF P94-051 and 052-1);

– mineralogical analysis (Standard NF P18-592);

– research on organic material content (Standard NF P94-055);

– measurement of the calcium carbonate content CO_3Ca (Standard NF P 94-048).

1.3.2. *Definition of parameters*

1.3.2.1. *Grading*

As shown in Figure 1.3, not all soil grains are the same size. In order to determine the distribution of soil grains, we proceed to the study of soil grading, in other words, to determine the distribution of grains according to their size. Particle size analysis (Standards NF P94-041 and 056) consists of superimposing a series of sieves with a decreasing diameter from top to bottom in order to let the grains (or "aggregates") pass. The large grains remain at the top of the series of sieves, while the small ones drop down to the lowest sieves.

For fine soils, this method is not valid and a procedure based on Stokes' law is used, which gives the limiting velocity of a soil particle falling under the action of gravity in a liquid. The particles with a large diameter fall faster than those with a small diameter.

The results of the particle size analysis are plotted on a *grading curve* (Figure 1.5), characterizing, as a function of grain size D, the percentage by weight of grains of size D_p (diameter D corresponding to percentage p) less than D. From this curve, we can define:

– The *uniformity coefficient* (called *Hazen's coefficient*) by the ratio:

$C_u = D_{60}/D_{10}$, which translates as:

- a uniform grading when C_u is less than 2; and

- a grading that is spread out if C_u is higher than 2.

– The curvature factor C_c, which is the ratio:

$$C_c = \frac{(D_{30})^2}{D_{10} \times D_{60}},$$

A factor describing well-graded embankments when it varies from 1 to 3 and $D_{60}/D_{10} < 4$.

COMMENTS.– When grain sizes are close (uniform grading: $C_u < 2$), the network of *Laboratoires des Ponts et Chaussées* adopts the following classification:

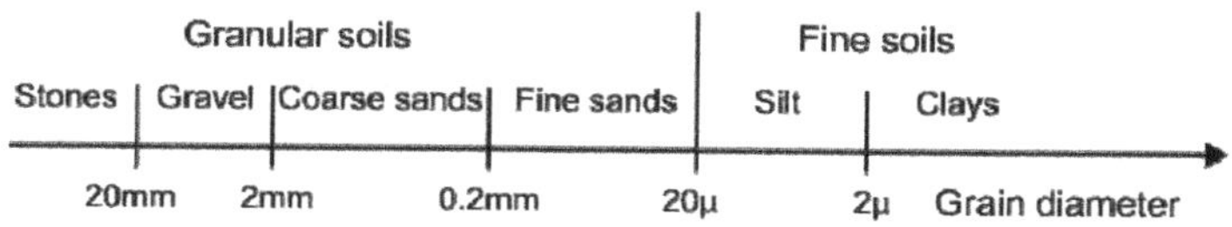

– In this classification, silts and clays differ only in their plastic potential, defined by the Atterberg limits (Figure 1.6).

1.3.2.2. *Methylene blue value test (Spot method)*

The VBS methylene blue test (standard NF P94-068) of a soil consists of injecting a solution of methylene blue until a blue peripheral aureole appears around the clay particles. The VBS methylene blue value is expressed in grams of methylene blue for 100 g of materials, and thus characterizes the degree of clay in the soil (Table 1.4).

1.3.2.3. *Sand equivalent test*

The sand equivalent test ES (standard XP P18-597 and 598) makes it possible to determine the relative proportion of fine soil and coarse soil in a soil. Carried out on elements of soil lower than 5 mm, it consists of dispersing the particles of the soil specimen, letting them rest and measuring the height (h_1) of the solid deposit of granular soil and the height (h_2) of the flocculant of fine soil, in order to define the sand equivalent by the following percentage:

$$ES = \frac{h_1}{h_2 - h_1} \times 100$$

This is the percentage with which we distinguish sands from clays (Table 1.4).

1.3.2.4. *Relative density (or Density index)*

The relative density D_r (also noted as density index, I_D) gives an idea of the density state a granular soil is in, through the two void indices e_{max} and e_{min} measured in the laboratory (equation [1.11]), as shown in Table 1.4.

1.3.2.5. *Organic soils*

Organic soil is soil that contains a significant proportion of more or less decomposed debris of animal or plant origin. It is often characterized by its high compressibility and a characteristic odor. Purely organic soil is peat, a vegetable debris form of deposit that has a fibrous structure, with a brown or blackish color. Peat is characterized by high levels of organic material, such as 55% for the peat of Bourgoin (Isère) and 45% for the peat of Lamentin (Martinique). Soils with organic content generally exceeding 12% are considered to be organic.

Figure 1.7 shows the results of tests for organic materials from a low-organic muddy silt and peat (organic).

1.3.2.6. *Soils containing calcium carbonate $CaCO_3$*

Soils containing $CaCO_3$ are classified in Table 1.4. This table presents orders of magnitude of some characteristic parameters from soil identification.

Parameter	Values	Soils
VBS	<0.2	Sand
	0.2–2.5	Silt
	2.5–6	Clay silt
	6–8	Clay soil
	>8	Very clayey soil
ES (%)	100	Pure and clean sand
	40	Non-plastic soil
	20	Plastic soil
	0	Clay
$D_r\ (I_D)\ (=\dfrac{e_{max}-e}{e_{max}-e_{min}})$	<20	Very loose sand
	20–40	Loose sand
	40–60	Medium sand
	60–80	Dense sand
	80–100	Very dense sand
Calcium carbonate content $CaCO_3$ (%)	0–10	Clay or silt
	10–30	Marly clay or marly silt
	30–70	Marl
	70–90	Marly limestone
	90–100	Limestone

Table 1.4. *Orders of magnitude of the characteristic parameters*

From the laboratory identification tests, we can proceed to the classification of loose soils and rocks, a classification presented in the tables provided in Appendix 1.1.

1.4. Examples of soil identification and applications

1.4.1. *Particle size analysis*

Remember that not all soil grains are the same size. For granular soils, particle size analysis is an important test because it makes it possible to determine the distribution of the grains. A series of sieves is set up with sieve openings of decreasing size. The soil sample is placed in the top sieve and the grains fall through the other sieves. This allows a grading curve to be drawn (Figure 1.5).

In this test, we define Hazen's coefficient, which is the ratio of the diameter corresponding to 60% of aggregates (D_{60}) through the sieves to the diameter corresponding to 10% of aggregates (D_{10}). This coefficient makes it possible to check if the grading of the grains is uniform or of different grain sizes.

Figure 1.5 shows two typical grading curves obtained in granular soils. From the dotted curve obtained in sandy-granular alluvium, it is apparent that:

– D_{60} = 6 mm;

– D_{10} = 0.02 mm; hence, C_u = 300 (>2): grading that is spread out.

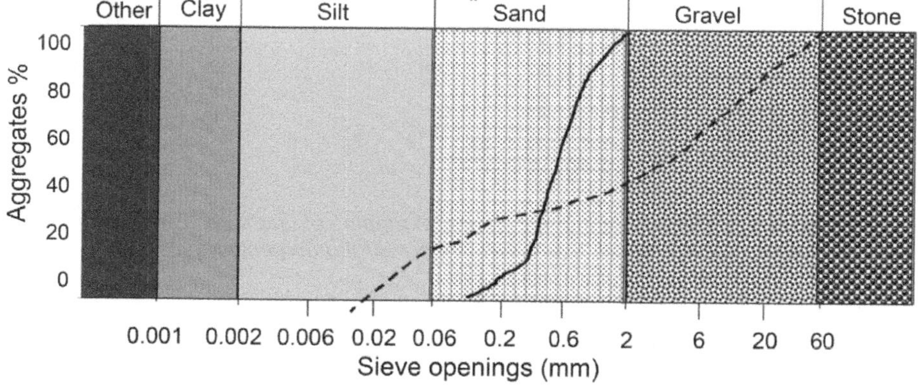

Figure 1.5. *Examples of soil particle size analysis*

The continuous curve corresponds to Fontainebleau sands with a dense and uniform grading, given that:

– $D_{60} = 0.35$ mm;

– $D_{10} = 0.20$ mm; hence, $C_u = 1.75$ (<2): uniform grading.

The percentage of aggregates beyond a sieve diameter of 0.08 m (80 mm) is called the *percentage of fines*.

Grading can help to understand the permeability of soils and the problems of liquefaction under earthquake, at least in qualitative terms.

1.4.2. *Characterization of soil plasticity*

Figure 1.6 shows the Casagrande plasticity diagram (Table 1.3), with the characterization of typical fine soils in France and the domain (green circles) of swelling and shrinking soils (see also section 2.4 of Volume 2, Chapter 2).

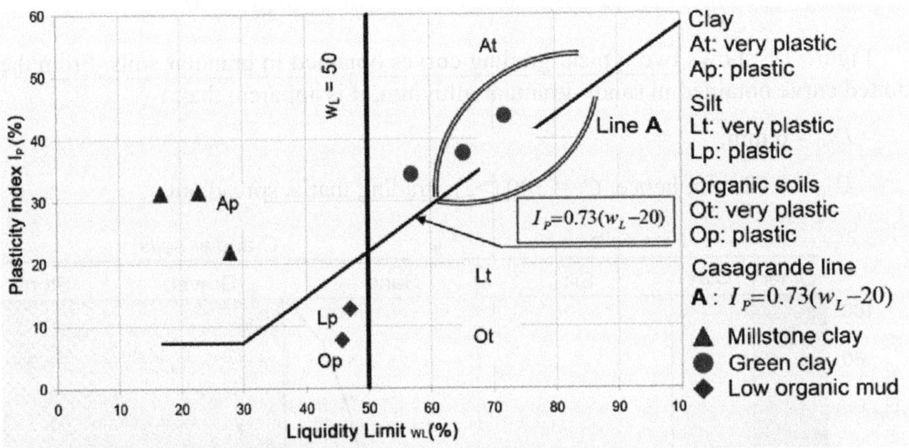

Figure 1.6. *Casagrande diagram and examples of fine soils. For a color version of this figure, see www.iste.co.uk/dhouib/geotechnics1.zip*

1.4.3. *Organic material content research*

The graph in Figure 1.7(a) shows the content of organic material measured in clay mud and peat (organic).

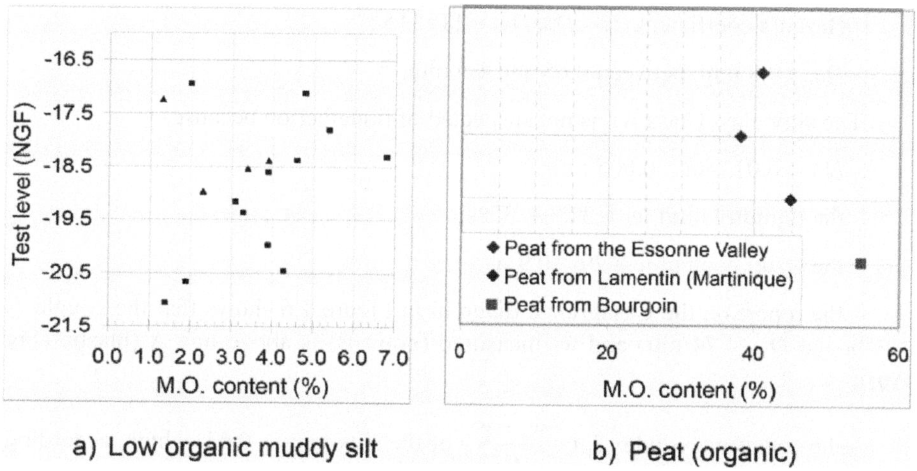

Figure 1.7. *Research of organic materials. For a color version of this figure, see www.iste.co.uk/dhouib/geotechnics1.zip*

The content of organic material measured in this silt is low and this type of material can be improved, contrary to peat, by "flexible" columns such as sand spikes or stone columns (see also section 2.6.4 of Volume 4, Chapter 2).

1.4.4. *Liquefaction risk analysis*

Table 1.5 summarizes laboratory test results at various depths (z) for analysis of the soil liquefaction risk under earthquake conditions. Soil liquefies when its grains disperse in water and lose resistance.

Soils	z **(m)**	w	w_L	I_P	S_r	C_u	D_{15}	D_{50}	**Class**
Silty sands	3.0	10	55	-	98	11	-	0.8	B_1
Silty clay	7.0	29	31	24	-	-	0.01	-	A_2

Table 1.5. *Laboratory test results (parameters defined and commented on below)*

From the data in Table 1.5, it can be concluded that:

– Silty sands, Class B_1 according to the tables in Appendix 1.1, are suspected of liquefaction because:

- the sands are quasi-saturated: degree of saturation $S_r = 98\%$;

- Hazen's coefficient $C_u = D_{60}/D_{10} = 11 < 15$;

- $D_{50} = 0.8$ mm, between 0.05 and 1.5 mm.

– The silty clay, Class A_2, is not suspected of liquefaction because:

- $D_{15} = 0.01$ mm > 0.005 mm;

- the liquidity limit $w_L = 31\% < 35\%$;

- the water content $w = 29 > 0.9\ w_L$;

- the report on the Casagrande diagram in Figure 1.6 shows that the couple I_P (>10% and D_{70} < 74 mμ) and w_L measured (point B) is above line A (plastic clay (Ap));

- the consistency index I_c ($= \frac{w_L - w}{I_P}$) of the silty clay is 0.63, which, according to Table 1.6, qualifies it as a clay of firm consistency.

I_c	**< 0.00 0.00**	**0.25**	**0.50**	**0.75**	**1.00**	**> 1.00**
Consistency	Liquid	Very soft	Soft	Firm	Very firm	Hard
State	Liquid		Plastic			Solid

Table 1.6. *Clay soil consistency class*

1.4.5. *Characterization of clay soils with respect to swelling and shrinkage*

The phenomena of swelling–shrinkage (a phenomenon identical to a sponge, which swells in water and shrinks once emptied of its water) of clayey soils are very widespread in France, as in other countries of the world (see also section 2.4 of Volume 2, Chapter 2). Justice experts are often confronted with these phenomena, which can sometimes be very expensive for the insurers. The traditional approach, adopted by geotechnicians and justice experts who are specialized in geotechnics (Heading C.125, according to usual court nomenclature), consists of having samples taken by specialized soil facilities for laboratory identification tests. These samples and laboratory tests must lead to specify:

– the nature of the soil through the plasticity index I_P, the percentage of fines (aggregates at 80 μm, noted as *f*%) and the methylene blue value (VBS): Table 1.7;

– the consistency range of clays through the consistency index I_c: Table 1.8.

Nature of the soil	I_P ($w_L - w_P$)	f% (aggregates at 80 μm)	VBS
Very clayey	>30	>90	>6
Moderately clayey	$15 < I_P < 30$	>50	2 < VBS < 6
Not very clayey	<15	<50	<2

Table 1.7. *Nature of the soil in terms of clay content*

Consistency range	$I_c = (w_L - w)/I_P$	Remarks
Very plastic consistency	<0.5	High water content
Plastic consistency	$0.5 < I_c < 1.0$	Medium water content
Firm consistency	>1.0	Low water content

Table 1.8. *Consistency range of clays*

The instability class of the clay according to parameters of plasticity and shrinkage (see also Table 2.10 and Table 2.11 of Volume 2, Chapter 2) makes it possible to determine the instability class of the clays with respect to the phenomena of shrinkage–swelling.

1.4.6. *Soil treatment for platforms and technical backfills*

1.4.6.1. *Proctor tests*

Backfills, labeled as "technical backfills" because they obey technical rules of installation and densification, are implemented by compaction. Compacting a soil means reducing its volume by a vibrating load, pressure or shocks. The degree of soil compaction depends on:

– the compaction energy;

– the compaction method; and

– the water content of the soil.

By the same compaction method used to demonstrate the importance of water content in fine and intermediate soils, in 1933, Proctor defined (AFNOR 1999) the variations of the specific weight of a dry soil γ_d (equation [1.6]) as a function of the water content w (equation [1.9]). Figure 1.8 represents the results of two Proctor tests on a clayey sand, classified as B5m, and a crushed and sifted marl-limestone at 20 mm, classified as R22/R23, and shows that the dry specific weight (γ_d) passes through a maximum value for a so-called "optimum" water content: w_{OPN}.

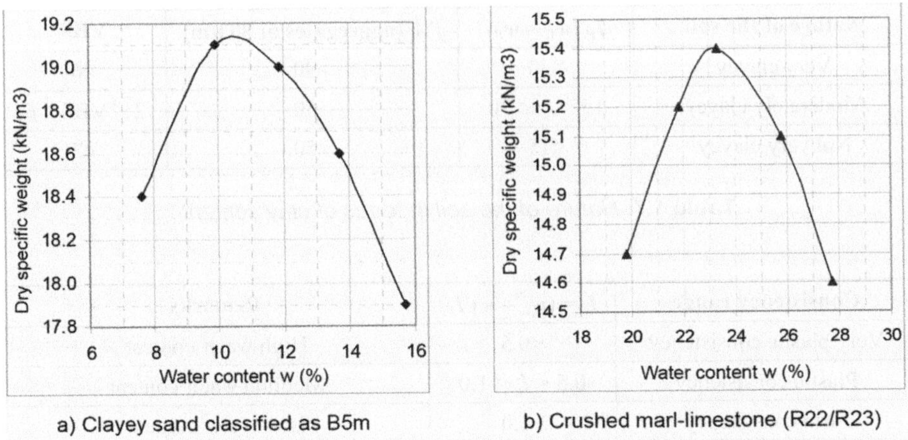

Figure 1.8. *Proctor tests on sand and marl. For a color version of this figure, see www.iste.co.uk/dhouib/geotechnics1.zip*

The Proctor test consists of compacting a soil sample in a mold in layers with a tamper. It includes two tests:

– the Standard Proctor test (called "OPN") for the study of backfill layers, where the weight of the tamper is 2.480 kg, the drop height is 305 mm and the number of soil layers to be compacted is 3;

– the Modified Proctor test (known as "OPM") for carriageway layers, with a tamper weight of 4.535 kg, and a drop height of 457 mm on five soil layers.

To characterize the bearing capacity of a compacted soil, a bearing capacity study is carried out by a specific test called a "Californian Bearing Ratio" (CBR), among others. This test consists of hammering a soil sample in a mold with a given state of density and water content, with a 19.3 cm^3 surface piston at a velocity of 1.27 mm/min, and of determining the CBR index conventionally defined by:

$$CBR = Max(\frac{P2.5}{0.7};\frac{P5}{1.05})$$

where $P2.5$ is the pressure at 2.5 mm of piston depression and $P5$ is the pressure at 5 mm of piston depression.

Controlling the compaction of the soils and the possibilities of their re-use in backfill can combine with the Proctor-CBR studies by drawing the curves:

– the specific weight of the dry soil γ_d as a function of the water content;

– the CBR index (ICBR) as a function of the density γ_d for two compaction energies constituting the energy of the standard Proctor;

– the CBR index (ICBR) as a function of the water content for each compaction energy.

To be able to appreciate the quality of the soils at their natural water content, the hammering is done on the natural soil received so as to define the Immediate Bearing Index, known as IPI.

In current practice, three types of indices are determined by mold tests in order to assess the bearing capacity of compacted backfills and layers, as well as their ability to support loads:

– the immediate CBR index (ICBRi), which makes it possible to check the bearing capacity of the prepared and compacted soil at different values of water content w (%);

– the CBRi index after immersion, which makes it possible to qualify the bearing capacity of the hammered soil in a soaked state (the soil and mold are immersed in water for 4 days, for example) in order to consider a soil under heavy rain;

– the IPI, which makes it possible to assess the bearing capacity of the soil received in its natural state (natural water content).

COMMENTS.– The IPI is obtained by way of hammering at a velocity of 1.27 mm and plotting a "stress-strain" curve for indentations of 1.25 mm, 2.0 mm, 2.5 mm, 5.0 mm, 7.5 mm and 10.0 mm.

– The ICBRi is conducted in the same way as the IPI, but is done by storing two overloads and after 4 days of immersion.

– In France, IPI and ICBR are more widely used than ICBRi, which is only rarely used.

Another test intended for the control of compacted platforms and backfills is the plate-bearing test, presented and studied in section 3.4.7 of Chapter 3.

In this application, we present two examples of soil treatment with lime and cement for railway embankments:

– example of sand, classified according to the GTR guide as B5m, and treated with 1%–3% lime;

– example of 20 mm crushed and sifted marl-limestone classified as R_{22}/R_{23} according to the GTR guide (classified as A_1 fine soil, once crushed), and treated like sand, with 1%–3% lime.

1.4.6.2. *Example 1: case of clayey sands used as technical backfill*

The material for this technical backfill project is a more or less brown clayey sand, with the following laboratory analysis and identification:

– Natural water content: $w = 10.0\%$;

– Maximum grain diameter: $D_{max} = 20$ mm;

– 2 mm aggregates: 89.0%;

– Aggregates <0.08 mm (fines): 19.0% (sand with fines);

– Methylene blue value: VBS = 1.30 (more or less clayey sand; sand is clayey when the methylene blue value VBS is higher than 1.5, Table 1.4).

Thus, these identification parameters make it possible to classify the more or less clayey sands as "B5m" according to the GTR classifications (SETRA-LCPC 2000). The sands, placed in successive layers that were 0.35 m thick and compacted to 98.5% of the Optimum Proctor Standard (OPN), were treated with lime (CaO) and load-bearing tests were carried out in order to measure the Immediate Proctor Index (IPI) according to the water content. Figure 1.9(a) shows the results obtained.

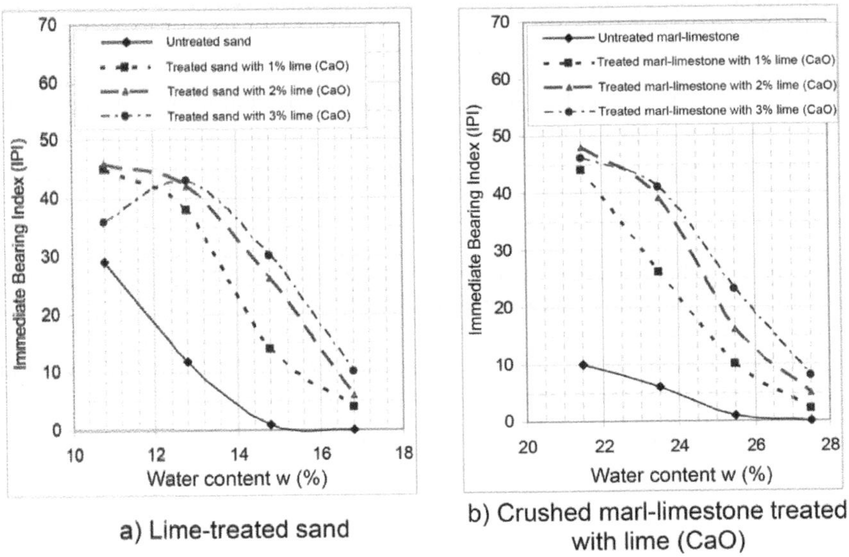

Figure 1.9. *Bearing capacity of soils treated with lime for use as technical backfill. For a color version of this figure, see www.iste.co.uk/dhouib/geotechnics1.zip*

1.4.6.3. *Example 2: case of crushed limestone*

Marl-limestone that is crushed and sifted to a maximum diameter of 20 mm is classified as "R22/R23" (Class A_1 when crushed) for an initial water content of 14.5%, a percentage of fines of 38% and a methylene blue value VBS of 1.15.

The lime treatment of these materials is conducted on samples with an initial water content of 21.5%, with lime treatment percentages ranging from 1% to 3%. The 21.5% water content corresponds to the maximum specific weight (w_{OPN} water content) of the marl-limestone without treatment. The tests were subsequently conducted at water contents: w_{OMN}, w_{OMN} +2, w_{OMN} +4 and w_{OMN} +6, in other words, at 21.5%, 23.5%, 25.5% and 27.5%.

Figure 1.9(b) shows the IPI as a function of water content, ranging from 23.5% to 27.5%. The two graphs in Figure 1.9 show that the 20 mm crushed and sifted marl-limestone has poor bearing capacity without treatment. With treatment, the sands and the marl-limestone have more or less similar bearing capacities.

It should be noted that the studies conducted on the sands and the marl-limestone treated with 1% and 2% of lime have made it possible to use them in technical blocks serving as a base for high embankments, implemented with a compaction energy which is, according to the GTR guide (LCPC-SETRA 1992a), Type q4 (material compacted to 95% of the OPN) or Type q3 (98.5% of the OPN), depending on the mechanical criteria conventionally required in compression resistance (R_c) for high embankments (h_r):

– $R_c > 0.5$ MPa for $h_r > 10$ m;

– $R_c > 0.8$ MPa for $h_r > 15$ m;

– $R_c > 1.0$ MPa for $h_r > 20$ m;

– $R_c > 1.2$ MPa for $h_r > 25$ m.

Laboratory shear tests were conducted on the lime- and cement-treated sands, and the lime-only treated marl-limestone to measure their shear characteristics (see Figure 2.55 in Volume 2, Chapter 2).

On the treated sands and marl-limestone put in technical backfills, pressuremeter tests were performed on the entire height of the technical backfills. Figure 3.60 in Chapter 3 represents the pressuremeter tests that were carried out on the technical backfills that were implemented from the treated sands and marl in Figure 1.9, in order to verify the compactness of these embankments and their capacity to support railway loads.

1.5. Soil classification

Before any work, especially earthwork, the identification and classification of soils are essential.

The classification of soils in France is described in the standard NF P94-300 (AFNOR 1992). Its use in current practice is detailed in the "*Guide Technique pour la Réalisation des remblais et des couches de forme*" (Technical Guide for the Construction of Embankments and Subgrades) (LCPC-SETRA 1992a).

This classification, referred to in this guide as the "GTR Classification", conventionally includes six major soil families (see Appendix 1.1), which are briefly described in Table 1.9 below.

Soil class	Sub-class	Soil type	Characteristics
A: Fine soils	A_1–A_4	Silt, loess*, fine sand, clay, marly clay	$D_{max} \leq$ 50 mm, and from 35% to 100% fines (1)
B: Sandy and granular soils with fines	B_1–B_6	Silty sand, not very clayey, clayey gravel, sandy, silty	$D_{max} \leq$ 50 mm and less than 35% fines
C: Heterogeneous soils with fine and coarse elements	C_1 or C_2 – A_1/B_1	Flint clay, millstone clay, sloping scree, moraine, coarse alluvium, flinty mills	D_{max} > 50 mm and with ± 12% fines
D: Clean sand and gravel resistant to water	D_1–D_3	Alluvial sand, dune sand, alluvial gravel	% of fines < 12 and VBS $\leq$ 0.1
R: Rocks	R_1–R_6	Chalk, limestone, marl, shale, sandstone, pudding, granite, gneiss, etc.	Density, hardness, fragmentability, degradability, etc. (2)
F: Natural organic soils and industrial by-products	F_1–F_9	Peat, mud, topsoil, coal shale, industrial products and waste	Organic content, water content and visual investigation
Remarks	(1) 0.076 mm aggregate (rounded to 80 μm). (2) Standard NF P11-300.		

* Loess is a very fine silt.

Table 1.9. *Soil classes according to the GTR guide in France*

Table 1.9 can be simplified into the synoptic form shown in Figure 1.10.

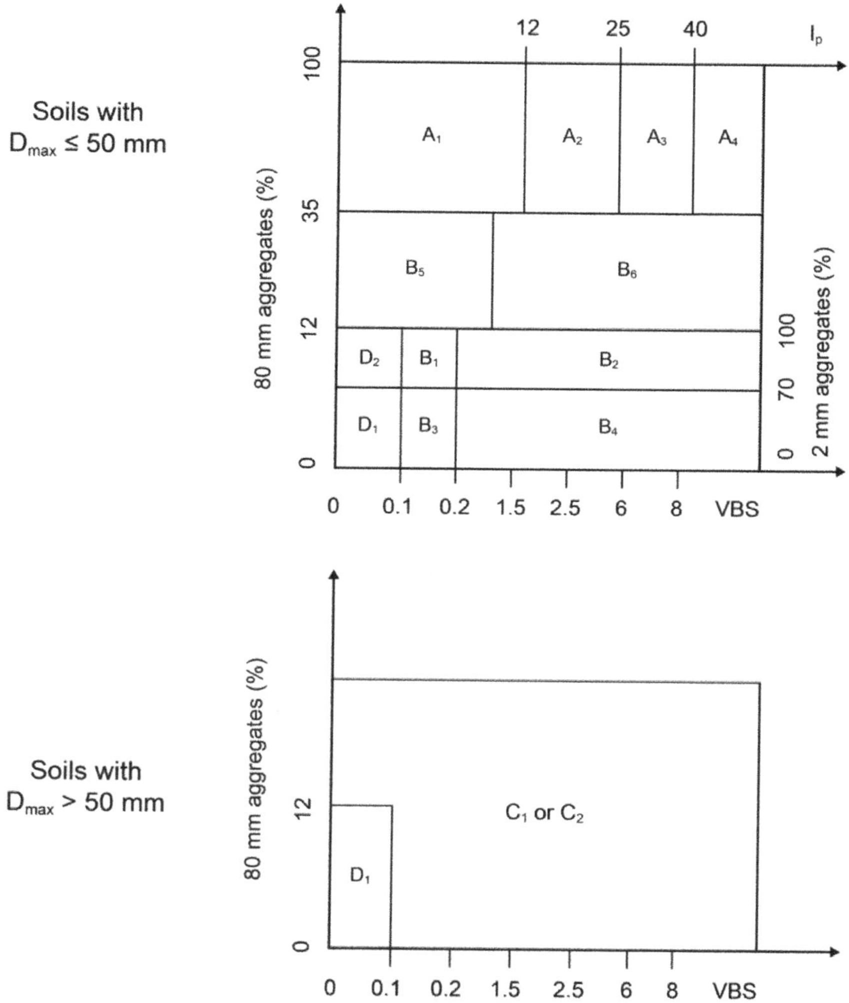

Figure 1.10. *Synoptic graph of soil classification*

Moreover, based on the particles sized between 2 and 20 μm, Demolon (1948) established a ternary diagram to classify the silts, sands and clays, as well as the intermediate soils, from these three soil layers.

Figure 1.11 shows the Demolon (1948) diagram that can be used, among other things, to classify intermediate soils according to the percentage of aggregates in the 2–20 µm particle size range.

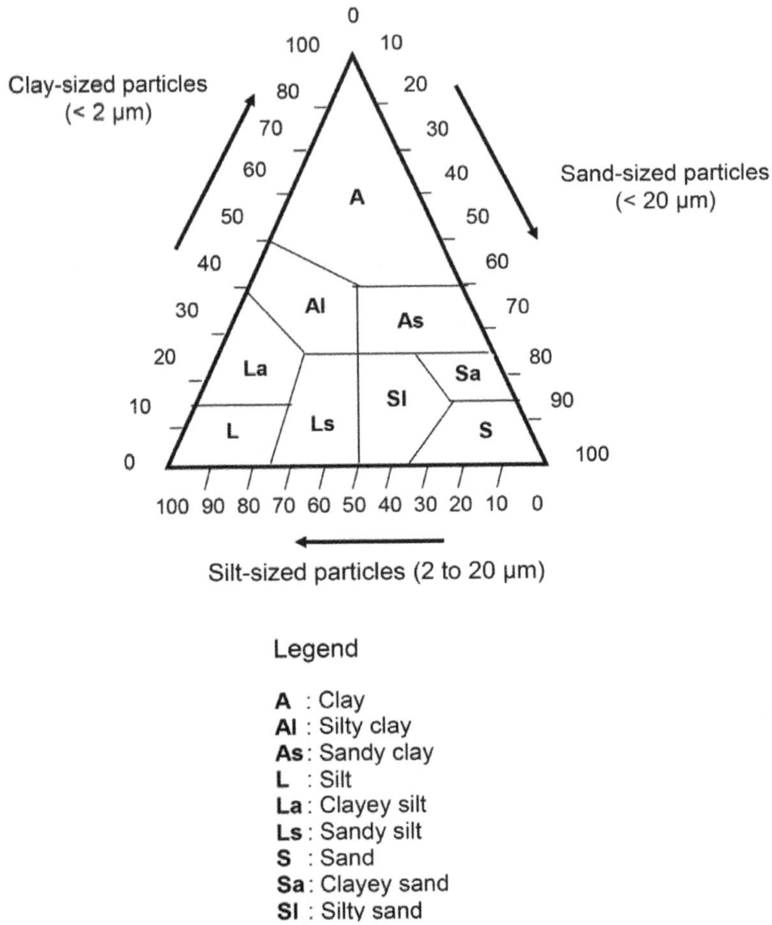

Figure 1.11. *Demolon ternary diagram (1948)*

Appendix 1.1 details the data in Table 1.8 relating to the classification of loose soil, rock, and organic soil and industrial waste, for practical application.

1.6. Illustrative samples

By way of illustration, Figure 1.12 shows a few samples of various soils.

Fontainebleau sand (Saclay Plateau)

Sands and gravels from the Seine (Paris basin)

Sands and gravels from Rhône (Lyon)

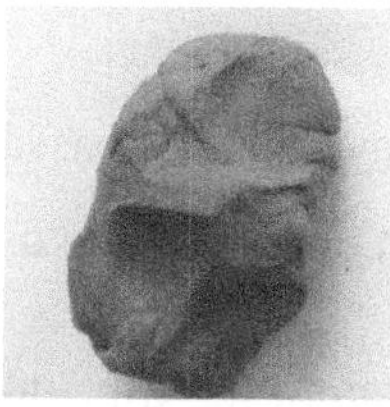

Green clay from the wet Sannoisian ("lump")

Silty alluvium from the Seine (dried up)

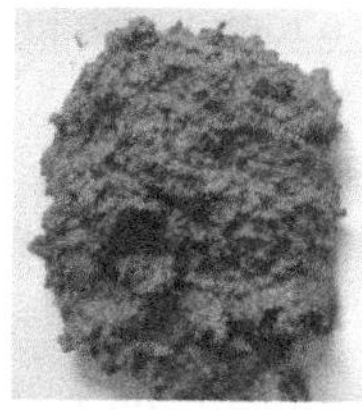

Sparnacian plastic clay

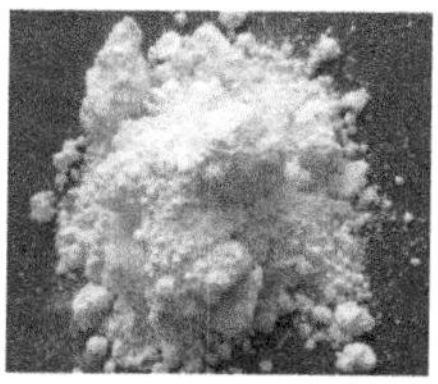

Saccharoid gypsum[2] (Parisian Basin)

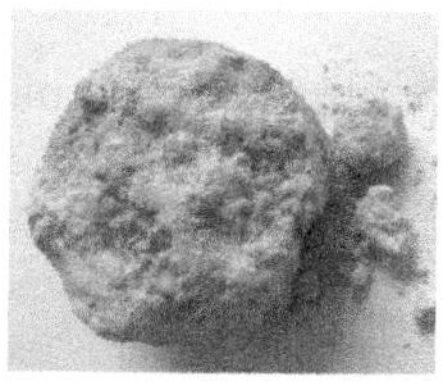

Marls and stones from the Lutetian (Paris Basin)

Coarse limestone from the Lutetian (Paris Basin)

Pelite from Tangier - Morocco (very swollen marl)

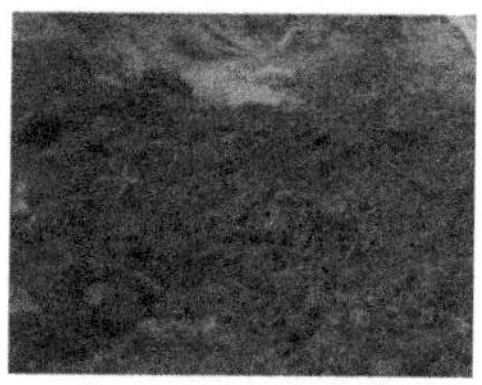

Peaty valley mud

Unclean stone (with fines)

Figure 1.12. *Various soil and rock samples. For a color version of this figure, see www.iste.co.uk/dhouib/geotechnics1.zip*

2 Gypsum which dissolves like sugar (hence the term "saccharoid gypsum") and leads to underground cavities, sometimes reaching 2 m in height (cases of dissolution encountered: 1.9 m at a depth of 39 m at the Porte d'Aubervilliers and 2.1 m at a depth of 60 m at the Goutte d'Or in Paris' 18th district).

1.7. Roads and carriageways fact sheet

A carriageway structure consists of several successive layers (Figure 1.13):

– an asphalt *surface layer* (1) with a wearing course and possibly a binder course;

– a base layer with:

- a base layer (2), in untreated gravel (UG) or in bitumen gravel,

- a foundation layer (3) in untreated gravel (UG).

These three layers constitute the structure of the carriageway, which is based on:

– a *sub-base* (4) made of granular materials or treated with lime and/or hydraulic binders;

– a *supporting soil* (5), or upper part of the earthwork (PST) of the soils that are in situ at the bottom of the excavations or at the top of the backfill. These are soils which can be applied, improved or treated with lime and/or hydraulic binders.

These two layers characterize the earthwork platform, whose sizing, "sub-base-PST" couple, and thickness and choice of materials, is presented in the SETRA-LCPC technical guides "Creation of embankments and sub-bases – GTR 2000", which is currently being revised, "Treatment of soils with lime and/or hydraulic binders – GTS January 2000" and in the recent European standards (series EN 16907-1 to 6).

Figure 1.13 shows the principle of design and execution of road and carriageway platforms.

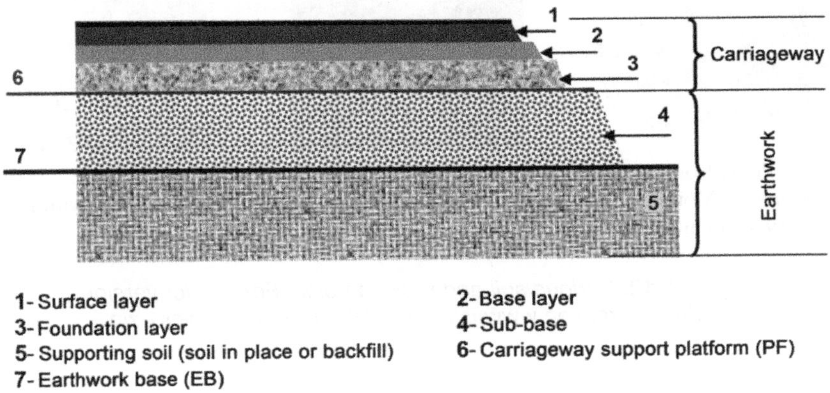

Figure 1.13. *Principle of implementation of road and carriageway layers. For a color version of this figure, see www.iste.co.uk/dhouib/geotechnics1.zip*

The principle of carriageway design is based on the characterization of the supporting soil (earthwork platform) and its bearing capacity (PF1 to PF4), the type, duration and intensity of traffic. The carriageway structure can be assessed from Peltier's empirical formula, which links the thickness (e) of the carriageway structure to the CBR by Figure (1.14).

$$e(cm) = \frac{482.4}{CBR + 5}$$

Example:
For a PF2 platform:
- EV2 = 50 MPa, (EV2/CBR=3.5 to 7)
- CBR=50/5=10,

Therefore, the carriageway structure has a thickness of e = 32 cm.

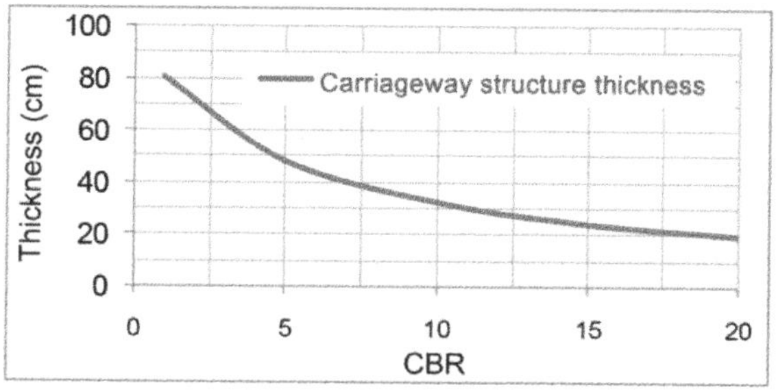

Figure 1.14. *Practical assessment of the thickness of the carriageway structure. For a color version of this figure, see www.iste.co.uk/dhouib/geotechnics1.zip*

Table 1.10 is a practical guide for the selection, design and execution of road and carriageway layers (Guide LCPC-SETRA 1998).

No.	Layer	Type		Thickness (cm)	Remarks
		Asphalt BBSG	(a)	6	-
1	Surface layer	Asphalt BBS BBSG	(b)	5	BBS for soft carriageway
		Asphalt BBM	(c)	4	-
		GB Class 3	(a)	8	-
2	Base layer	GNT 0/20	(b)	20	-
		GB Class 3	(c)	9	-
		GNT 0/20 or 0/31.5	(a)	12	Compacted
3	Foundation layer	GNT 0/20 or 0/31.5	(b)	25	-
		GNT 0/20 or 0/31.5	(c)	35	-
		Treated in situ	(a)	35 (PF3)*	With 5% binders
4	Sub-base	Stones 0/80	(b)	(PF2)*	E_{v2} = 50 MPa
		Stones 0/80	(c)	(PF2)*	E_{v2} = 50 MPa

a): GB3 carriageway on GNT for TC4-20 traffic (1.5–2.5 million heavy vehicles over 20 years).
b): GNT/GNT carriageway for TC2-20 traffic (0.0–0.5 million heavy vehicles over 20 years).
c): GB3 carriageway on GNT for TC2-20 traffic (1.5–2.5 million heavy vehicles over 20 years).
BBSG: Semi-granular asphalt concrete; BBS: Flexible asphalt concrete for soft carriageway; GNT: Untreated gravel; BBM: Thin bituminous concrete.
* See Table 3.10 in Chapter 3.

Table 1.10. *Design and execution of roads and carriageways*

1.8. Railway platforms fact sheet

The design and execution of railway platforms are more complex than those of roads and carriageways, as their function is to:

– carry the trains and direct the forces toward the railway platform through the track frame and the stone layer, to ensure good functioning in terms of stiffness and maintenance of geometry and leveling, thus guaranteeing the safety of the trains;

– absorb the vertical and horizontal stresses of the trains and guide them so that they do not derail.

Indeed, a railroad track is composed of several elements that are part of the "rolling stock/rail" system and successive layers (Figure 1.15):

– the track frame or reinforcement: rails, rail fasteners, sleepers (wood or concrete) and rail to rail assembly (1);

– stone layer (2);

– sub-layer (3);

– sub-base (4);

– support layer, either of the soil in situ or in the form of a backfill (5).

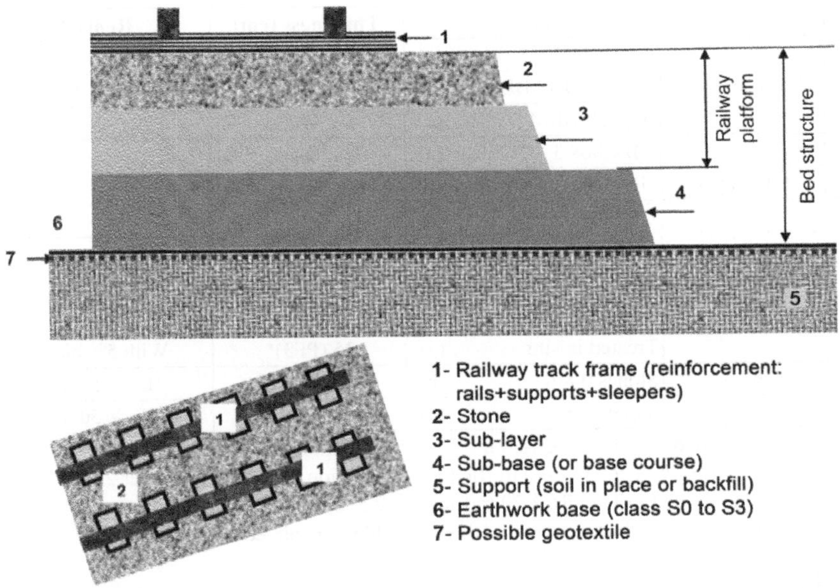

Figure 1.15. *Principle of implementation of typical railway platforms. For a color version of this figure, see www.iste.co.uk/dhouib/geotechnics1.zip*

COMMENT ON FIGURE 1.15.–

– The three layers (1), (2) and (3) constitute the track bed structure.

– Railroad tracks can be laid on a slab, which must be specifically designed.

The sizing of the track bed structure depends on:

– the type of traffic;

– the type of track reinforcement (wooden or concrete sleepers);

– the bearing capacity class of the platform (PF1 to PF3) and the base class (S1 to S3).

By way of illustration, Table 1.11 presents a fact sheet for the selection, design and execution of the thicknesses of the successive layers of Figure 1.15 that constitute the railway platform of new high-speed lines (TGV), according to the combinations of the Technical Reference IN 3278 (2006).

Base classes	Stone (cm)	Sub-layer (cm)	Sub-base	
			Thickness (cm)	Nature
S1 (30 MPa)*	35**	20	50	Report (granular)
S1 (30 MPa)*	35**	20	35	Treated (Class 4)***
S1 (30 MPa)*	35**	20	40	Treated (Class 5)***
S2 (50 MPa)*	35**	35	Integrated over 30 cm	
S2 (50 MPa)*	35**	20	35	Report (granular)
S3 (80 MPa)*	35**	20	Integrated over 30 cm	

* Reference bearing capacity.

** 31.5/50 or 31.5/63 stone (on 15–30 cm for classic lines).

*** Mechanical class according to GTR (Fascicle I).

Table 1.11. *Example of the thickness of the layers of railway platforms of new high-speed lines (TGV)*

1.9. Appendix 1.1: Classification of soft soil and rock

Classification by nature				Classification by water status				Reuse in backfill
	Sub-class	Second level of classification	Soil type		IPI or	w_n or	I_c	
Class A: ***Fine soils*** *First level of classification:*	A_1	I_P: ≤ 12 or VBS: ≤ 2.5	Silt, loess, fine sand	A_1th A_1h A_1m A_1s A_1ts	≤ 3 3–8 8–25 - -	$\geq 1.25\ w_{OPN}$ 1.10–1.25 w_{OPN} 0.90–1.10 w_{OPN} 0.70–0.90 w_{OPN} $<0.70\ w_{OPN}$	- - - - -	Yes
→ D_{max} $\leq$ 50 mm and with:	A_2	I_P: 12–25 or VBS: 2.5–6	Fine clayey sands, silts, low plastic clays	A_2th A_2h A_2m A_2s A_2ts	≤ 2 2–5 5–15 - -	$\geq 1.30\ w_{OPN}$ 1.10–1.30 w_{OPN} 0.90–1.10 w_{OPN} 0.70–0.90 w_{OPN} $<0.70\ w_{OPN}$	≤ 0.90 0.90–1.05 1.05–1.20 1.20–1.40 >1.40	Yes
→ from 35% to 100% fines (aggregates at 80 µm)	A_3	I_P: 25–40 or VBS: 6–8	Clays, marly clays, very plastic silts	A_3th A_3h A_3m A_3s A_3ts	≤ 1 1–3 3–10 - -	$\geq 1.4\ w_{OPN}$ 1.20–1.40 w_{OPN} 0.90–1.20 w_{OPN} 0.70–0.90 w_{OPN} $<0.70\ w_{OPN}$	≤ 0.80 0.80–1.00 1.00–1.15 1.15–1.30 >1.30	(1)
	A_4	I_P: >40 or VBS: >8	Clays and very plastic marly clays	A_4th A_4h A_4m A_4s	The threshold values are to be defined with the support of an appropriate study.			No (2)
(1): Lime or lime-cement treatment. (2): Unless specific study supported by full-scale tests.								

D_{max}: Maximum grain diameter; IPI: Immediate Proctor Index; w_n: Natural water content; I_c: Consistency index; I_P: Plasticity index; VBS: Methylene blue value; w_{OPN}: Water content at Optimum Proctor Standard.

Classification level required for use in backfill									
Classification level required for use in sub-base									
Classification by nature					Classification by water status			Classification according to behavior	
	Sub-class	Second level of classification	Soil type	Main characteristics	Sub-class	IPI or	w_n	Sub-class	Threshold parameters
Class B: Sandy and granular soils with fines	B_1	80 µm ≤ 12% 2 mm > 70% **VBS** = 0.1–0.2 or ES > 35	Silty sand	Sandy soils that are generally resistant to water. However, this resistance must be confirmed by specific studies. Their use in the sub-base requires the measurement of their mechanical resistance (Friability of Sand: FS)				B_{11}	FS ≤ 60
								B_{12}	FS >60
First level of classification:	B_2	80 µm ≤ 12% 2 mm > 70% **VBS** > 0.2 or ES ≤ 35	Clayey sand (not very clayey)	Plastic and water-sensitive soils. Their water status does not change very much. Their use in the sub-base without treatment requires the measurement of FS.	B_2th	≤ 4	≥ 1.25 w_{OPN}	$B_{21}th$	FS ≤ 60
								$B_{22}th$	FS >60
					B_2h	4–8	1.10–1.25 w_{OPN}	$B_{21}h$	FS ≤ 60
								$B_{22}h$	FS >60
					B_2m	-	0.90–1.10 w_{OPN}	$B_{21}m$	FS ≤ 60
								$B_{22}m$	FS >60
					B_2s	-	0.50–0.90 w_{OPN}	$B_{21}s$	FS ≤ 60
								$B_{22}s$	FS >60
D_{max} ≤ 50 mm and with less than 35% fines					B_2ts	-	<0.50 w_{OPN}	$B_{21}ts$	FS ≤ 60
								$B_{22}ts$	FS >60
	B_3	80 µm ≤ 12% 2 mm ≤ 70% **VBS** = 0.1–0.2 or ES > 25	Silty gravel	Granular soils that are generally resistant to water. However, this resistance must be confirmed by specific studies. Their use in the sub-base without treatment requires the measurement of their mechanical resistance (LA, MDE in presence of water)				B_{31}	LA ≤ 45 and MDE ≤ 45
								B_{32}	LA >45 or MDE >45

Classification by nature					Classification by water status			Classification according to behavior	
	Sub-class	Second level of classification	Soil type	Main characteristics	Sub-class	IPI	w_n	Sub-class	Threshold parameters
Class B: Sandy and granular soils with fines *First level of classification:* $D_{max} \leq 50$ mm and with less than 35% fines	B_4	80 µm ≤ 12% 2 mm ≤ 70% **VBS** > 0.2 or ES ≤ 25	Clayey gravel (not very clayey)	Water-sensitive soil. Permeable. Their water status does not change very much. LA and MDE measurement before use (CF) without treatment.	B_4th	≤ 7	≥ 1.25 w_{OPN}	$B_{41}th$	LA ≤ 45 and MDE ≤ 45
								$B_{42}th$	LA >45 or MDE >45
					B_4h	7–15	1.10–1.25 w_{OPN}	$B_{41}h$	LA ≤ 45 and MDE ≤ 45
								$B_{42}h$	LA >45 or MDE <45
					B_4m	-	0.90–1.10 w_{OPN}	$B_{41}m$	LA ≤ 45 and MDE ≤ 45
								$B_{42}m$	LA >45 or MDE >45
					B_4s	-	0.60–0.90 w_{OPN}	$B_{41}s$	LA ≤ 45 and MDE ≤ 45
								$B_{42}s$	LA >45 or MDE >45
					B_4ts	-	<0.6 w_{OPN}	$B_{41}ts$	LA ≤ 45 and MDE ≤ 45
								$B_{42}ts$	LA >45 or MDE >45

Class B: Sandy and granular soils with fines *First level of classification:* $D_{max} \leq 50$ mm and with less than 35% fines	B_5	80 µm between 12% and 35% **VBS** ≤ 1.5 or $I_P \leq 12$	Very silty sand and gravel	Close to Class A_1 fine soils. VBS is the preferred criterion for their identification. Measurement of LA and MDE before use (CF) without treatment.	B_5th	≤ 5	$\geq 1.25\ w_{OPN}$	B_{51}th	LA ≤ 45 and MDE ≤ 45
								B_{52}th	LA >45 or MDE >45
					B_5h	5–12	1.10–1.25 w_{OPN}	B_{51}h	LA ≤ 45 and MDE ≤ 45
								B_{52}h	LA >45 or MDE >45
					B_5m	12–30	0.90–1.10 w_{OPN}	B_{51}m	LA ≤ 45 and MDE ≤ 45
								B_{52}m	LA >45 or MDE >45
					B_5s	-	0.60–0.90 w_{OPN}	B_{51}s	LA ≤ 45 and MDE ≤ 45
								B_{52}s	LA >45 and MDE >45
					B_5ts	-	<0.6 w_{OPN}	B_{51}ts	LA ≤ 45 and MDE ≤ 45
								B_{52}ts	LA >45 or MDE >45
	B_6	80 µm between 12% and 35% **VBS** > 1.5 or $I_P > 12$	Clayey to very clayey sand and gravel	Water-sensitive (large sandy fraction).	B_6th	≤ 4	$\geq 1.30\ w_{OPN}$ or $I_c \leq 0.8$		
					B_6h	4–10	1.10–1.30 w_{OPN} or $I_c = 0.8–1.0$		
					B_6m	10–25	0.90–1.10 w_{OPN} or $I_c = 1.0–1.2$		
					B_6s	-	0.70–0.90 w_{OPN} or $I_c = 1.2–1.3$		
					B_6ts	-	< 0.7 w_{OPN} or $I_c \geq 1.3$		

D_{max}: Maximum grain diameter; IPI: Immediate Proctor Index; w_n: Natural water content; VBS: Methylene blue value; w_{OPN}: Water content at Optimum Proctor Standard; FS: Friability of Sand; I_c: Consistency index; LA: Los Angeles (Standard NF EN 1097-2); MDE: Micro-Deval (Standard NF EN 1097-1).

Classification by nature					Classification according to water status and behavior		
	Sub-class	Second level of classification	Soil type	Main characteristics			
Class C: Soils with fine and coarse elements	C_1A_1	Angular materials: 0/50 mm fraction > 60%– 80%. Wearing materials: 0/50 mm fraction of Class A.	Flint clay, millstone clay, scree, moraine, coarse alluvium	The 0/50 mm fraction is essential for the classification of these soils (especially for angular materials). The classification of these soils is always valued by a double symbol: $A_1(C_1)$ or $A_1(B_1)$, where A_1 and B_1 are the classes of the fraction 0/50 mm.	Sub-classification according to the 0/50 mm fraction of Class A or B:		
					C_1A_1 C_1A_2 C_1A_3 C_1A_4	C_2A_1 C_2A_2 C_2A_3 C_2A_4	Water status: th, h, m, s or ts
First level of classification:	C_1B_1	Angular materials: 0/50 mm fraction > 60%–80%. Wearing materials: 0/50 mm fraction of Class B.			C_1B_{11} C_1B_{12} C_1B_{31} C_1B_{32}	C_2B_{11} C_2B_{12} C_2B_{31} C_2B_{32}	Soils generally resistant to water
D_{max} > 50 mm and with more or less than 12% fines. VBS > at 0.1	C_2A_1	Angular materials: 0/50 mm fraction ≤ 60%80%. 0/50 mm fraction of Class A.	Flint clay, millstone clay, scree, flint mills	Difficulty in practice to evaluate the 50/D mm fraction where D is >50 mm. Full-scale measurements are often useful to assess the 0/50 mm fraction.	C_1B_{21} C_1B_{22} C_1B_{41} C_1B_{42} C_1B_{51} C_1B_{52} C_1B_6	C_2B_{21} C_2B_{22} C_2B_{41} C_2B_{42} C_2B_{51} C_2B_{52} C_2B_6	Water status: th, h, m, s or ts
	C_2B_1	Angular materials: 0/50 mm fraction ≤ 60%–80%. 0/50 mm fraction is Class B.					

D_{max}: Maximum grain diameter; VBS: Methylene blue value.

<table>
<tr><td colspan="10">Classification level required for use in backfill ←→</td></tr>
<tr><td colspan="10">Classification level required for use in sub-base ←→</td></tr>
<tr><td rowspan="13">Class D:
Water-resistant soils

First level of classification:

With less than 12% fines
and
VBS ≤ at 0.1</td><th colspan="4">Classification by nature</th><th colspan="5">Classification according to behavior</th></tr>
<tr><th rowspan="2">Sub-class</th><th rowspan="2">Second level of classification</th><th rowspan="2">Soil type</th><th rowspan="2">Main characteristics</th><th rowspan="2">Sub-class</th><th colspan="3">Threshold parameters retained</th><th rowspan="2">Remarks</th></tr>
<tr><th>FS</th><th>LA</th><th>MDE</th></tr>
<tr><td rowspan="2">D_1</td><td rowspan="2">$D_{max} \leq$ 50 mm
and
2 mm > 70%</td><td rowspan="2">Clean alluvial sands, dune sands</td><td rowspan="2">Soils without cohesion and permeable.
Badly scaled grading → erodible and difficult trafficability.</td><td>D_{11}</td><td>≤ 60</td><td>-</td><td>-</td><td rowspan="6">Their use in the sub-base without treatment requires the measurement of their mechanical resistance (LA, MDE in the presence of water) or FS</td></tr>
<tr><td>D_{12}</td><td>>60</td><td>-</td><td>-</td></tr>
<tr><td rowspan="2">D_2</td><td rowspan="2">$D_{max} \leq$ 50 mm
and
2 mm ≤ 70%</td><td rowspan="2">Clean alluvial gravel, sand</td><td rowspan="2">Soils without cohesion and permeable.
Less erodible and can withstand traffic after compaction.</td><td>D_{21}</td><td>-</td><td>≤ 45
and</td><td>≤ 45</td></tr>
<tr><td>D_{22}</td><td>-</td><td>>45
or</td><td>>45</td></tr>
<tr><td rowspan="2">D_3</td><td rowspan="2">D_{max} > 50 mm</td><td rowspan="2">Clean alluvial gravel</td><td rowspan="2">Soils without cohesion and permeable.
Unsuitable for mixing.</td><td>D_{31}</td><td>-</td><td>≤ 45
and</td><td>≤ 45</td></tr>
<tr><td>D_{32}</td><td>-</td><td>>45
or</td><td>>45</td></tr>
</table>

D_{max}: Maximum grain diameter; VBS: Methylene blue value; FS: Friability of Sand; LA: Los Angeles (Standard NF EN 1097-2); MDE: Micro-Deval (Standard NF EN 1097-1).

Sedimentary rocks											
Classification by nature				**Classification according to water status and behavior**							
						Parameters and threshold values retained					
	Sub-class	**Rock type**	**Main characteristics**	**Sub-class**	**Rock quality**	**Water or mechanical status**					
							th	h	m	s	ts
Carbonate rocks	R_1	Chalk: Calcite particles from 1 to 10 µm.	Fragile and porous structure, pasty in the saturated state. Stiff and difficult to compact when dry.	R_{11}	Dense	ρ_d	>1.7				
				R_{12}	Medium density	ρ_d	-	1.5–1.7			
						w_n	-	≥ 27	22–27	18–22	<18
				R_{13}	Low density	ρ_d	≤ 1.5				
						w_n	≥ 31	26–31	21–26	16–21	<16
	R_2	Rocky limestone: Coarse limestone, Travertine, Tuffs	Friable and gel-like when fragmentable. Formation of fines → water-sensitive.	R_{21}	Hard	MDE ≤ 45					
				R_{22}	Medium density	MDE >45 and ρ_d >1.8					
				R_{23}	Fragmentable	ρ_d ≤ 1.8					
Clayey rocks	R_3	Marls, Sedimentary shales, Argillites, Pelites	Swelling clay minerals. Fine plastic and water-sensitive. R_{34}: fragmentable → to be characterized by the 0/50 mm fraction.	R_{31}	Very degradable	FR ≤ 7 and DG >20					
				R_{32}	Med. degradable	FR ≤ 7 and DG = 5–20					
				R_{33}	Not very degradable	FR ≤ 7 and DG ≤ 5					
				R_{34}	Fragmentable *or: IPI < 2 ** IPI = 2–5		th	h	m	s	ts
						FR	> 7 and				
						w_n	≥ 1.3w_{OPN}*	1.1–1.3 w_{OPN}**	0.9–1.1 w_{OPN}	0.7–0.9 w_{OPN}	<1.3 w_{OPN}

Silica rocks R_4	Sandstone, Conglomerates, Breccias	Conglomerate of sand grains cemented with silica or calcite.	R_{41}	Hard	LA $\leq$ 45 and MDE $\leq$ 45
			R_{42}	Medium hard	LA >45 or MDE >45 and FR $\leq$ 7
			R_{43}	Fragmentable	FR >7
Salt rocks R_5	Gypsum, Salt, Gem*, Anhydrite	Solubility → risk of disruptions on works.	R_{51}	Not very soluble	T_{ss} $\leq$ 5%–10% (gem) and $\leq$ 30%–50% (gypsum)
			R_{52}	Very soluble	T_{ss} >5%–10% (gem) and >30%–50% (gypsum)
Magmatic and metamorphic rocks					
R_6	Granite, basalt, trachyte, andesite, gneiss, shale	Very different fragmentability and friability.	R_{61}	Hard	LA $\leq$ 45 and MDE $\leq$ 45
			R_{62}	Medium hard	LA >45 or MDE >45 and FR $\leq$ 7
			R_{63}	Fragmentable or altered	FR >7

ρ_d: Rock density; IPI: Immediate Proctor Index; w_n: Natural water content; FR: Friability; DG: Rock degradability (NF P 94-067); T_{ss}: Soluble salt content; w_{OPN}: Water content at Optimum Proctor Standard; LA: Los Angeles (Standard NF EN 1097-2); MDE: Micro-Deval (Standard NF EN 1097-1 and 2).

Organic soils and industrial by-products				
Family of materials	**Main characteristics**	**Classification of materials**		
		Parameters	**Threshold values**	**Sub-class**
F_1 Natural soils with organic matter	Organic material content (MO %)	MO (%)	$3 \leq$ % MO ≤ 10	F_{11} Low organic content: topsoil, mud
			% MO > 10	F_{12} High organic content: peat, forest humus
F_2 Fly ash Silica-alumina from thermal power plants	60% fines Non-plastic	Ratio $\frac{w_n}{w_{OPN}}$	**IPI** ≤ 4 or $w_n \geq 1.3\ w_{OPN}$	Fly ash: F_2**th** (very humid)
			$4 <$ **IPI** ≤ 10 $1.2\ w_{OPN} \leq w_n < 1.3\ w_{OPN}$	F_2**h** (humid)
			$0.85\ w_{OPN} \leq w_n < 1.2\ w_{OPN}$	F_2**m** (w_n medium)
			$0.75\ w_{OPN} \leq w_n < 0.85\ w_{OPN}$	F_2**s** (dry)
			$w_n < 0.75\ w_{OPN}$	F_2**ts** (very dry)
F_3 Coal shales	Coal mine heaps	Combustion	Red to purple color	F_{31} (burnt)
			Black to orange color	F_{32} (not burnt)
F_4 Potash mine shales	Tailings from potash mining	Sodium chloride (NaCi) content	**%** NaCi ≤ 10	F_{41}
			% NaCi >10	F_{42}

F_5 Phosphogypsum	Artificial gypsum (by-product of phosphate fertilizers)	D_{50} w_n	$D_{50} > 80$ µm $w_n \geq 1.2\ w_{OPN}$	**F_{51}h** (coarse, neutralized with lime)
			$D_{50} > 80$ µm $w_n < 1.2\ w_{OPN}$	**F_{51} m** and **s** (ditto, less humid)
			$D_{50} \leq 80$ µm	**F_{52}** (fine and coarse, not neutralized with lime)
F_6 Incineration of household waste	Variable compositions	Degree of incineration measured by loss of fire (PF) at 500°C	PF $\leq$ 5%	**F_{61}** (old, well incinerated)
			PF $\leq$ 5%	**F_{62}** (fresh, well incinerated)
			PF >5%	**F_{62}** (poorly incinerated)
F_7 Demolition materials	Variable compositions	Presence of: - plaster - putrescible elements	Visual assessment	**F_{71}** (without plaster and putrescible elements)
				F_{72} (ditto with plaster)
				F_{73} (with putrescible elements)
F_8 Blast furnace slag	By-products of ironmaking	Cooling process	Grading, cleanliness	-
F_9 Other industrial wastes and by-products	Steel mill slag, foundry sands	Specific study		-

D_{50}: Diameter corresponding to 50% of aggregates; IPI: Immediate Proctor Index; w_n: Natural water content; w_{OPN}: Water content at Optimum Proctor Standard; %MO: Percentage of organic materials; NaCi: Sodium chloride content; PF: Loss of fire.

1.10. References

AFNOR (1999). NF P94-093. Détermination des références de compactage d'un matériau – essai Proctor normal. Essai Proctor modifié, AFNOR, Paris.

AFBOR (2019). NF P 98-086: Dimensionnement structurel des chaussées routières (neuves). Norme française, October.

Demolon, A. (1948). *Principes d'agronomie. Tome I : Dynamique du sol*, 4th edition. Dunod, Paris.

IN 3278 (EF1C1) (2006). Référentiel Technique pour la réalisation des LGV-Partie Génie Civil. Document, SNCF.

LCPC-SETRA (1992a). Guide Technique pour la réalisation des remblais et couches de forme. SETRA, Paris.

LCPC-SETRA (1992b). Guide Technique pour la réalisation des remblais et des couches de forme – GTR Fascicule I : principes généraux. SETRA, Paris.

LCPC-SETRA (1992c). Guide Technique pour la réalisation des remblais et des couches de forme – Fascicule II : annexes techniques. SETRA, Paris.

LCPC-SETRA (1998). Guide Catalogue des structures types de chaussés neuves. SETRA, Paris.

Legrand, J. (1974). Description, identification et classification des sols. *Techniques de l'ingénieur*, C208.

Philipponnat, G. and Hubert, B. (2008) *Fondations et ouvrages en terre*. Edition Eyrolles, Paris.

Pierre, C. (2012). Voie ferrée – composants, construction et maintenance. *Techniques de l'ingénieur*, TRP3050.

Schlosser, F. (2006). *Eléments de mécanique des sols*. Edition des Presses de l'Ecole Nationale des Ponts et Chaussées, Paris.

SETRA-LCPC (2000). GTR – Réalisation des remblais et des couches de forme – GTR Fascicule II : annexes techniques. SETRA, Paris.

2

Stresses, Deformations and Behavioral Laws of Soils[1]

Between rest and stress, man ends up wearing himself out.
Soil eternally resists the directed pressure that forces it to deform.

Ammar DHOUIB

2.1. Basic concepts of mechanics of continuous media

2.1.1. *Concept of stress*

Let us consider a wire that is subjected to a tensile force T (Figure 2.1(a)) at its two ends, A and B. At point M in a hypothetical perpendicular plane, the wire is subjected to a tensile stress (σ), defined as the ratio of tension T to the cross-section (S) of the wire; in other words:

$$\sigma = \frac{T}{S} \tag{2.1}$$

To apply this concept of stress to the soil, let us take the case of a walker's foot. The trace of their foot on the soil constitutes a support surface (S). The weight of the walker (W), related to their footprint (S), translates (Figure 2.1(b)) into a vertical compressive stress that is supposedly uniform (σ), formulated by:

$$\sigma = \frac{W}{S} \tag{2.2}$$

---

1 Attention: In spite of the simplifications made by the author, this chapter requires knowledge in continuum mechanics, numerical analysis and soil rheology.

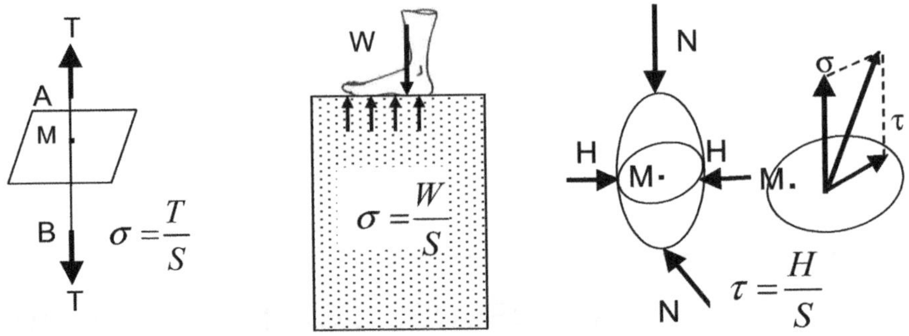

a) Tensile stress (wire stretched at both ends) b) Compressive stress under a walker's foot c) Normal and shear stress (solid under any load)

Figure 2.1. *Concept of stress in soil*

COMMENT ON FIGURE 2.1.–

1) *In the two cases presented in Figure 2.1(a) and Figure 2.1(b), the applied stress σ is in a plane that is normal to the stressed section. It is a normal tensile stress in the wire and compressive stress in the soil.*

2) *When the forces applied to a solid are arbitrary ("multidirectional" forces), tangential shear stresses develop in addition to normal stresses. Indeed, force N in Figure 2.1(c) creates normal stresses (σ) at point M, and under the horizontal force H, shear stresses (τ) are mobilized. The two stresses can act separately.*

2.1.2. Concept of deformation

Let us return to the example of the stretched wire. Under tension T, the wire undergoes an elongation (Δl) and its length goes from l_0 to ($l_0 + \Delta l$), and point A moves to A'. The deformation (ε) of the wire, which is dimensionless, is the ratio of this elongation (Δl) to the initial length l_0, that is to say (Figure 2.2(a)):

$$\varepsilon = \frac{\Delta l}{l_0} \quad [2.3]$$

Under the foot of the walker (Figure 2.2(b)), as well as on a soil sample subjected to the action of the piston (Figure 2.2(c)), the soil undergoes a settlement

(Δh). The deformation (ε) of the soil of thickness h is the ratio of the settlement (Δh) of the soil to its thickness (h), and is expressed by:

$$\varepsilon = \frac{\Delta h}{h} \quad [2.4]$$

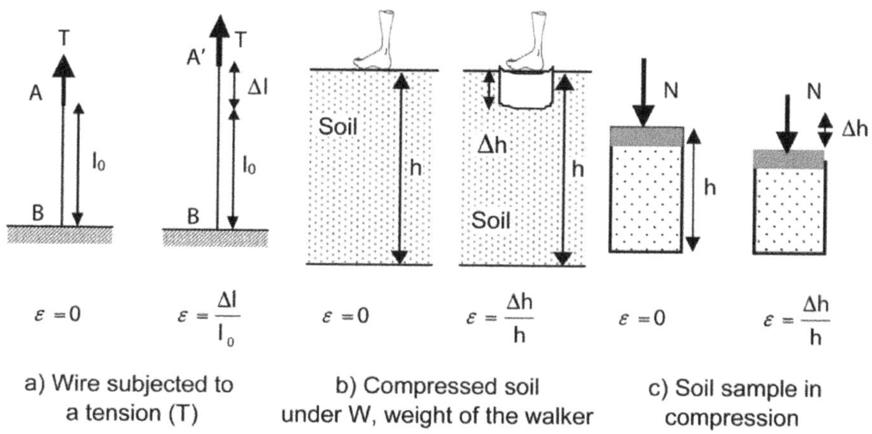

Figure 2.2. *Concept of deformation in soil*

COMMENT ON FIGURE 2.2.– *It is important not to confuse displacement (a quantity with dimensions in meters or its derivatives) with deformation (a dimensionless quantity given in percent). In the example of the wire, it is an elongation. In the other two cases, the compressed soil undergoes a settlement.*

– The deformation defined in Figure 2.2(a) reflects the increase in length (l_0) of the wire. Those in Figure 2.2(b) and Figure 2.2(c) represent the decrease in the height of the mass or soil sample. In all three cases, the deformation is said to be linear because it corresponds to the variation of the length of the wire and the height of the soil mass and sample, without any change in direction.

– When a change of direction (distortion) occurs during the deformation in the xyz "reference mark" (Figure 2.3(a)), the deformation is called "angular deformation". Indeed, solid S (Figure 2.3(b)) deforms and becomes S', the segment AB undergoes an elongation and becomes A'B'. This elongation is accompanied by a change of direction of angle θ_a in this reference mark. The angular deformation is given by this angle, or:

$$\theta_a = (\overrightarrow{AB}, \overrightarrow{A'B'}) \quad [2.5]$$

The angular deformation is a very small-value "rotation".

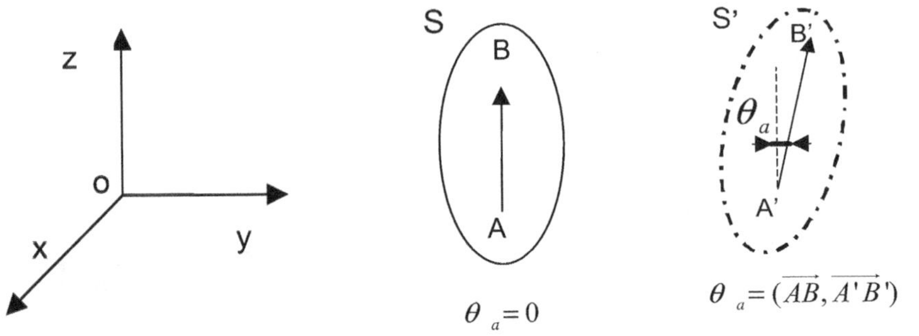

a) Reference mark

b) Solid S before angular deformation

c) Solid S' after angular deformation

Figure 2.3. *Definition of angular deformation*

2.1.3. *Generalization of the state of stresses*

2.1.3.1. *Stress tensor*

Let us examine the cubic sample of Figure 2.4(a), taken in a mass subjected to external and internal forces (gravity, for example), attached to the coordinates system with three dimensions, Oxyz (Figure 2.4(b)). On the six facets of the cube, the following stresses are exerted:

– normal stresses perpendicular to each facet: σ_{xx} (on the facet perpendicular to the ox axis), σ_{yy} (on the facet perpendicular to the oy axis) and σ_{zz} (on the facet perpendicular to the oz axis);

– tangential stresses in the plane of each facet: τ_{xy} and τ_{xz} (on the facet perpendicular to the ox axis), τ_{yx} and τ_{yz} (on the facet perpendicular to the oy axis) and τ_{zx} and τ_{zy} (on the facet perpendicular to the oz axis).

The distribution of normal and tangential stresses on the different facets is given in Figure 2.4(b).

The generalized state of stresses can be put in a matrix form called "stress tensor", generally noted as $\bar{\sigma}$, and written in the Oxyz reference mark:

$$\bar{\sigma} = \begin{bmatrix} \sigma_{xx} & \tau_{xy} & \tau_{xz} \\ \tau_{yx} & \sigma_{yy} & \tau_{yz} \\ \tau_{zx} & \tau_{zy} & \sigma_{zz} \end{bmatrix} \quad [2.6]$$

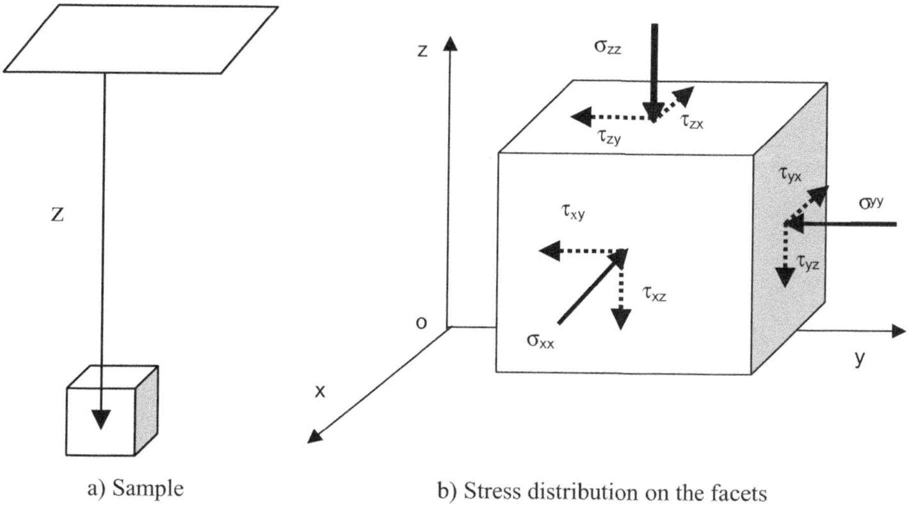

Figure 2.4. *Stress distribution on the facets of a sample within the soil*

COMMENT ON FIGURE 2.4.–

1) *Due to symmetry, the shear stresses are identical; in other words:*

$$\begin{cases} \tau_{xy} = \tau_{yx}, \\ \tau_{yz} = \tau_{zy}, \\ \tau_{zx} = \tau_{xz} \end{cases} \qquad [2.7]$$

2) *The normal stresses (located on the diagonal axis: σ_{xx}, σ_{yy}, σ_{zz}) constitute the "spherical tensor", they are called "spherical stresses".*

3) *The shear stresses (τ_{xy}, τ_{xz} and τ_{yz}, as well as their symmetries) correspond to the "deviatoric tensor", they are called "deviatoric stresses".*

4) *The stress tensor ($\bar{\sigma}$) is sometimes written as σ_{ij}, as in equation [2.6], where i is the number of rows (i = 1–3) and j is the number of columns (j = 1–3). In this case, the stresses are attached to the "1-2-3" reference mark instead of "xyz".*

2.1.3.2. *Principal stresses*

When there are no shear stresses ($\tau = 0$ at any point in the solid), only normal stresses (σ_1, σ_2 and σ_3) develop in the solid body. Each plane, that is perpendicular to each other, is considered a "principal plane" and the stresses that apply to it are called "principal stresses" (Figure 2.5).

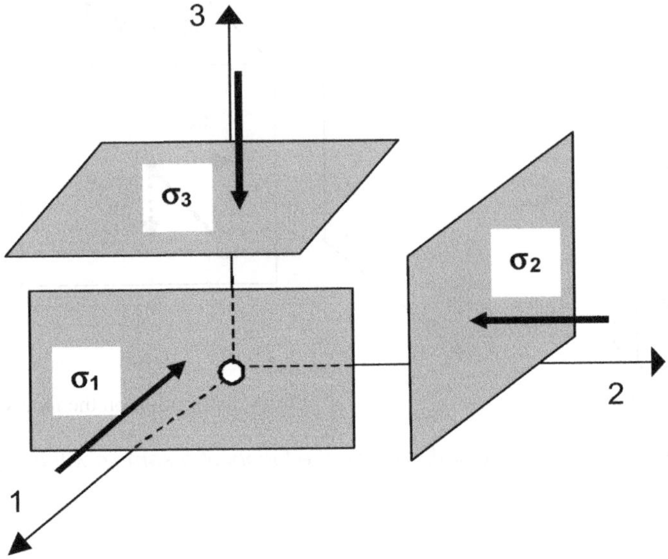

Figure 2.5. *Principal planes and stresses*

COMMENT ON FIGURE 2.5.– *When the three principal stresses differ,* as in *the following*:

$$\sigma_3 < \sigma_2 < \sigma_1,$$

then, these stresses are called:

σ_3*: minor stress;*

σ_2*: intermediate stress;*

σ_1*: major stress.*

2.1.3.3. *Mohr's circle*

To better understand the state of stresses around a point, it is easier to replace the analytical formulation by a graphical representation, which in this case is the Mohr diagram. This is an often-used diagram that consists of plotting vector $\vec{f}$ (Figure 2.6(a)) in a system of axes (τ, σ), as shown in Figure 2.6(b).

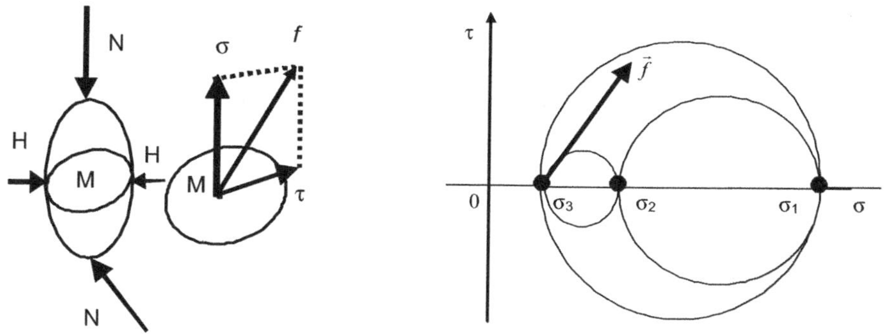

a) Stresses around point M

b) Mohr's circle

Figure 2.6. *Mohr's circle: graphical representation of stresses*

2.1.3.4. *Special case of a liquid*

In a liquid there is no shear stress (τ = 0). Only pressure (p) prevails in the liquid. The normal stresses (σ_1, σ_2 and σ_3) are identical (equation [2.8]) to the pressure p (Figure 2.7(a)). Mohr's circle is reduced to a point on the axis of the normal stresses σ (Figure 2.7(b)):

$$\sigma_1 = \sigma_2 = \sigma_3 = \mathrm{p} \qquad [2.8]$$

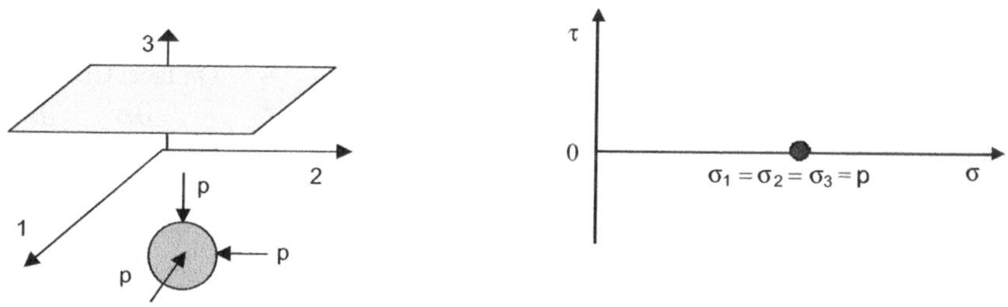

a) Pressure (p) in a liquid

b) Mohr's circle representative of the liquid

Figure 2.7. *Graphical representation of stresses in a liquid. For a color version of this figure, see www.iste.co.uk/dhouib/geotechnics1.zip*

COMMENT ON FIGURE 2.7.– *The state of stresses in a liquid is purely spherical. Indeed, at a given depth, a diver in the sea or in a swimming pool undergoes a uniform water pressure (p) in all directions.*

2.1.3.5. *Equilibrium equations*

The components of the stress tensor in Figure 2.4 (equation [2.6]) are variable and depend on the x, y and z coordinates of the point M. We will establish the equilibrium equations in a two- and three-dimensional setting.

Two-dimensional setting (two-dimensional medium: 2D)

For simplicity, let us consider a two-dimensional setting and an element represented by a square with center M, attached to the xz reference mark. In this aspect, forces per unit area are exerted at point M: f_x (along ox) and f_z (along oz), and four stresses are mobilized; two normal stresses (σ_{xx} and σ_{zz}) and two shear stresses (τ_{xz} and τ_{zx}), which are identical ($\tau_{xz} = \tau_{xy}$). The state of the stresses on each facet of the square is shown in Figure 2.8.

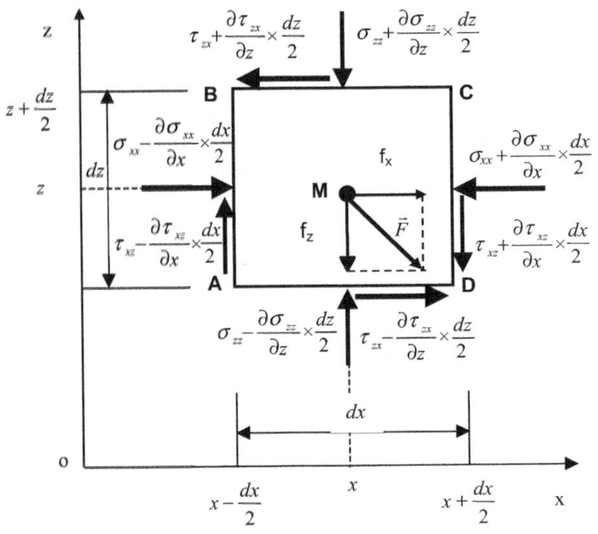

a) State of stresses on the facets of the square

On facet AB:

$$\begin{cases} (\sigma_{xx} - \frac{\partial \sigma_{xx}}{\partial x} \times \frac{dx}{2}).dz \\ (\tau_{xz} - \frac{\partial \tau_{xz}}{\partial x} \times \frac{dx}{2}).dz \end{cases}$$

On facet BC:

$$\begin{cases} -(\tau_{zx} + \frac{\partial \tau_{zx}}{\partial z} \times \frac{dz}{2}).dx \\ -(\sigma_{zz} + \frac{\partial \sigma_{zz}}{\partial z} \times \frac{dz}{2}).dx \end{cases}$$

On facet CD:

$$\begin{cases} -(\sigma_{xx} + \frac{\partial \sigma_{xx}}{\partial x} \times \frac{dx}{2}).dz \\ -(\tau_{xz} + \frac{\partial \tau_{xz}}{\partial x} \times \frac{dx}{2}).dz \end{cases}$$

On facet AD:

$$\begin{cases} (\tau_{zx} - \frac{\partial \tau_{zx}}{\partial z} \times \frac{dz}{2}).dx \\ (\sigma_{zz} - \frac{\partial \sigma_{zz}}{\partial z} \times \frac{dz}{2}).dx \end{cases}$$

b) Components of the forces

Figure 2.8. *Stresses and force components on the facets of the 2D medium*

By writing the balance of forces on ox and oz (resultants of zero force in both directions, Figure 2.8(b)) and dividing by the quantity (*dx.dz*), we get:

$$\begin{cases} \dfrac{\partial \sigma_{xx}}{\partial x} + \dfrac{\partial \tau_{xz}}{\partial z} = f_x \quad (on\, ox) \\ \dfrac{\partial \tau_{zx}}{\partial x} + \dfrac{\partial \sigma_{zz}}{\partial z} = f_z \, (on\, oz) \end{cases} \qquad [2.9]$$

COMMENT.– The two equations (equation [2.9]) constitute the equilibrium equations of a two-dimensional solid body (two-dimensional medium) that is subjected to purely static forces.

Three-dimensional setting ("three-dimensional medium")

The three-dimensional medium is represented by the cube of Figure 2.4(b), attached to the xyz reference mark. At point M, the state of stresses and forces are respectively defined by:

– the three normal stresses σ_{xx}, σ_{yy} and σ_{zz}, as well as the three tangential stresses τ_{xy}, τ_{yz} and τ_{xz} and their symmetries;

– the three components of the force $\vec{F}$: $\begin{cases} f_x \\ f_y \\ f_z \end{cases}$ [2.10]

The same approach, previously undertaken for the two-dimensional medium, leads to the following equilibrium equations in the three-dimensional setting:

$$\begin{cases} \dfrac{\partial \sigma_{xx}}{\partial x} + \dfrac{\partial \tau_{yx}}{\partial y} + \dfrac{\partial \tau_{zx}}{\partial z} = f_x \\ \dfrac{\partial \tau_{xy}}{\partial x} + \dfrac{\partial \sigma_{yy}}{\partial y} + \dfrac{\partial \tau_{zy}}{\partial z} = f_y \\ \dfrac{\partial \tau_{zx}}{\partial x} + \dfrac{\partial \tau_{yz}}{\partial y} + \dfrac{\partial \sigma_{zz}}{\partial z} = f_z \end{cases} \qquad [2.11]$$

COMMENTS.– The equilibrium equations obtained in equation [2.9] and equation [2.11] result from the equilibrium of the forces (assumed to be zero at any point of the solid).

– The void of the moments in any point of the solid leads to the deduction of the equalities between the symmetrical tangential stresses given by the relations in equation [2.7].

– Most often, the volume forces are reduced to the gravity forces, where, in the representation of Figure 2.4 (ascending vertical axis oz), we obtain:

$$\vec{F}: \begin{cases} f_x = 0 \\ f_y = 0 \\ f_z = -\gamma \end{cases} \qquad [2.12]$$

2.1.3.6. *General formulation of equilibrium equations*

Under dynamic loads where the solid, of homogeneous density ρ is subjected to an acceleration γ_i, the equilibrium equations (also called "equations of motion") can be written in the following general form:

$$\sigma_{ij,j} + f_i = \rho \times \ddot{u} \qquad [2.13]$$

with:

– $\sigma_{ij,j}$: designates the tensor of the six stresses considered in a 1-2-3 reference mark (instead of xyz) and derived from j (j = 1–3), as in the equilibrium equations of the three-dimensional medium given by the equalities [2.11];

– f_i: represents the three components (i = 1–3) of the volume force vector $\vec{F}$ (equation [2.10]);

– $\ddot{u}$: these are the three components of the acceleration vector $\vec{\gamma}$: $\begin{cases} \gamma_1 \\ \gamma_2 \\ \gamma_3 \end{cases}$ (defined in equation [2.24]).

COMMENT.– Equation [2.13] is strictly identical to the fundamental relation of dynamics taught in high school, where the force (F) is expressed as:

$$\vec{F} = M \times \vec{\gamma}$$

where M is the mass of the solid subjected to the acceleration $\vec{\gamma}$.

2.1.4. *Generalization of the deformation state*

Under the stress state of Figure 2.4, a deformation state is developed, which is represented by the deformation tensor, usually noted ($\overline{\varepsilon}$), and is written as:

$$\overline{\varepsilon} = \begin{bmatrix} \varepsilon_{xx} & \gamma_{xy} & \gamma_{xz} \\ \gamma_{yx} & \varepsilon_{yy} & \gamma_{yz} \\ \gamma_{zx} & \gamma_{zy} & \varepsilon_{zz} \end{bmatrix} \qquad [2.14]$$

COMMENTS.–

1) As for the stress tensor, let us also distinguish the following:

– volume deformations, called "volumetric strain" located on the diagonal line (ε_{xx}, ε_{yy} and ε_{zz});

– deformations due to shear stress, located on both sides of the diagonal, called "angular deformations" or "distortions".

2) The volumetric strain corresponds to the volume variation (ΔV) of the volume element (V), that is to say:

$$\frac{\Delta V}{V} = \varepsilon_{xx} + \varepsilon_{yy} + \varepsilon_{zz} \qquad [2.15]$$

3) Due to symmetry, the angular deformations are identical; in other words:

$$\begin{cases} \gamma_{xy} = \gamma_{yx} \\ \gamma_{yz} = \gamma_{zy} \\ \gamma_{xz} = \gamma_{zx} \end{cases} \qquad [2.16]$$

4) As for the stresses, there are also principal planes where principal deformations are exerted when the angular deformations are zero: ε_1, ε_2 and ε_3.

5) The principal directions of the stresses coincide with the principal directions of the deformations.

2.1.5. *Deformation–displacement relations*

Let us consider the solid S attached to the xyz reference mark of Figure 2.9. Point M of coordinates (x, y, z) undergoes displacements (point M') in the three directions represented by the vector:

$$\vec{U}:\begin{cases} u \\ v \\ w \end{cases} \qquad [2.17]$$

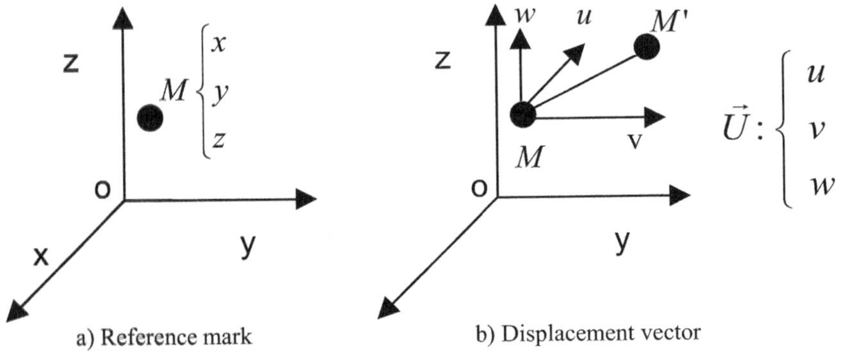

Figure 2.9. *Displacement vector of a solid*

In the general xyz reference mark, the principal deformations are given by the following equations, in terms of displacements:

$$\begin{cases} \varepsilon_{xx} = \dfrac{\partial u}{\partial x} \\ \varepsilon_{yy} = \dfrac{\partial v}{\partial y} \\ \varepsilon_{zz} = \dfrac{\partial w}{\partial z} \end{cases} \qquad [2.18]$$

And the angular deformations are given by:

$$\begin{cases} \gamma_{xy} = \dfrac{1}{2}(\dfrac{\partial u}{\partial y} + \dfrac{\partial v}{\partial x}) \\ \gamma_{yz} = \dfrac{1}{2}(\dfrac{\partial v}{\partial z} + \dfrac{\partial w}{\partial y}) \\ \gamma_{zx} = \dfrac{1}{2}(\dfrac{\partial w}{\partial x} + \dfrac{\partial u}{\partial z}) \end{cases} \qquad [2.19]$$

COMMENTS.–

1) Deformations are defined in the scope of small deformations.

2) Under this assumption, the deformation tensor can be expressed as a function of the displacements in a three-dimensional $x_1x_2x_3$ reference mark (instead of xyz), in the generalized form known as the Cauchy equations, that is to say:

$$\varepsilon_{ij} = \frac{1}{2}\left(\frac{\partial u_i}{\partial x_j} + \frac{\partial u_j}{\partial x_i}\right) \qquad [2.20a]$$

3) In the scope of large displacements, these are the cross-referenced terms that add to the quantities in equation [2.20a], namely:

$$\varepsilon_{ij} = \frac{1}{2}\left(\frac{\partial u_i}{\partial x_j} + \frac{\partial u_j}{\partial x_i}\right) + \frac{\partial u_i}{\partial u_j} \cdot \frac{\partial u_j}{\partial x_i} \qquad [2.20b]$$

2.1.6. *Deformation rate*

Considering a 1-2-3 reference mark, where the displacement vector, analogous to that of Identity [2.17], has the following components:

$$\vec{U} : \begin{cases} u_1 \ (u) \\ u_2 \ (v) \\ u_3 \ (w) \end{cases} \qquad [2.20c]$$

The components of the deformation rate vector ($\bar{\dot{\varepsilon}}$) are defined in the same reference mark by:

$$\varepsilon_i = \frac{\partial u_i}{\partial t} \qquad [2.21]$$

or, by:

$$\vec{\dot{\varepsilon}} : \begin{cases} \dot{\varepsilon}_1 = \dfrac{\partial u_1}{\partial t} \\ \dot{\varepsilon}_2 = \dfrac{\partial u_2}{\partial t} \\ \dot{\varepsilon}_3 = \dfrac{\partial u_3}{\partial t} \end{cases} \qquad [2.22]$$

2.1.7. *Acceleration*

The components of the acceleration vector are given by the second derivatives of the displacements, which are expressed in the general form by:

$$\gamma_i = \frac{\partial^2 u_i}{\partial^2 t} \tag{2.23}$$

Or, attached to the reference mark, by:

$$\vec{\gamma} : \begin{cases} \gamma_1 = \dfrac{\partial^2 u_1}{\partial^2 t} \\ \gamma_2 = \dfrac{\partial^2 u_2}{\partial^2 t} \\ \gamma_3 = \dfrac{\partial^2 u_3}{\partial^2 t} \end{cases} \tag{2.24}$$

In soils, the relations between stresses and deformations are complex and the solution for problems in soil mechanics often requires the use of behavioral laws. We are interested here in the behavioral models that are most commonly used in geotechnical problems and programmed in geotechnical calculation software: PLAXIS, CESAR-LPC, FLAC, etc.

2.2. Behavioral laws

The behavior of a material is described by a mathematical relationship between the stresses caused by the loading and the resulting deformations. This relationship translates the "response" of the material to the applied stress. It is a "behavioral law" defined in the framework of a soil "behavior model".

The relationship between the applied stresses and the resulting deformations depends on the behavior of the material. Indeed, there are materials that can be reasonably associated with elastic behavior (steel, for example) and other materials with more or less complex behavior, such as soil that can be loosely attributed to several behavior laws:

– elasto-plastic behavior;

– viscoelastic behavior;

– creep behavior.

2.2.1. *Elastic behavior model*

There are two schools of thought in this domain:

– For the first, the soil is linear elastic and isotropic in a very small range of deformations ($10^{-6}/10^{-5}$). The relationship between the stresses applied to it and the resulting deformations is linear, and the deformations are reversible. That is to say, once the stress is eliminated, there are no more deformations and the soil returns to its initial position.

– The second school of thought refutes the hypothesis of the first and never considers the soil to be linear elastic. We share the postulate of the second: the soil is not linear elastic and Young's modulus (E) improperly attributed to the soil does not make sense. It is simply a "conventional" deformation modulus, determined from on-site and/or laboratory measurements.

A linear and isotropic elastic medium is characterized by the following modules:

– modulus of elastic deformation "Young's modulus": E;

– modulus of compressibility: K;

– shear modulus: G.

2.2.1.1. *Case of a one-dimensional setting ("one-dimensional" medium)*

To simplify the concepts and understand the notion of behavioral law, let us consider the case of a "one-dimensional" sample on which a normal stress (σ) is applied, creating a shortening (Δh) of the sample and involving a deformation (ε), as shown in Figure 2.10.

In the assumed linear elastic and isotropic medium, the relationship between stress (σ) and deformation (ε) is as follows:

$$\sigma = E.\varepsilon \quad [2.25]$$

This is the law of linear elasticity ("Hooke's law"), applied to the one-dimensional medium. This law characterizes a model with just one parameter, which is Young's modulus E. This modulus represents the elastic sample assimilated to a spring. In soil mechanics, the conventional deformation modulus E (noted elsewhere as E_s and which is not a Young's modulus because the soil is not elastic) is determined experimentally, either specifically in a laboratory or more generally from on-site tests.

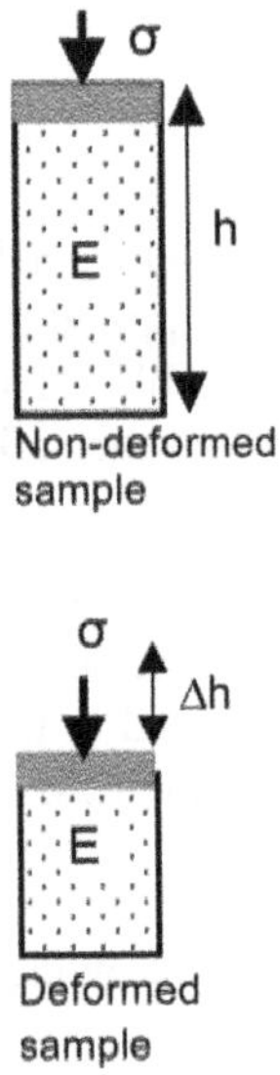

Experimentally, the soil deformation modulus E_s can be assessed by:

$$E_s = \frac{\sigma}{\frac{\Delta h}{h}}$$

The deformation is:

$$\varepsilon = \frac{\Delta h}{h}$$

The stress-deformation relationship is:

$$\sigma = E_s \varepsilon$$

a) Soil sample b) Stress-deformation relationship

Figure 2.10. *Behavioral law in a one-dimensional medium*

2.2.1.2. *Octahedral stresses and deformations*

In a linear and isotropic elastic medium, the stress-deformation relations are expressed by linear law. We can also write:

$$\sigma_{oct} = K \times \varepsilon_{oct} \quad [2.26]$$

and

$$\tau_{oct} = G.\gamma_{oct} \quad [2.27]$$

where, in a 1-2-3 reference mark:

– σ_{oct} is the octahedral normal stress, also called “mean or spherical stress”, given by:

$$\sigma_{oct} = \frac{\sigma_1 + \sigma_2 + \sigma_3}{3} \quad [2.28]$$

– τ_{oct} is the octahedral tangential stress, also called "deviatoric stress", expressed by:

$$\tau_{oct} = \frac{1}{3}\sqrt{(\sigma_1 - \sigma_2)^2 + (\sigma_2 - \sigma_3)^2 + (\sigma_3 - \sigma_1)^2} \qquad [2.29]$$

COMMENTS.– Similarly, the octahedral components (ε_{oct} and γ_{oct}) of the deformation state are defined as follows:

$$\varepsilon_{oct} = \frac{\varepsilon_1 + \varepsilon_2 + \varepsilon_3}{3} \qquad [2.30]$$

and $\gamma_{oct} = \frac{1}{3}\sqrt{(\varepsilon_1 - \varepsilon_2)^2 + (\varepsilon_2 - \varepsilon_3)^2 + (\varepsilon_3 - \varepsilon_1)^2}$ [2.31]

2.2.1.3. *Case of a two-dimensional setting ("two-dimensional medium")*

In a two-dimensional medium (2D medium), a distinction must be made between the state of plane stresses (Figure 2.11(a)) and the state of plane deformations (Figure 2.11(b)).

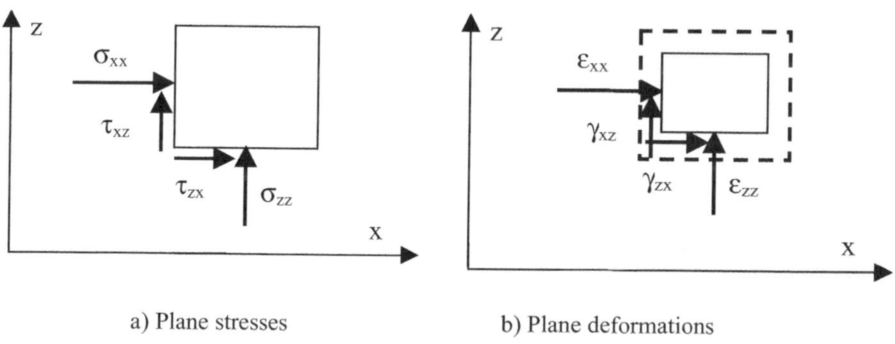

Figure 2.11. *State of stresses and deformations in a 2D medium*

Plane stresses

In an isotropic linear elastic medium (Poisson's ratio ν and E modulus), the plane stresses are related to the deformations in an xz reference mark by:

$$\begin{cases} \sigma_{xx} = \dfrac{E}{(1-\nu^2)}(\varepsilon_{xx} + \nu.\varepsilon_{zz}) \\ \sigma_{zz} = \dfrac{E}{(1-\nu^2)}(\varepsilon_{zz} + \nu.\varepsilon_{xx}) \\ \tau_{zx} = \dfrac{E}{(1+\nu)} \times \gamma_{zx} \end{cases} \quad [2.32]$$

COMMENTS.–

1) In the plane stress state, the following stresses and deformations are zero in the chosen xz reference mark:

$$\begin{cases} \sigma_{yy} = \tau_{yx} = \tau_{zy} = 0 \\ \text{and} \\ \gamma_{yx} = \gamma_{yz} = 0 \end{cases} \quad [2.33]$$

2) In the principal direction that is perpendicular to the (oxz) plane, the principal deformation (ε_{yy}) is given by:

$$\varepsilon_{yy} = \frac{\nu}{E}(\sigma_{xx} + \sigma_{zz}) \quad [2.34]$$

Plane deformations

In an isotropic linear elastic medium, plane deformations are related to stresses, in the same xz reference mark, by:

$$\begin{cases} \varepsilon_{xx} = \dfrac{1+\nu}{E}[(1-\nu) \times \sigma_{xx} - \nu \times \sigma_{zz}] \\ \varepsilon_{zz} = \dfrac{1+\nu}{E}[(1-\nu) \times \sigma_{zz} - \nu \times \sigma_{xx}] \\ \gamma_{zx} = \dfrac{1+\nu}{E} \times \tau_{zx} \end{cases} \quad [2.35]$$

COMMENTS.–

1) In the plane deformation state, the following deformations and stresses are zero in the xz reference mark:

$$\begin{cases} \varepsilon_{yy} = \gamma_{yx} = \gamma_{zy} = 0 \\ \text{and} \\ \tau_{yx} = \tau_{zy} = 0 \end{cases} \qquad [2.36]$$

2) In the principal direction that is perpendicular to the (oxz) plane, the principal stress (σ_{yy}) is given by:

$$\sigma_{yy} = \nu \times (\sigma_{xx} + \sigma_{zz}) \qquad [2.37]$$

2.2.1.4. *Case of a three-dimensional setting (three-dimensional medium)*

In a linear and isotropic three-dimensional elastic medium, the stresses (equation [2.6]) are related to the deformations (equation [2.14]) by:

$$\begin{cases} \sigma_{xx} = \dfrac{E}{(1+\nu)\times(1-2\nu)}\left[(1-\nu)\varepsilon_{xx} - \nu\times\left(\varepsilon_{yy} + \varepsilon_{zz}\right)\right] \\ \sigma_{yy} = \dfrac{E}{(1+\nu)\times(1-2\nu)}\left[(1-\nu)\varepsilon_{yy} - \nu\times\left(\varepsilon_{zz} + \varepsilon_{xx}\right)\right] \\ \sigma_{zz} = \dfrac{E}{(1+\nu)\times(1-2\nu)}\left[(1-\nu)\varepsilon_{zz} - \nu\times\left(\varepsilon_{yy} + \varepsilon_{xx}\right)\right] \end{cases} \qquad [2.38]$$

and

$$\begin{cases} \tau_{xy} = \dfrac{E}{(1+\nu)}\times\gamma_{xy} \\ \tau_{zy} = \dfrac{E}{(1+\nu)}\times\gamma_{zy} \\ \tau_{zx} = \dfrac{E}{(1+\nu)}\times\gamma_{zx} \end{cases} \qquad [2.39]$$

And conversely, deformations are related to stresses by:

$$\begin{cases} \varepsilon_{xx} = \dfrac{1}{E}\left[\sigma_{xx} - \nu\times\left(\sigma_{yy} + \sigma_{zz}\right)\right] \\ \varepsilon_{yy} = \dfrac{1}{E}\left[\sigma_{yy} - \nu\times\left(\sigma_{xx} + \sigma_{zz}\right)\right] \\ \varepsilon_{zz} = \dfrac{1}{E}\left[\sigma_{zz} - \nu\times\left(\sigma_{xx} + \sigma_{yy}\right)\right] \end{cases} \qquad [2.40]$$

and

$$\begin{cases} \gamma_{xy} = \dfrac{2(1+\nu)}{E} \times \tau_{xy} \\ \gamma_{yz} = \dfrac{2(1+\nu)}{E} \times \tau_{yz} \\ \gamma_{zx} = \dfrac{2(1+\nu)}{E} \times \tau_{zx} \end{cases} \qquad [2.41]$$

In a 1-2-3 Cartesian reference mark (instead of xyz), the stress tensor (σ_{ij}) can be given (equations [2.38] and [2.39]) in the following general form:

$$\sigma_{ij} = \frac{E}{(1+\nu)(1-2\nu)} \times \left[(1-2\nu)\varepsilon_{ij} - \nu \times \varepsilon_{kk} \times \delta_{ij} \right] \qquad [2.42]$$

and the deformation tensor (ε_{ij}) by (equations [2.40] and [2.41]):

$$\varepsilon_{ij} = \frac{1}{E} \times \left[(1+\nu)\sigma_{ij} - \nu \times \sigma_{kk} \times \delta_{ij} \right] \qquad [2.43]$$

where:

- $i = 1$–3 and $j = 1$–3;
- $\delta_{ij} = 1$ when $i = j$ and is worth 0 when $i \neq j$.

2.2.1.5. *Matrix formulation of stress-deformation relations*

The numerical resolution of problems in soil mechanics, such as in finite elements, requires linking the stresses and deformations by a matrix form that describes the behavioral law of the soil. In general terms, we can thus write that the stress tensor (σ_{ij}) is related to the deformation tensor (ε_{ij}) by:

$$\{\sigma_{ij}\} = [D] \times \{\varepsilon_{ij}\} \qquad [2.44]$$

where $[D]$ designates the matrix of elasticity parameters of the medium (assumed to be linear elastic), which is involved in the relations between stresses and deformations. It is called a "constitutive", as well as a "rheological" matrix, which describes the behavioral law of the soil in the applied state of stress or deformation.

In a two-dimensional medium, stresses are related to deformations in an xz reference mark by:

– plane stresses (Figure 2.11(a) and equations [2.32]–[2.34]):

$$\begin{Bmatrix} \sigma_{xx} \\ \sigma_{zz} \\ \tau_{zx} \end{Bmatrix} = \frac{E}{1-\nu^2} \begin{bmatrix} 1 & \nu & 0 \\ \nu & 1 & 0 \\ 0 & 0 & (1-\nu) \end{bmatrix} \times \begin{Bmatrix} \varepsilon_{xx} \\ \varepsilon_{zz} \\ \gamma_{zx} \end{Bmatrix} \qquad [2.45]$$

– plane deformations (Figure 2.11(b) and equations [2.35]–[2.37]):

$$\begin{Bmatrix} \sigma_{xx} \\ \sigma_{zz} \\ \tau_{zx} \end{Bmatrix} = \frac{E}{(1+\nu)(1-2\nu)} \begin{bmatrix} \nu & 1-\nu & 0 \\ 1-\nu & \nu & 0 \\ 0 & 0 & (1-2\nu) \end{bmatrix} \times \begin{Bmatrix} \varepsilon_{xx} \\ \varepsilon_{zz} \\ \gamma_{zx} \end{Bmatrix} \qquad [2.46]$$

In many practical geotechnical problems, the stresses applied to the soil are symmetrical with respect to the axis of the sample. These are the "axisymmetric condition". Such is the case of a cylinder filled with liquid exerting a pressure p on the walls of the cylinder, and which is axially compressed (Figure 2.12(a)), or a sample that is laterally confined by a uniform pressure p and compressed at the top by a vertical stress σ_{zz} (Figure 2.12(b)). In both solids, the stress state is described in a reference mark in cylindrical coordinates (r, z, θ), as shown in Figure 2.12(c).

The constitutive law that describes the relations between stresses and deformations under axisymmetric conditions is given by:

$$\begin{Bmatrix} \sigma_{rr} \\ \sigma_{zz} \\ \tau_{rz} \\ \sigma_{\theta\theta} \end{Bmatrix} = \frac{E(1-\nu)}{(1+\nu)(1-2\nu)} \begin{bmatrix} 1 & \frac{\nu}{1-\nu} & 0 & \frac{\nu}{1-\nu} \\ \frac{\nu}{1-\nu} & 1 & 0 & \frac{\nu}{1-\nu} \\ 0 & 0 & \frac{1-2\nu}{2(1-\nu)} & 0 \\ \frac{\nu}{1-\nu} & \frac{\nu}{1-\nu} & 0 & 1 \end{bmatrix} \times \begin{Bmatrix} \varepsilon_{rr} \\ \varepsilon_{zz} \\ \gamma_{rz} \\ \varepsilon_{\theta\theta} \end{Bmatrix} \qquad [2.47a]$$

The equilibrium equation in cylindrical coordinates is given, in a $(r - \theta)$ plane, by:

$$\frac{\sigma_\theta - \sigma_r}{r} - \frac{d\sigma_r}{dr} = 0 \qquad [2.47b]$$

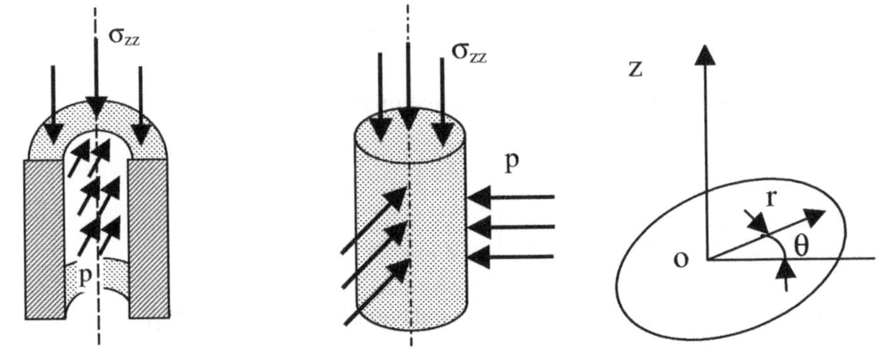

a) Action of a liquid on a compressed hollow cylinder

b) Compressed and laterally confined sample

c) Reference mark in cylindrical coordinates

Figure 2.12. *Axisymmetric conditions*

This equation describes the expansion of a cylindrical cavity, as in Louis Menard's pressuremeter test where a "soft" rubber probe is inflated in a small diameter (63–66 mm) preliminary boring until the ground breaks. We will come back to this in the presentation of the pressuremeter test in section 3.4.5 of Chapter 3, devoted to the determination of geotechnical parameters for the design and justification of geotechnical construction projects.

In a three-dimensional medium, the stress-deformation relation is as follows:

$$\begin{Bmatrix} \sigma_{xx} \\ \sigma_{yy} \\ \sigma_{zz} \\ \tau_{xy} \\ \tau_{yz} \\ \tau_{zx} \end{Bmatrix} = \frac{E(1-\nu)}{(1+\nu)(1-2\nu)} \begin{bmatrix} 1 & \frac{\nu}{1-\nu} & \frac{\nu}{1-\nu} & 0 & 0 & 0 \\ \frac{\nu}{1-\nu} & 1 & \frac{\nu}{1-\nu} & 0 & 0 & 0 \\ \frac{\nu}{1-\nu} & \frac{\nu}{1-\nu} & 1 & 0 & 0 & 0 \\ 0 & 0 & 0 & \frac{1-2\nu}{2(1-\nu)} & 0 & 0 \\ 0 & 0 & 0 & 0 & \frac{1-2\nu}{2(1-\nu)} & 0 \\ 0 & 0 & 0 & 0 & 0 & \frac{1-2\nu}{2(1-\nu)} \end{bmatrix} \times \begin{Bmatrix} \varepsilon_{xx} \\ \varepsilon_{yy} \\ \varepsilon_{zz} \\ \gamma_{xy} \\ \gamma_{yz} \\ \gamma_{zx} \end{Bmatrix}$$

[2.48]

2.2.1.6. *Relationship between elasticity parameters*

The main relations between the elasticity parameters are as follows.

– volumetric deformation modulus *K*:

$$K = \frac{E}{3(1-2\nu)} \tag{2.49}$$

– shear modulus *G*:

$$G = \frac{E}{2(1+\nu)} \tag{2.50}$$

The λ and μ parameters, called “Lamé parameters”, are given by:

$$\lambda = \frac{\nu \times E}{(1-2\nu)(1+\nu)} \tag{2.51}$$

and

$$\mu = G = \frac{E}{2(1+\nu)} \tag{2.52}$$

to which Poisson’s ν ratio is related by:

$$\nu = \frac{\lambda}{2(\lambda+\mu)} \tag{2.53}$$

and Young’s modulus *E* by:

$$E = \frac{\mu \times (3\lambda + 2\mu)}{(\lambda+\mu)} \tag{2.54}$$

2.2.2. *Elasto-plastic behavioral models*

2.2.2.1. Concept of plasticity

The plastic behavior of a solid is characterized by residual deformations. That is to say, unlike elastic behavior, a solid does not return to its initial position when the force that deformed it is removed. Before explaining the main models used in soil mechanics, let us look at the simple example of a beam-bracket of an average section (S) that is subjected to a one-dimensional compression N at its extremity (Figure 2.13(a)). Under the force N, a normal compressive stress (σ) is mobilized, given by a relation analogous to the equations in equation [2.2] and described by Figure 2.36(a), which will be presented later. This stress (σ) generates a deformation (ε) in the solid body – which is supposedly linear elastic – that is related to the stress by Hooke's law (equation [2.25]). Plasticity appears when the deformation (ε) developed under the stress (σ) reaches a deformation threshold (ε_e), corresponding to a stress level (σ_e) in the beam. We can then distinguish three possible cases (Figure 2.13(b)):

Case 1

The current deformation (ε) remains below the deformation threshold (ε_e), the stress σ is lower than σ_e (line 0A on the curve in Figure 2.13(c)), giving us:

$$\sigma < \sigma_e \rightarrow \sigma - \sigma_e < 0 \qquad [2.55]$$

the beam then behaves elastically and returns to its initial position ($\varepsilon = 0$), as soon as the force N ($\sigma = 0$) is removed.

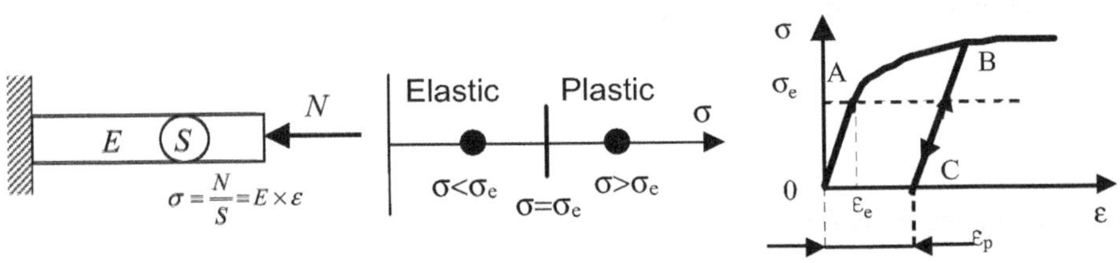

a) Compressed beam b) Boundaries between domains C) Stress-deformation curve

Figure 2.13. *Concept of plasticity in a one-dimensional medium*

Case 2

The current deformation (ε) reaches the deformation threshold (ε_e), the stress σ is identical to σ_e (point A on the curve in Figure 2.13(c)) and gives the following:

$$\sigma = \sigma_e \rightarrow \sigma - \sigma_e = 0 \quad [2.56]$$

the beam is at the yield point and plastic deformations will appear as soon as ε exceeds ε_e.

Case 3

The current deformation (ε) is larger than the deformation ε_e, the stress σ increases beyond the value of σ_e (path AB on the curve in Figure 2.13(c)), which leads to:

$$\sigma > \sigma_e \rightarrow \sigma - \sigma_e > 0 \quad [2.57]$$

the beam then plasticizes and the residual deformations will cause the beam to gradually break.

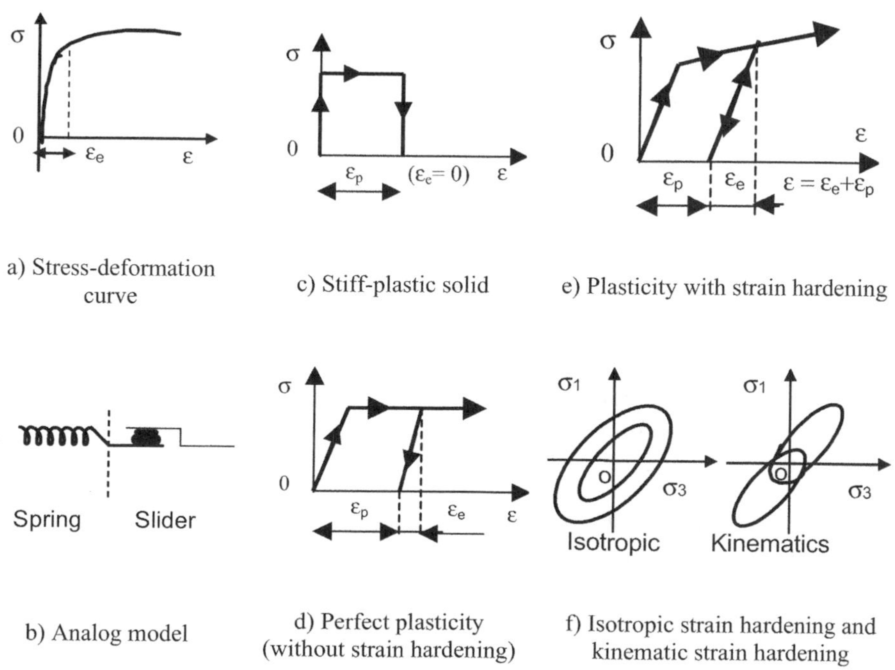

Figure 2.14. *Elasto-plastic models of a solid*

COMMENT ON FIGURE 2.14.–

1) *Plastic deformations* (ε_p) *are called "irreversible", "permanent" or "remanent" deformations.*

2) *These deformations appear beyond a certain deformation threshold, characterized by elastic deformations* (ε_e)*, as shown in Figure 2.13(c).*

3) *The elasto-plastic behavior of Figure 2.14(a) can be compared to an analogical model, represented by a spring and a slider (Figure 2.14(b)).*

4) *Like Figure 2.13(c), the stress-deformation curve of Figure 2.14(a) describes the behavior of a deformable medium, where the elastic deformations* (ε_e) *can be mobilized, unlike the soil, which is not elastic.*

5) *A solid is stiff-plastic when the elastic deformations* (ε_e) *are equal to zero (Figure 2.14(c)), like glass which breaks under very small deformations.*

6) *The plasticity of a solid is said to be "perfect" when the plastic deformations are not accompanied by viscosity or strain hardening; in other words, the elastic domain remains fixed (Figure 2.14(d)). In the opposite case (Figure 2.14(e)), the strain hardening can be "isotropic" (Taylor and Quinney 1931), that is to say, the elastic domain inside the plasticity criterion is transformed by the homothetic center during the deformations, that are "kinematic" (Prager 1955, 1958), which corresponds to a translation of the elastic domain in the stress space (Figure 2.14(f)).*

7) *The total deformation* (ε) *is the sum of the two deformations* (ε_e) *and* (ε_p)*.*

2.2.2.2. *Soil behavior*

Soil is not elastic and, contrary to an elastic solid (Figure 2.15(a)), the stress-deformation curve follows a shape that can be represented by Figure 2.15(b). Indeed, after deformation (ε), the elastic solid subjected to a given stress (σ) returns to its initial position according to a linear curve, that is to say, as soon as the stress is removed ($\sigma = 0$) the deformation is equal to zero ($\varepsilon = 0$). This is exactly the case for an elastic thread (Figure 2.15(a)).

As described in Chapter 1, soil is composed of a solid skeleton where the discontinuities created by the presence of water and gas between grains make the soil behavior, and the relations between the stresses and deformations in it, complex. Soil is not very reversible and deformations are mainly of plastic origin.

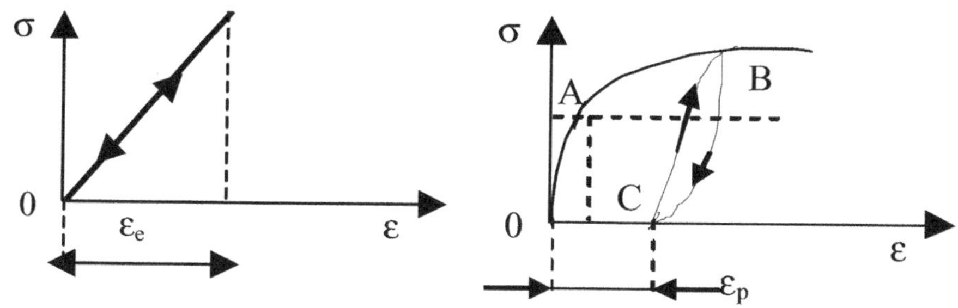

a) Behavior of an elastic solid

b) Soil behavior

Figure 2.15. *Behavior of an elastic solid and soil*

The following should be noted when reading the stress paths of a classical stress-deformation diagram.

– In the low deformation range (path 0A in Figure 2.15(b)), the deformations are not very reversible; in other words, the stress-deformation curve is not linear and the deformations (ε) do not cancel out immediately when the stress (σ) is removed.

– Beyond this threshold of stress and deformation, the medium plasticizes (path AB in Figure 2.15(b)) and the deformations are irreversible, that is to say, when the stress (σ) is cancelled, a residual deformation (ε_p) remains in the soil (path 0C in Figure 2.15(b)).

– When unloading the soil again (zero σ stress), the stress-deformation curve follows a path that is more or less analogous to the BC path (Figure 2.15(c)). On reloading, the stress-deformation curve follows the CB path, which is very close to the unloading path BC.

The nonlinearities and the irreversible character of the soil deformations result from the discontinuity and the complexity of the laws of contact between the grains, which constitute the skeleton of the soil. Depending on the state of compactness of the grains, the soil can be loose or dense. Soil is in a dense state when the grains are tightly packed; it is in a loose state in the opposite case. In both states, when soil undergoes a shear stress τ, it deforms and its volume changes (volume variation ΔV).

In granular soil (sand, for example), the soil shearing results in a volume variation which, depending on the state of the grains, involves the following.

– In the case of a soil in a loose initial state (Figure 2.16(a)), a re-entanglement of solid grains with a decrease in volume (Figure 2.16(b)), represented by the following (equation [2.15]):

$$\frac{\Delta V}{V} < 0$$

In this case, the soil is called "contracting soil", where the stress-deformation curve follows a curve that stabilizes according to the angular deformations (γ) on a shear stress level (τ), called "critical state" (Figure 2.16(c.1)), and the deformations of volumetric order (ΔV) decrease (Figure 2.16(c.2)). This is the behavior of loose sand, for example.

– In the case of disentanglement of soil grains when the initial soil state is dense (Figure 2.17(a)), with an increase in volume (Figure 2.17(b)), represented by (equation [2.15]):

$$\frac{\Delta V}{V} > 0$$

In this case, the soil is referred to as "dilatant". The stress-deformation curve increases toward a peak stress and the stresses "relax" to stabilize at the critical state level (Figure 2.17(c.1)). The deformations of volumetric order (ΔV) increase with distortion (Figure 2.17(c.2)). This is the case with a dense sand.

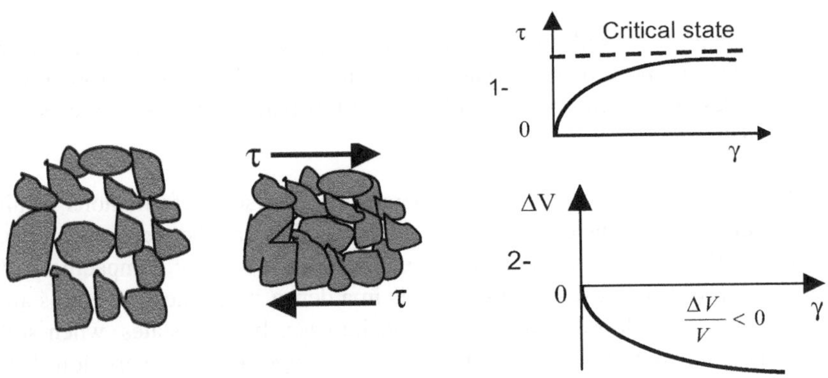

a) Loose initial state b) Re-entanglement of the grains ($\frac{\Delta V}{V}<0$) c) Contracting soil (case of loose sand)

Figure 2.16. *Behavior of a soil sample subjected to shear stress (τ); case of loose soil (loose particles)*

On the other hand, the state of compactness of a saturated clay is indicated by its degree of consolidation. Thus, we can have under-consolidated clays, normally consolidated clays and over-consolidated clays. In practice, there is a similar behavior between sands and clays (Figure 2.17).

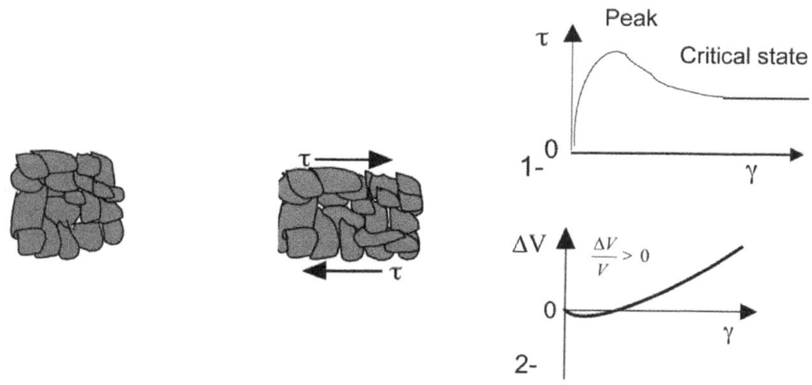

a) Dense initial state b) Disentanglement of the grains ($\frac{\Delta V}{V} > 0$) c) Contracting soil (case of a dense sand)

Figure 2.17. *Behavior of a soil sample submitted to shear stress (τ); case of dense sand (tightly packed grains)*

COMMENT ON FIGURE 2.17.–

1) *The behavior of saturated clays is similar to that of sands. Indeed, clays can be either under-consolidated or over-consolidated. In an over-consolidated clay that is sheared under a mean stress and is lower than the pre-consolidation stress, an increase in volume can develop, as in a dense sand ("dilatancy" phenomenon), and the stress-deformation curve (curve I in Figure 2.18(a)) passes through a peak level and stabilizes at the critical state level, toward which the curve II of the under-consolidated clay also tends.*

2) *The dilatant behavior of saturated clay results in a decrease in pore pressures (curve I on Figure 2.18(b)), whereas when the soil contracts, pore pressures increase due to tightening of the grains (curve II on the same figure).*

3) *These peak and critical state resistance levels differ from the shear resistance, which, after developing following large displacements, decreases to reach a "residual" resistance level. At the peak level, the grains are tightly packed. At the*

critical state level, the adhesion between grains begins to decrease until they break up at the residual resistance level (Figure 2.18(c)).

4) *The distortions (γ_I) developed in the contracting soil sample are much larger than the deformations (γ_{II}) that are mobilized in the dilatant soil sample. Generally speaking, in dilatant soils, only a few percent (1%–5%) is needed to reach peak deformation level, whereas in contracting soils, four to five times more is needed to reach the limit (around 20%).*

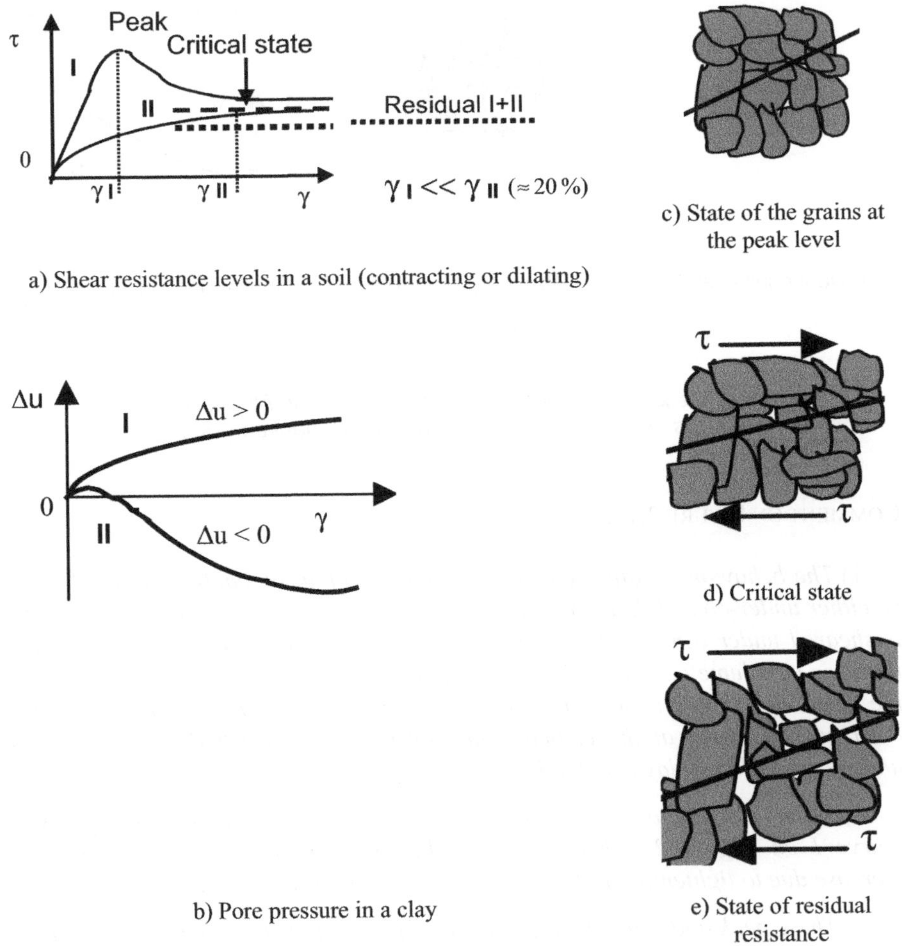

Figure 2.18. *Behavior of a soil sample subjected to shear stress τ*

2.2.2.3. *Behavioral models*

Let us return to the example of the beam-bracket shown in Figure 2.13, illustrating its compression behavior. The diagram in Figure 2.13(b) shows that there is a boundary that separates the elastic domain from the plastic domain. This boundary is described by equation [2.56], which can be written in the following general form:

$$F(\sigma) = 0 \quad \text{(Boundary)} \qquad [2.58]$$

When the deformations are elastic (path 0A on the curve shown in Figure 2.13(c)), we can write (equation [2.55]):

$$F(\sigma) < 0 \ \text{(Elasticity domain)} \qquad [2.59]$$

Similarly, in the plastic domain (path AB on the curve in Figure 2.13(c)), we can write (equation [2.57]):

$$F(\sigma) > 0 \quad \text{(Failure domain)} \qquad [2.60]$$

Indeed, the function $F(\sigma)$ defines a plasticity criterion named "yield function", which distinguishes the elastic domain from the plastic domain.

In an isotropic medium that is only governed by principal stresses σ_1, σ_2 and σ_3 (Figures 2.5 and 2.6(b)), the load function limiting the elastic domain is:

$$F(\sigma_1, \sigma_2, \sigma_3) < 0 \quad \text{(Elasticity domain)} \qquad [2.61]$$

In the Mohr graphical representation (Figure 2.6), there is an envelope curve known as the "intrinsic curve" (Figure 2.19) that separates stress domain that can be supported by the soil (interior of the curve) from the stress domain causing its failure (shaded outer part of the intrinsic curve in Figure 2.19).

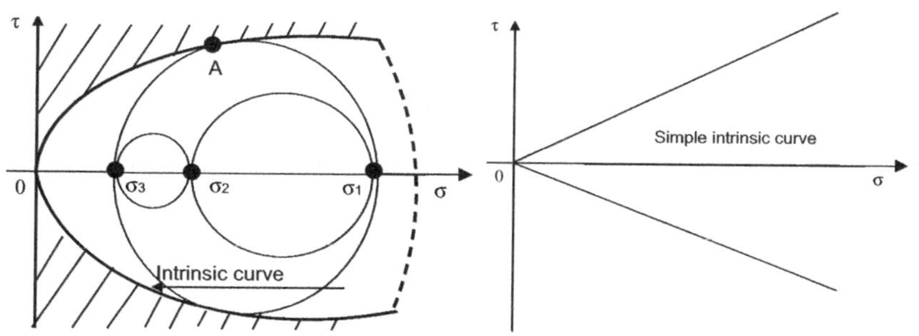

Figure 2.19. *Intrinsic soil curve in the Mohr graphical representation*

COMMENT ON FIGURE 2.19.–

1) *When Mohr's circle is tangent to the intrinsic curve (point A), soil failure occurs by sliding on the facet that corresponds to point A, which establishes the contact between the circle and the curve which envelops it.*

2) *Experimental studies and the development of digital tools show that the intrinsic curve presented in Figure 2.19 cannot accurately describe the behavior of the soil under the main stresses (σ_1, σ_2 and σ_3), and that it is necessary to close the elastic domain on the axis of these stresses (dashed line).*

3) *Indeed, a solid cannot be indefinitely stressed by the principal stresses. This can be likened to the case of a balloon that eventually bursts when it is over-inflated (spherical stresses).*

4) *This observation has led several authors (Cambridge Models 1958–1969; Di Maggio and Sendler 1971; Di Maggio et al. 1976; Wild 1977a, 1977b, 1979; Vermer 1978; Nova and Wood 1979) to propose models where the elasticity domain is closed.*

In soil mechanics, several behavioral models have been developed over the last few decades. For the most part, these are models developed in academic institutions and calibrated in the laboratory on homogeneous and isotropic samples (Hostun and Fontainebleau sand which is very clean in France). These models are often complex and introduce a large number of parameters, determined under ideal conditions and far from the realities of the soil in situ and its heterogeneities. In current practice, some simpler models are programmed in finite element codes ("PLAXIS", "CESAR-LPC", "FLAC", etc.) and are applied to geotechnical structures. We briefly present those that are most used by practitioners.

Mohr–Coulomb model

This model is based on a simplified approach to the intrinsic curve (Figure 2.19). Let us consider the stress diagram of Figure 2.20, the so-called Lambe representation, equivalent to Mohr's circle. By applying the theorem of corresponding states (section 2.3.5 and Figure 2.19 of Volume 2, Chapter 2), dedicated to the shear resistance of soils, we can relate the stresses to each other by the following general relation:

$$F(\sigma) = q - p \times \sin\phi - c \times \cos\phi = 0 \qquad [2.62]$$

with:

– q: deviatoric stress given by:

$$q = \frac{\sigma_1 - \sigma_3}{2} \text{ (Radius of Mohr's circle)} \quad [2.63]$$

– p: mean stress expressed by:

$$p = \frac{\sigma_1 + \sigma_3}{2} \text{ (Center of Mohr's circle)} \quad [2.64]$$

This is the cohesion of the soil and φ its angle of friction.

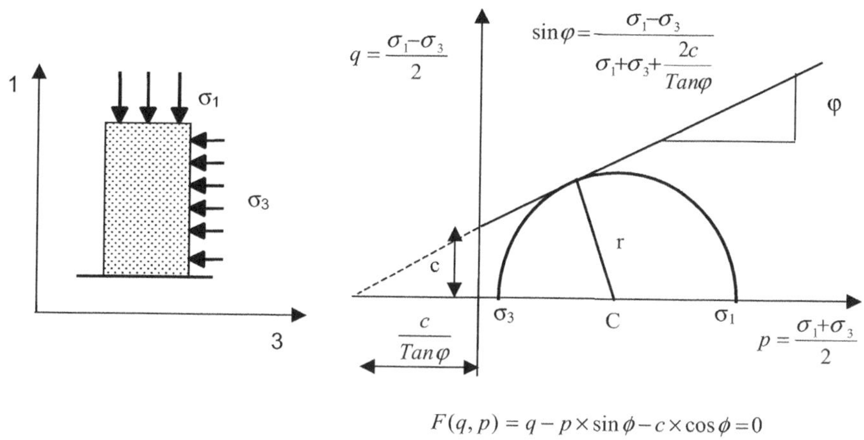

a) State of stresses

b) Intrinsic curve (Lambe diagram)

Figure 2.20. *Mohr-Coulomb soil model*

In dilatant soils (dense sand, over-consolidated clay), some authors define the dilatancy (d) of the material. Rowe (1962) expresses it by the ratio of the volumetric plastic deformation rates ε_{pv} and shear rates ε_{pc}; in other words:

$$d = \frac{\dot{\varepsilon}_{pv}}{\dot{\varepsilon}_{pc}} \quad [2.65]$$

From plane deformation tests on sand, Stroud (1971) and Smith (1972) express the dilatancy as a function of the q/p ratio and of two material constants, k_1 and k_2, that is to say:

$$d = k_1 - k_2 \times \frac{q}{p} \quad [2.66]$$

An angle of dilatancy (ψ) is usually defined by:

$$\sin\psi = \frac{\dot{\varepsilon}_{pv}}{\dot{\varepsilon}_{pc}} \quad [2.67]$$

The plastic potential is defined by a relation analogous to equality [2.62] and involving the angle of dilatancy ψ:

$$F(\sigma) = q - p \times \sin\psi - c \times \cos\psi = 0 \quad [2.68]$$

In practice, the angle of dilatancy is given empirically by:

$$\psi = \phi - 30 \text{ degrees} \quad [2.69]$$

Note that when the angle of dilatancy is equal to the angle of friction ($\psi = \varphi$), the material is called "standard".

Tresca criterion

Under undrained conditions, with the soil governed by undrained cohesion (c_u), the Tresca (1864) criterion is widely used; the yield function is written as a function of the deviator q (equation [2.63]) as (Figure 2.21):

$$F(\sigma) = q - c_u = 0 \quad [2.70]$$

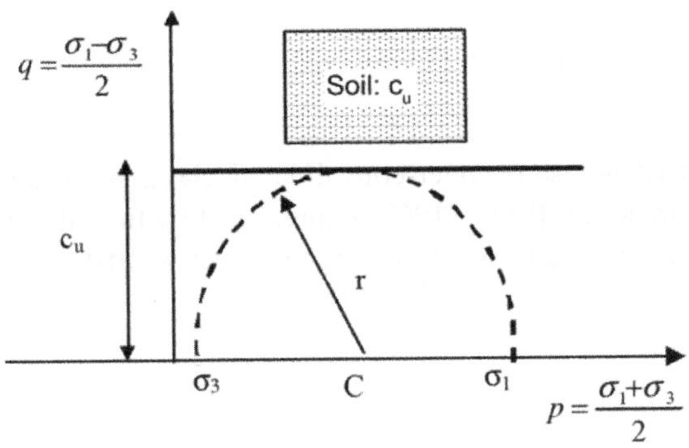

Figure 2.21. *Tresca criterion*

Parabolic criterion

In practice, the foundation material (concrete, steel) is considered to be a linear elastic material. Sometimes, a parabolic criterion is introduced in order to be able to understand the possible plasticization zones of the material. This criterion is written as a function of the spherical (J_1) and deviatoric (J'_2) stress invariants, and the tensile (R_t) and compressive (R_c) resistance of the foundation material; in other words (Figure 2.22):

$$F(\sigma) = J'_2 - \frac{R_c - R_t}{3} \times J_1 - \frac{R_c \times R_t}{3} = 0 \qquad [2.71]$$

with:

$$\begin{cases} J_1 = \sigma_{oct} \\ J'_2 = 2\tau_{oct} \end{cases} \qquad [2.72]$$

σ_{oct} and τ_{oct} are the mean octahedral and deviatoric stresses, respectively, given by equation [2.28] and equation [2.29].

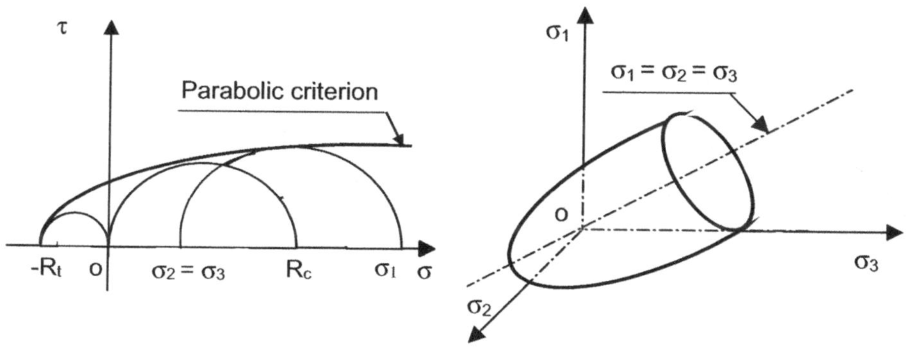

a) In a plane (τ,σ): two-dimensional medium b) In space: three-dimensional medium

Figure 2.22. *Parabolic criterion*

Cam-Clay model

This model, developed between 1958 and 1968 at the University of Cambridge (Soil Mechanics Team: Roscoe, Schofield, Poorooshasb, Wroth and Thurairajah), has had several versions. The last one, named "modified version", introduces a plasticity criterion (limit of the elasticity domain) called "Limit State Curve: LSC" and links the mean stress p, the deviatoric stress (q) and the pre-consolidation pressure (p_0) by (Figure 2.23(a)):

$$F(\sigma) = (\frac{q}{p})^2 - M^2 \times (\frac{p_0}{p} - 1) = 0 \quad [2.73]$$

where:

q, in the Cambridge representation, is twice the radius r of Mohr's circle (Figure 2.20, equation [2.63]); in other words:

$$q = \sigma_1 - \sigma_3 \quad [2.74]$$

p is the mean stress (analogous to equation [2.28]), and M designates a dimensionless parameter given by (Figure 2.23(b)):

$$M = Arc\tan\,(\frac{6\sin\phi}{3-\sin\phi}) \quad [2.75]$$

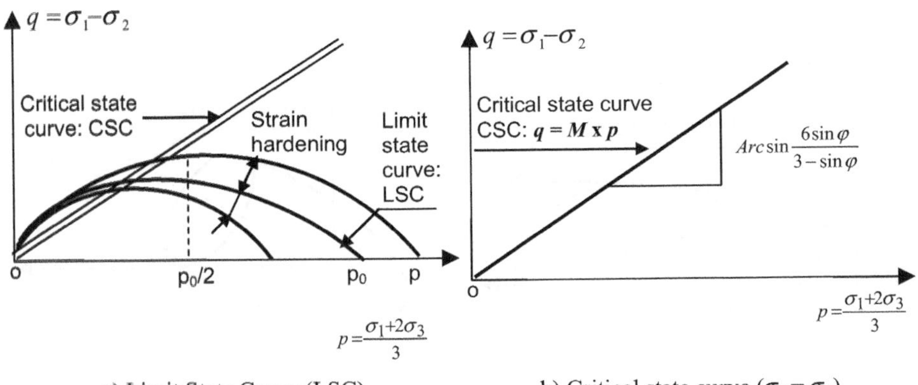

a) Limit State Curve (LSC) b) Critical state curve ($\sigma_2 = \sigma_3$)

Figure 2.23. *Cam-Clay model; plasticity criterion (modified version)*

COMMENT ON FIGURE 2.23.–

1) *The Cam-Clay models are based on the concept of the critical state, represented in the (q, p) plane by a straight line called the "critical state curve: CSC", translating the fact that the soil flows like a liquid when it is continuously*

sheared. This curve is a straight line, represented by the following equation (Figure 2.23(b)):

$$q = M \times p \quad [2.76]$$

2) *They introduce the concept of normality, which stipulates that plastic deformations are carried by a normal force to the load surface ($F(\sigma) = 0$), directed outward. This concept is disputed because it is not generally verified for soils.*

2.2.3. *Viscoelastic behavior models*

The behavior is viscoelastic when it is viscous and linear. This behavior can be represented by the combination of a spring and a damper (of viscosity η). When the spring and damper are in a series, it is the Maxwell model (Figure 2.24(a)). When they are parallel, it is the Kelvin–Voigt model (Figure 2.24(b)). The serial assembly of the two models leads to the Burgers model.

In Maxwell's model, the stress (σ) is related to the elastic strain (ε_e) and the viscous strain rate ($\dot{\varepsilon}_{vs}$) by (Figure 2.24(a)):

$$\sigma = E \times \varepsilon_e = \eta \times \dot{\varepsilon}_{vs} \quad [2.77]$$

and the total deformation (ε) by:

$$\varepsilon = \varepsilon_e + \varepsilon_{vs} \quad [2.78]$$

In the Kelvin–Voigt model, the stress is the sum of the two terms (Figure 2.24(b)), which is:

$$\sigma = E \times \varepsilon + \eta \times \dot{\varepsilon} \quad [2.79]$$

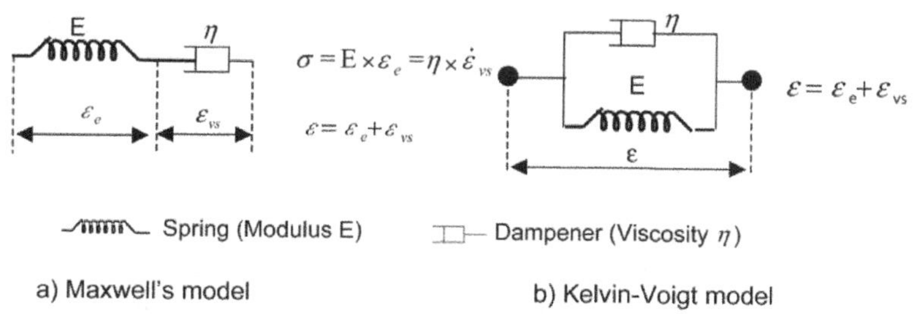

Figure 2.24. *Viscoelastic models*

2.2.4. *Dynamic behavior of soils*

2.2.4.1. *Dynamic excitations*

The behavior of soil under dynamic actions differs significantly from the behavior of soil under static loads or monotonic loads. Under a static load, the load is monotonous, that is to say, it is constant over a given time interval. Under dynamic action, the load fluctuates in time; in other words, it is variable over a given time interval.

In dynamic actions, it is necessary to distinguish:

– the periodic dynamic excitations;

– the non-periodic dynamic excitations.

Figure 2.25 illustrates the case of a simple harmonic dynamic excitation $S(t)$ (induced for example by vibrating machines, Figure 2.25(a)) and that of any periodic excitation (such as the action of a wave on a boat, Figure 2.25(b)).

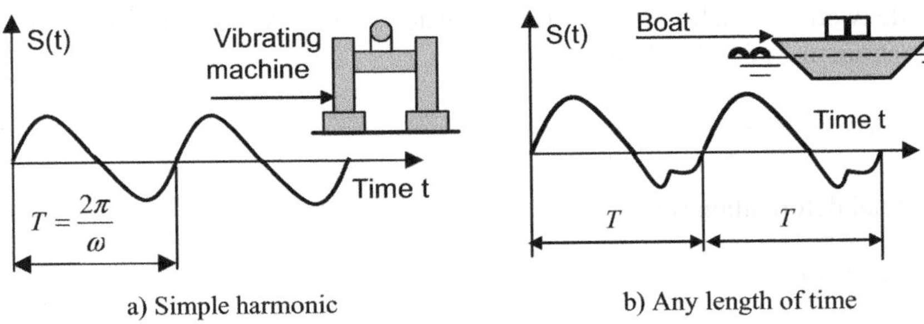

Figure 2.25. *Periodic dynamic excitations*

Figure 2.26 shows other types of non-periodic dynamic excitations, such as impulsive actions. An example of this would be the case of a bomb explosion in the vicinity of a building (Figure 2.26(a)) or non-periodic loads of long duration, such as the force induced on the surface of the soil by an earthquake (Figure 2.26(b)).

2.2.4.2. *Characteristic parameters in soil dynamics*

Consider a soil sample subjected to cyclic shear stress $\tau(t)$. The sample is subjected to closed loading-unloading cycles centered at the origin (Figure 2.27(a)). For a closed cycle, the soil behavior is characterized by a so-called "hysteresis loop", whose area and inclination depend on the magnitude of the distortion (γ_c) during the cycle. The hysteresis loop is characterized by two fundamental parameters:

– the shear modulus, called the "secant modulus" and noted as G (or G_s), which designates the slope of the straight line connecting the two ends of the loop (Figure 2.27(b));

– the damping coefficient, noted as β, which represents the energy dissipated by the soil during the cycle, thus the area of the loop (Figure 2.27(b)).

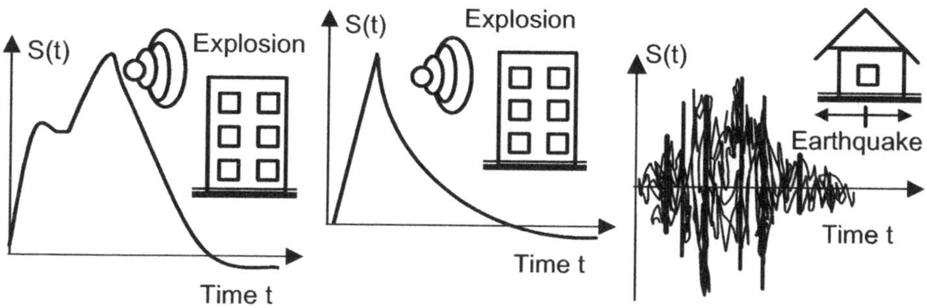

a) Any impulse (explosion in the vicinity of a building)

b) Surface force due to an earthquake

Figure 2.26. *Non-periodic dynamic excitations*

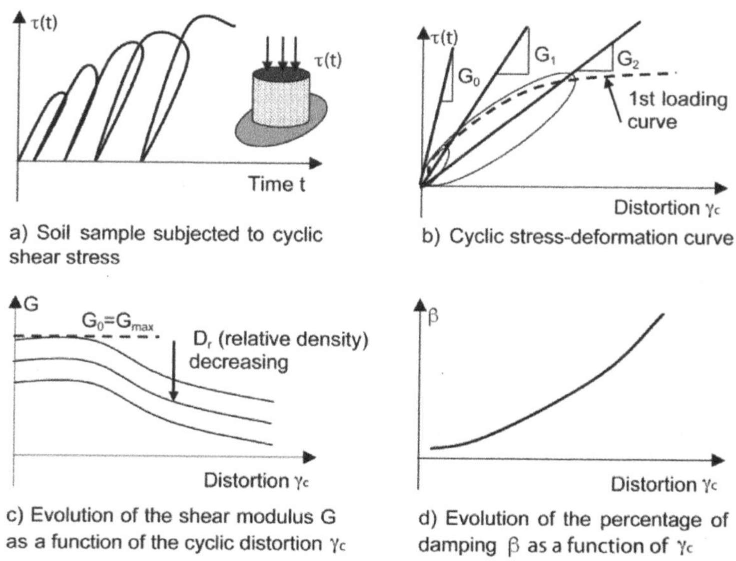

a) Soil sample subjected to cyclic shear stress

b) Cyclic stress-deformation curve

c) Evolution of the shear modulus G as a function of the cyclic distortion γ_c

d) Evolution of the percentage of damping β as a function of γ_c

Figure 2.27. *Behavior of a soil sample subjected to cyclic shear stress: characteristic parameters (G and β)*

COMMENT ON FIGURE 2.27.–

1) *As the distortion* γ_c *increases, the soil deforms and the shear modulus (G) decreases.*

2) *The shear modulus* (G_0), *which is at a maximum at the origin of the distortions* ($\gamma_c = 0$), *corresponds to the elastic behavior of the soil (equation [2.50]).*

3) *The decrease of the shear modulus as a function of distortions is accentuated for soils when the relative density* D_r *decreases (Table 1.4, Chapter 1) (Figure 2.27(c), case of loose sand, for example).*

4) *The damping* (β) *percentage increases as the soil distortion* (γ_c) *increases (Figure 2.27(d)).*

The shear modulus G (G_s) and the damping β percentage are the two main characteristic parameters that depend on the distortion γ_c. Several authors have proposed relations relating G and β to γ_c. Seed and Idriss (1970) presented average experimental curves for sands and clays. Kondner (1963) developed a hyperbolic model where G and β are related to distortion by the following relations:

$$G = \frac{G_{max}}{1+\gamma_c} \quad [2.80]$$

and

$$\beta = \frac{\gamma_c}{1+\gamma_c} = \beta_{max} \times \left(1 - \frac{G}{G_{max}}\right) \quad [2.81]$$

In the Kondner (1963) model, the shear stress cyclic deformation curve is represented by the following hyperbolic relation (Figure 2.28(a)):

$$\tau = \frac{G_{max}}{1+\frac{\gamma_c}{\gamma_r}} \quad [2.82]$$

where γ_r is the reference distortion given by:

$$\gamma_r = \frac{\tau_{max}}{G_{max}} \quad [2.83]$$

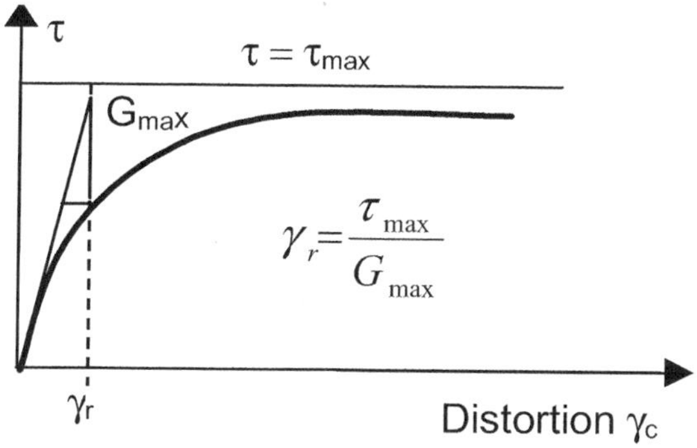

a) Kondner's hyperbolic model (1963): shear stress-cyclic deformation relation

Dynamic parameters :

– Deformation modulus Edyn:

$$E_{dyn} = 2\rho \times v_s^2 \times (1+\nu)$$

– Shear modulus Gdyn:

$$G_{dyn} = \rho \times v_s^2$$

– Compressibility modulus Kdyn:

$$K_{dyn} = \rho \times (v_p^2 - 1 \times 33 \times v_s^2)$$

ρ: volume mass,
vp and vs: compression and shear velocities.

Figure 2.28. *Dynamic data*

As with static loading, the elastic deformations are very small in the soil when it is subjected to repetitive actions. Indeed, depending on the level of deformation reached, the soil behavior can be represented in one of the following behavioral domains (Figure 2.29):

– linear behavior domain, noted as D^L, where deformations are elastic for levels remaining below 10^{-5} (Hujeux 1985), an assumption that can be disputed (the soil is not elastic);

– stabilized hysteretic domain, noted as D^S, where the variations of deformation are supposed to be "elastic" and the permanent deformations are considered negligible, the deformations remaining in a range of 10^{-5}–10^{-4} (Hujeux 1985);

– non-stabilized hysteretic domain, noted as D^{nS}, in which the deformations exceed 10^{-4} (Hujeux 1985) are not linear and result from the elastic component and the plastic component;

– non-linear domain (or domain of incremental laws), noted as D^l, where, beyond a certain threshold of 10^{-3} (Hujeux 1985), the variations of the permanent deformations become significant and increase during the application of the repetitive action until infinity, where soil failure occurs.

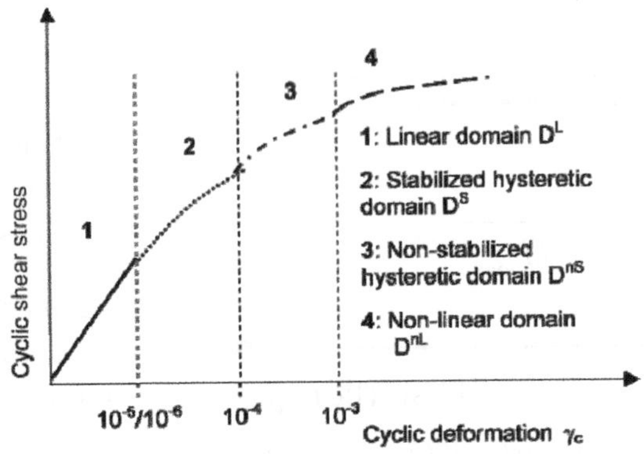

Figure 2.29. *Domains of soil behavior under cyclic shear stress (Dhouib 1987)*

2.3. Soil creep phenomenon

2.3.1. *Creep mechanism*

Some soils, such as clays, organic soils (peat, for example) or certain plastic marls, creep over time. That is to say, under a constant stress, the soil continues to deform over time. The soil sample (Figure 2.30(a)), subjected to a constant stress in time (Figure 2.30(b)), undergoes deformations (ε) which increase in time according to three directions of deformation (Figure 2.30(c)).

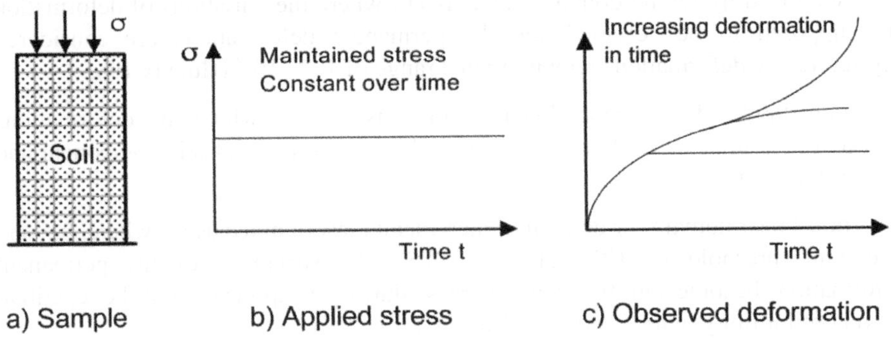

Figure 2.30. *Soil creep (one-dimensional behavior)*

2.3.2. *Creep stresses and deformations*

When the soil is subjected to several states of stress (e.g. rise of backfill in phases), deformations similar to those given in Figure 2.31 follow, if the supporting soil is fluid. The stresses are represented by a "stepped" function (Figure 2.31(a)) and the deformations by an "echelon" function (Figure 2.31(b)):

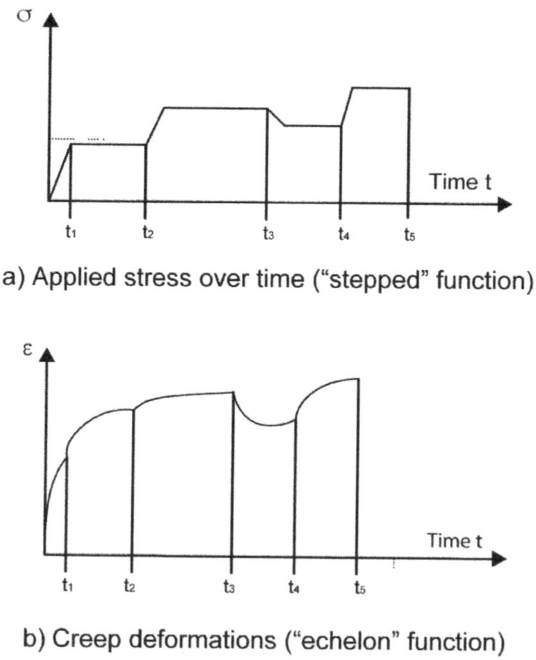

Figure 2.31. *Creeping of the supporting soil under various stress conditions*

2.4. Mechanics of continuous media applied to soils

2.4.1. *Terzaghi's fundamental relation*

We stated in Chapter 1 (section 1.1) that soil is a mixture of water, air and solid grains. Like air, water occupies the voids that exist between the grains. These voids are called "pores". The water flowing through these interstices exerts, as noted in section 2.1.3.4 of this chapter, identical normal stresses in all directions, that reduce to an isotropic pressure p (Figure 2.7). This water pressure is referred to as the "*pore pressure*" and is generally noted as u.

On each facet of a soil sample, normal σ' and tangential τ' stresses are applied to the solid grains. These stresses are called "*effective stresses*" because they act on the solid skeleton of the soil. When this facet is globally subjected to normal σ and tangential τ stresses, then these stresses are called "total stresses" because they act on both the water and the soil grains (see Figure 2.4 and Table 2.1 in Volume 2, Chapter 2).

The total (σ and τ) and effective (σ' and τ') stresses are related to the pore pressure (u) by Terzaghi's relation, which is the basis of soil mechanics and is given by:

$$\sigma = \sigma' + u \quad (\tau = \tau') \qquad [2.84]$$

(Terzaghi's fundamental relation).

2.4.2. *Concepts of short term and long term*

As mentioned above, soil is composed of a solid skeleton, formed by grains and pore water. In soils, water can either evacuate quickly or very slowly.

Indeed, in granular soils (sands, sands and gravels, and stones), water bypasses the grains and evacuates more or less quickly because these soils are permeable (fine sands) to very permeable (coarse sand, gravels and pebbles). On the other hand, fine soils (clays, silts, consistent and plastic marls) are not very permeable, and the water evacuates very slowly (a few months for silts and years for clays) because of the strong viscosity generated in the small interstices between the microscopic layers of the soil.

It is therefore possible to distinguish two distinct behaviors in fine soils; short-term behavior and long-term behavior:

– In the short term, the water under pressure has not started to escape when the load is applied, and such behavior involves both the behavior of the water and the solid skeleton.

– In the long term, the water escapes and this behavior only involves the behavior of the soil grains.

COMMENT.– The concept of short and long term can be represented by the analogous mechanical model consisting of a cylinder filled with water, subjected to the action of a piston, equipped with a water drain valve, and in which a spring R is embedded (see Figure 2.4 of Volume 2, Chapter 2 on soil consolidation).

2.4.3. *Equilibrium equations*

Generally speaking, and with the exception of a few special cases, problems in soil mechanics are always solved in two dimensions (plane deformations, plane stresses or in axisymmetric states).

Let us examine the equilibrium equations (see Figure 2.8 and equation [2.9]) for general configurations that are specific to the soil, and that can be reasonably considered as an undefined medium in terms of depth and lateral extensions.

2.4.3.1. *Undefined homogeneous soil with free or overloaded horizontal surface*

In an Oxz reference mark (z descending, see Figure 2.32(a)), let us consider an undefined homogeneous soil mass with a free horizontal surface, that is only subjected to the action of gravity (soil with total specific gravity γ, see Figure 2.32(b)). By symmetry, the stresses σ_{xx} and σ_{zz} are principal (Figure 2.5), and the equilibrium equations, in a two-dimensional medium, are reduced to:

$$\begin{cases} \dfrac{\partial \sigma_{xx}}{\partial x} = 0 \quad (on\, ox) \\ \dfrac{\partial \sigma_{zz}}{\partial z} = \gamma \quad (on\, oz) \end{cases} \qquad [2.85]$$

After integration, this leads to:

$$\begin{cases} \sigma_{xx} = f(z) \\ \sigma_{zz} = \gamma \times z + C \end{cases} \qquad [2.86]$$

where:

– *C* is the integration constant; and

– *f(z)* depends on the depth and the state of the lateral displacement of the soil.

COMMENTS.–

1) When the soil surface is free (Figure 2.32(b)), the constant C is zero. In the opposite case (soil subjected to an overload q on its free surface, see Figure 2.32(c)), the constant C is q.

2) In what follows, we are interested in the principal vertical stress (σ_{zz}) that prevails in the soil at a depth z. For the sake of simplification, this stress will be noted as σ_{zz} or σ_z.

3) The principal horizontal stress (σ_{xx}) will be studied in Volume 4, Chapter 1, which is on the fundamental states of horizontal stress in soil: resting soil state, active earth pressure and passive earth pressure states.

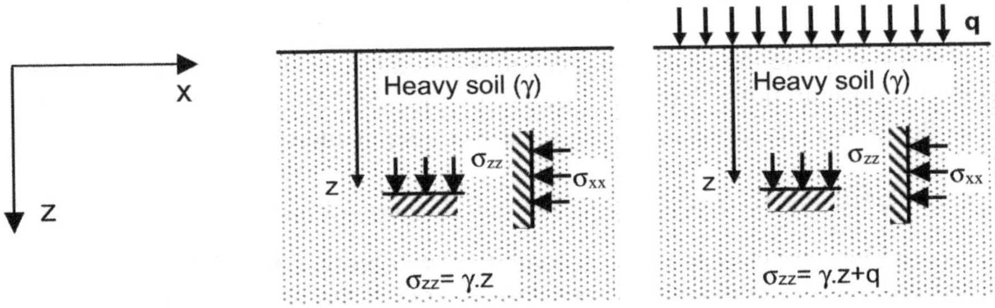

a) Reference mark b) Homogeneous soil with free horizontal surface (q=0) c) Homogeneous soil with a loaded horizontal surface (q)

Figure 2.32. *Stress states in an undefined homogeneous soil*

2.4.3.2. *Undefined heterogeneous soil with free or overloaded horizontal surface*

When, in the Oxz reference mark (Figure 2.33(a)), the soil is heterogeneous and composed of *n* horizontal layers of total specific gravity γ_i (Figure 2.33(b)), which is often the case, then the vertical stress σ_{zz} is written as:

$$\sigma_{zz} = \sum_{i=1}^{n} \gamma_i \times h_i \quad [2.87]$$

where γ_i and h_i are the unit volume weight and thickness of soil layer *i* among the *n* soil layers, respectively.

COMMENT.– When the soil surface is subjected to an overload q, then the constant C (equation [2.86]) is q; in other words (Figure 2.33(c)):

$$\sigma_{zz} = (\sum_{i=1}^{n} \gamma_i \times h_i) + q \quad [2.88]$$

In the two configurations shown in Figure 2.32 and Figure 2.33, the soil is dry. Let us examine the case of soil masses moistened by a water table, where Terzaghi's fundamental relation of equation [2.84] can be directly applied.

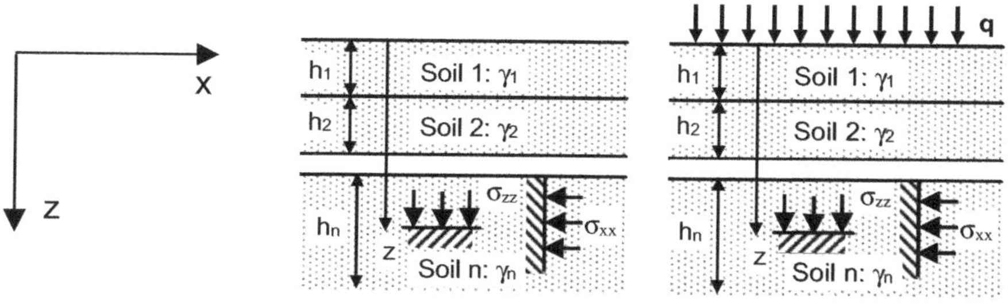

$\sigma_{zz} = \sum_{i=1}^{n} \gamma_i \times h_i$ $\qquad$ $\sigma_{zz} = (\sum_{i=1}^{n} \gamma_i \times h_i) + q$

a) Reference mark

b) Heterogeneous soil with free horizontal surface (q=0)

c) Heterogeneous soil with a loaded horizontal surface (q)

Figure 2.33. *Stress states in an undefined heterogeneous dry soil*

2.4.3.3. *Undefined soil with a horizontal surface moistened by a free water table*

Figure 2.34 illustrates the case of an undefined heavy soil with a horizontal surface moistened by a free water table; in other words, the water stagnates without any flow in the lateral direction. The soil mass is submerged, the water overcomes the assumed homogeneous (Figure 2.34(b)) or heterogeneous (Figure 2.34(c)) soil mass by a height h_W.

By applying Terzaghi's fundamental relation (equation [2.84]), the total vertical stress (σ_{zz}) is the sum of the effective stress (σ'_{zz}) and the pore pressure (u). In this framework, two cases can be distinguished to express the vertical stress (σ_{zz}):

– Case of a homogeneous soil (Figure 2.34(b)):

$$\sigma_{zz} = \gamma' \times z + (h_w + z) \times \gamma_w \qquad [2.89]$$

– Case of a heterogeneous soil (Figure 2.34(c)):

$$\sigma_{zz} = (\sum_{i=1}^{n} \gamma'_i \times h_i) + (h_w + z) \times \gamma_w \quad [2.90]$$

with:

γ': submerged unit weight of the soil (see equation [1.8], Chapter 1);

γ_w: unit volume weight of the water;

z: depth at which the stress σ_{zz} is calculated;

h_W: height of the water table;

h_i: height of the soil layer i.

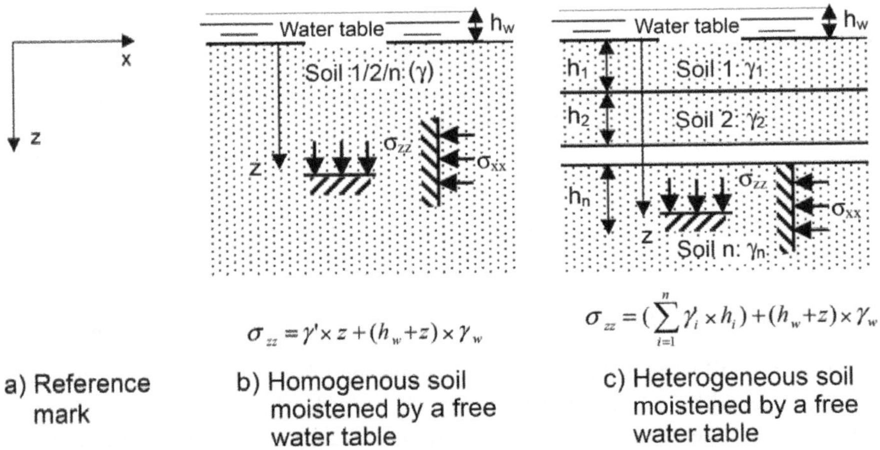

Figure 2.34. *Stress states in an undefined soil moistened by a free water table (without flow)*

2.4.3.4. *Undefined homogeneous soil with free or overloaded inclined surface*

In an Oxz reference mark (z descending, see Figure 2.35(a)), let us consider an undefined homogeneous soil mass with an inclined surface (inclination β with respect to the horizontal surface) that is free and only subject to the action of gravity (soil with a total specific weight γ, see Figure 2.35(b)).

In the Oxz reference mark, inclined by an angle β to the horizontal surface, the equilibrium equations in a two-dimensional medium (see Figure 2.8 and equation [2.9]) are expressed by the following equalities:

$$\begin{cases} \dfrac{\partial \sigma_{xx}}{\partial x} + \dfrac{\partial \tau_{xz}}{\partial z} = -\gamma \times \sin\beta \quad (on\, ox) \\ \dfrac{\partial \tau_{zx}}{\partial x} + \dfrac{\partial \sigma_{zz}}{\partial z} = \gamma \times \cos\beta \quad (on\, oz) \end{cases} \qquad [2.91]$$

This is a plane problem, where the state of stress at a given point does not depend on x. After integration, this leads to:

$$\begin{cases} \tau_{xz} = -\gamma \times z \times \sin\beta + C_1 \\ \sigma_{zz} = \gamma \times z \times \cos\beta + C_2 \end{cases} \qquad [2.92]$$

where C_1 and C_2 are the integration constants.

COMMENT.– When the soil surface is free (Figure 2.35(b)), constants C_1 and C_2 are equal to zero. In the opposite case (soil subjected to an overload q on its free surface, Figure 2.35(c)), then constant C_2 is q and constant C_1 refers to the horizontal stress due to the overload q.

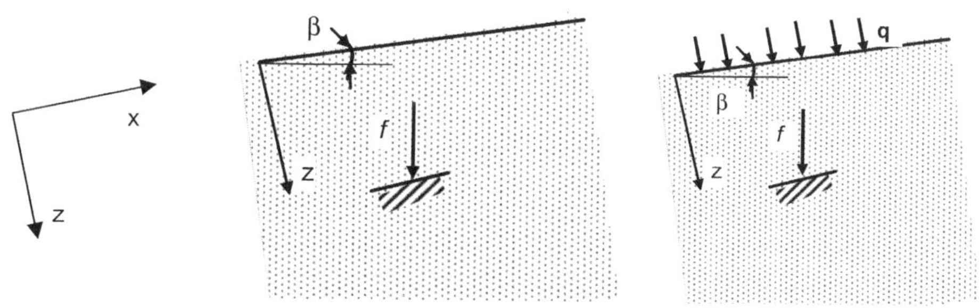

a) Reference mark

b) Homogeneous soil with free inclined surface (q=0)

c) Homogeneous soil with loaded inclined surface (q)

Figure 2.35. *Stress state in a homogeneous undefined soil with inclined surface*

2.4.4. *Experimental determination of elastic parameters of soil*

In the field of soil mechanics, experimentation is of considerable interest to determine the physical, hydraulic and mechanical properties of soils.

The various parameters are either determined in a laboratory or directly on site. In Chapter 3, we will deal with all of the tests and measurements on site, in order to determine the physical, hydraulic and mechanical properties of the soils. Here, we briefly remind you of the principles of the fundamental tests carried out in the laboratory, for the determination of the principal elasticity parameters that allow stresses and deformations to be linked. These tests are:

– the simple compression test;

– the isotropic compression test;

– the compression test with zero lateral deformation: oedometer test.

We limit ourselves here to the representation of the state of stresses and deformations (Figure 2.36), so as to explain the way in which these tests determine the elasticity parameters involved in the relations between the stresses applied to a soil sample and the resulting deformations.

COMMENTS.–

1) In the simple compression test (Figure 2.36(a)), the vertical force N applied to the straight section sample (*S*) results in a normal stress (σ), which is the ratio of the force N to the sample section (*S*); in other words, in the reference mark:

$$\sigma = \frac{N}{S} = \sigma_1 \quad [2.93]$$

Under stress σ_1, the sample undergoes an axial deformation ε_1 and lateral deformations ε_2 and ε_3 ("barrel effect"); when the crushed sample is homogeneous and isotropic, it is possible to consider that both deformations, ε_2 and ε_3, are identical:

$$\varepsilon_2 = \varepsilon_3 = \varepsilon \quad [2.94]$$

The simple compression test assumes that the sample has some cohesion. It is widely used in rock mechanics to determine the compressive resistance (R_C) and deformation modulus (E) of the rock.

2) Isotropic compression, performed with the triaxial apparatus, consists of applying a homogeneous state of stress (σ) around the triaxial sample (Figure 2.36(b)); this state of stress is analogous to that of hydrostatic stresses (Figure 2.7). The deformations generated by these stresses are purely volumetric. When the soil sample is homogeneous and isotropic, the deformations are considered to be identical in all three directions; in other words:

$$\varepsilon_1 = \varepsilon_2 = \varepsilon_3 = \varepsilon \qquad [2.95]$$

The change in volume (ΔV) relative to the volume (V) of the sample (equation [2.15]) is then reduced to:

$$\frac{\Delta V}{V} = 3\varepsilon \qquad [2.96]$$

3) In oedometric compression (Figure 2.36(c)), the stiff walls of the box prevent any lateral deformation ($\varepsilon_2 = \varepsilon_3 = 0$). The soil sample is compressed and the deformation is strictly axial (ε_1).

4) The experimentally determined deformation parameters from these tests are:

– the deformation modulus E (E_s);

– Poisson's ν ratio;

– the oedometric modulus E_{oed}.

5) The soil sample of radius r and height h shown in Figure 2.36(a) undergoes, under normal stress σ, a lateral displacement Δr and a vertical displacement h. The modulus of deformation (E) is given by:

$$E = \frac{\sigma}{\frac{\Delta h}{h}} = \frac{\sigma_1}{\varepsilon_2} \qquad [2.97]$$

and Poisson's ν ratio (ratio of radial and horizontal deformations) by:

$$\nu = \frac{\Delta r}{\Delta h} \qquad [2.98]$$

The oedometric compression (Figure 2.37(c)) makes it possible to determine the oedometric modulus E_{oed} by the following relation:

$$E_{oed} = \frac{\sigma_3}{\frac{\Delta h}{h}} = \frac{\sigma_1}{\varepsilon_1} \qquad [2.99]$$

In soil mechanics, the oedometric deformation modulus (E_{oed}) is related to the vertical deformation modulus (E_s) and Poisson's ratio (ν) by the following relation:

$$E_{oed} = \frac{(1-\nu)}{(1+\nu)(1-2\nu)} \times E_s \qquad [2.100]$$

for $\nu = 0.33$ (a value that is usually retained in the long term), it comes to:

$$E_{oed} = 1 \times 48\, E_s \qquad [2.101]$$

6) In fine soils, a distinction must be made between undrained behavior (short term), governed by parameters measured on undrained tests (E_u and ν_u), and drained behavior (long term), where the parameters, noted as E' and ν', are determined by drained tests with:

In the short term:

$$\begin{cases} E = E_u \\ \nu = \nu_u \end{cases} \qquad [2.102]$$

In the long term:

$$\begin{cases} E = E' \\ \nu = \nu' \end{cases} \qquad [2.103]$$

7) In permeable granular soils (such as sands and gravels), the short term and long term are mixed up, and the elasticity parameters are as follows:

$$\begin{cases} E = E' = E_u \\ \nu = \nu' = \nu_u \end{cases} \qquad [2.104]$$

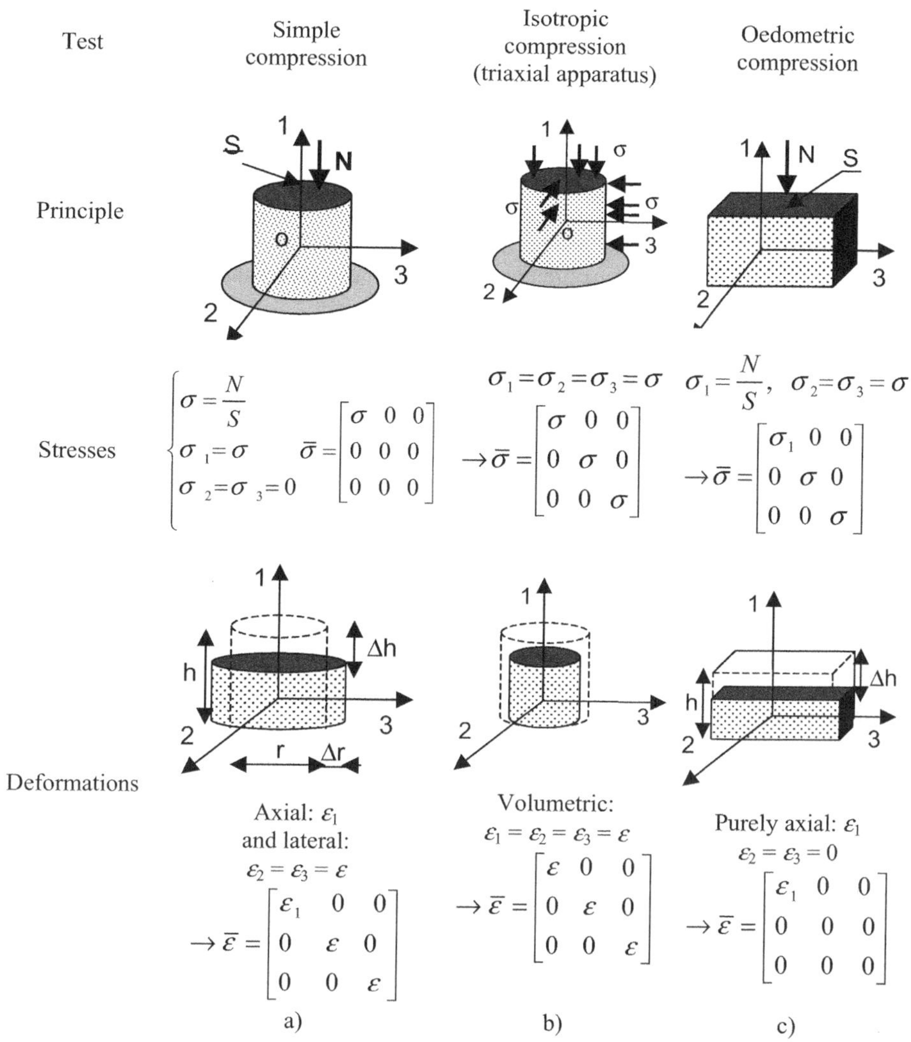

Figure 2.36. *Principal compression tests (in a laboratory) on a soil sample for the determination of deformation parameters*

2.4.5. *Order of magnitude of soil parameters*

Before closing this chapter with practical applications, we are going to give a few values of soil deformation parameters. These parameters are given as an indication so that students, engineers and practitioners can have a certain order of magnitude of the parameters in this book. Indeed, it is difficult to give precise values of the deformation parameters of the various types of soils, because each soil has its own "history". This history depends on the load that the soil has undergone in its "life", therefore on its state of consolidation in time, either under its own weight, or under the loads and overloads that the soil has been subjected to. The structure of the soil is very complex and the consolidation of the soil essentially depends on its grading, that is to say, the sizes of the grains that compose it, the loads that are applied to it and the duration of the application of these loads and overloads:

– For example, a saturated granular soil (sand and gravel) settles quickly under the applied loads and overloads, because the water drains quickly, therefore, allowing the soil to consolidate quickly.

– A fine saturated soil (silt and clay) settles slowly, because the incompressible water takes a very long time to evacuate.

In the context of geotechnical projects, on-site and/or laboratory measurements are required when structures are subjected to static loads (dead weights and overloads), or dynamic forces need to be integrated (vibrations, wave action and earthquakes). For comparison, the deformation parameters of other typical materials (steel and concrete) are also provided in Table 2.1(a).

There are two kinds of loads:

– static forces: gravity forces, flow forces, overloads, wind, etc.;

– dynamic actions due to vibrating machines, earthquakes, etc.

When the forces are small, the resulting deformations are small and the material can remain in an elastic domain. However, this domain is very small and absent in soils. When the forces applied to the material are high, the resulting deformations become plastic, and the material fails under a state of limit stress.

Contrary to other materials (steel for example), soil is very complex (due to its tri-phase components: grains, water and air), heterogeneous (compilations of layers, like a sandwich) and anisotropic (different mechanical properties in the three directions). It is within this complexity of the soil that the soil mechanics engineer must find the best adapted solutions to "combine" the soil and the foundations, by building various projects; projects that have been presented, studied and commented on in the various chapters related to foundations (Volume 3, Chapters 1 to 3), retaining walls (Volume 4, Chapter 1), reinforcement and improvement of soils (Volume 4, Chapter 2) and underground structures (Volume 4, Chapter 3).

In soil mechanics, the relations between the various deformation modulus (undrained E_u and drained E'), in Table 2.1 are summarized, and corresponding equations and graphs (Leroueil *et al.* 1985) can be found in Table 2.2.

SOILS					
Soils	**State of Compactness**	**Short term**		**Long term**	
		E_u (MPa)	**ν_u**	**E' (MPa)**	**ν'**
Sands	Loose	-[1]	-[1]	3–8	0.33
	Medium	-[1]	-[1]	8–12	
	Dense	-[1]	-[1]	12–30	
Sands and gravels	Dense	-[1]	-[1]	10–40	0.33
Compacted gravel	Dense	-[1]	-[1]	100–500	0.25
Silt	Soft	12–25	0.33–0.5[2]	8–12	0.33
Clays	Soft	8–15	0.33–0.50[2]	4–8	0.33
	Hard	15–60		10–30	
Chalk	Soft	4–8	0.33–0.50[2]	4–8	0.33
	Altered	10–20		10–20	
	Hard	20–400[3]		20–400[3]	0.25
Marl	Soft	20–40	0.33–0.50[2]	40–70	0.33
	Compact	50–400[3]	0.25[3]	50–400[3]	0.25
Rock	Soft	500–10,000[3]	0.25[3]	500–10,000[3]	0.25
OTHER MATERIALS					
Concrete[4]	Hard	32,000–37,000	0.20	10,800–12,600	0.20
Steel	Very hard	210,000	0.15	210,000	0.15

1) The short term (ST) and long term (LT) are mixed up due to the draining nature of granular soils.

2) When the soil is saturated, ν_u is exactly 0.5 (constant volume deformations).

3) There is no distinction between E_u and E' due to grain cementation.

4) Short term (ST): $E_u = 11000 \times \sqrt[3]{f_{c28}}$, fc28 is the compressive resistance of concrete at 28 days. Long term (LT): $E' = 3700 \times \sqrt[3]{f_{c28}}$ for fc28 going from 25–40 MPa.

Table 2.1(a). *Order of magnitude of deformation parameters: soil and other materials*

$$\frac{E_u}{E'} = \frac{1.5}{1+\nu'}$$

$$\frac{E'_{oed}}{E'} = \frac{1-\nu'}{1-\nu-2\nu'^2}$$

$$\frac{E_u}{E'_{oed}} = 1.5 \times \frac{1-\nu''-2\nu'^2}{1-\nu'^2}$$

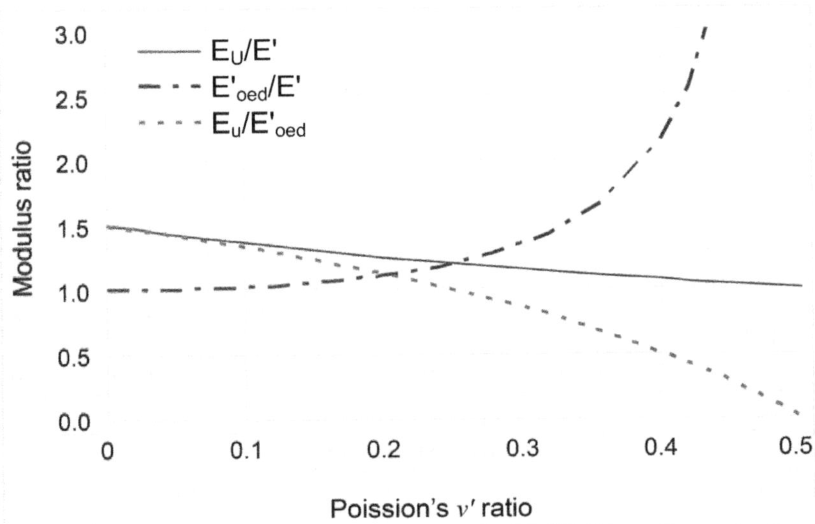

Table 2.1(b). *Soil deformation modules (Leroueil* et al. *1985)*

2.5. Practical applications to construction projects

2.5.1. *Calculation of vertical stresses in a homogeneous soil*

Consider a soil mass that is composed of loose, homogeneous sand with a horizontal surface and moistened by a water table of height h_W above the free soil surface (Figure 2.37(a)). The total density (γ) of the sand is (Table 1.1, Chapter 1): 18 kN/m^3 (saturated loose sand). According to equation [1.8], its submerged unit weight is 8 kN/m^3. The density of the water is 10 kN/m^3.

2.5.1.1. *Geotechnical data*

– γ = 18 kN/m^3;

– γ' = 8 kN/m^3 (Saturated sands[2]);

– γ_w = 10 kN/m^3;

– h_W = 2 m.

2.5.1.2. *Calculation of vertical stresses*

At a depth z of 10 m, the vertical stress state (σ_{zz}) is as follows (Figure 2.37(b)):

– $z = 0\,\text{m} \rightarrow \sigma_{z=0\,m} = u = \gamma_w \times h_w = 10 \times 2 = 20\,\text{kPa}$;

– $z = 10\,\text{m} \rightarrow \sigma_{z=10\,m} = \sigma'_{z=10\,m} + u = \gamma' \times z + \gamma_{w.} \times (h_w + z) = 8 \times 10 + 10 \times (2 + 10) = 200\,\text{kPa}$.

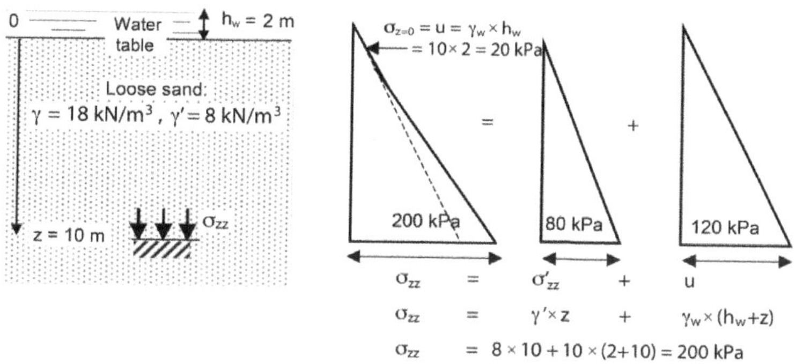

a) Sandy soil mass moistened by a water table

b) Total and effective stress states at 10 m depth (z = 10 m).

Figure 2.37. *Calculation of vertical stresses in a homogeneous sandy mass moistened by a free water table*

COMMENT ON FIGURE 2.37.–

1) *When the height h_W is zero, the pore pressure (u) is equal to zero for z = 0 m. The total vertical stress at 10 m depth is then:*

- $z = 10\,\text{m} \rightarrow \sigma_{z=10\,\text{m}} = \rightleftarrows \sigma'_{z=10\,\text{m}} + u = \gamma' \times z + \gamma_w \times z = 8 \times 10 + 10 \times 10 = 180\,\text{kPa}$.

2 When the soil is not saturated, the γ' value can be higher.

– *For a dry soil (no water table), we obtain the following*:

- $u = 0\,\text{kPa}$, *and*

- $z = 10\,\text{m} \rightarrow \sigma_{z=10\,\text{m}} = \rightleftarrows \sigma'_{z=10\,\text{m}} + 0 = \gamma \times z = 18 \times 10 = 180\,\text{kPa}.$

2.5.2. *Steps to solving finite element calculations: simple cases*

Since the 1970s, the academic world has continued to develop rheological soil models (section 2.2.2 of this chapter), without "revolutionizing" the world of geotechnics that has been practiced by engineers in design offices and companies. However, these models, often associated with ideal laboratory tests, have made it possible to deepen certain phenomena related to soils, such as the concept of expansion and strain hardening (isotropic and kinematic), which have been previously dealt with in this chapter.

The complexity of such models and the rapid growth of computing resources have led academics and research laboratories to develop numerical calculation codes based on finite element, finite difference and boundary calculation methods. Up until one or two decades ago, few industrialists had followed this movement, perhaps because of "mistrust" or "disinterest". Nowadays, finite element calculations with PLAXIS, FLAC and CESAR software are flourishing, and the graphical outputs embellish geotechnical reports.

2.5.2.1. *Reminder: analysis (calculation) at failure and deformation analysis*

Theoretical formulations

Before presenting, analyzing and commenting on the two chosen examples, let us recall that finite element calculation is a deformation calculation, as opposed to traditional break calculation.

To understand the two nuances, consider a one-dimensional beam element AB, subjected to a tension T at B, of:

– full section S;

– length l;

– deformation modulus E (Table 2.1(a));

– elastic limit σ_e (Figure 2.38(a)).

Figure 2.38 shows the beam element studied.

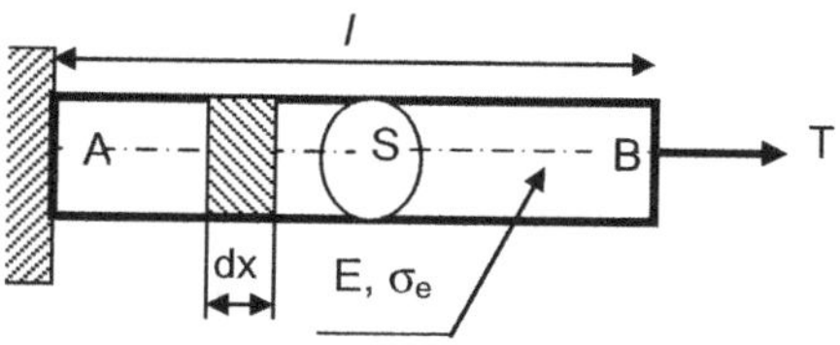

a) Beam element under tension T

- Break calculation:

$$\sigma = \frac{T}{S} < \sigma_e$$

- Deformation calculation:

$$\sigma = E.\frac{\Delta l}{l}(=\frac{T}{S})$$

b) Putting it into equations

Figure 2.38. *Analysis (calculation) at failure and deformation calculation*

The stress which prevails in any section S of the beam element AB is (see equation [2.1] and Figure 2.1(a)):

$$\sigma = \frac{T}{S} \qquad [2.105a]$$

The break calculation stops at the stress calculation (σ) by means of the relation (equation [2.105a]), and makes it possible to compare it with the elastic limit σ_e of the material constituting the beam element, whose stability against failure (Figure 2.13) is ensured when:

$$\sigma = \frac{T}{S} < \sigma_e \qquad [2.105b]$$

By introducing a one-dimensional behavioral law (see equation [2.25] and Figure 2.13) for the assumed isotropic elastic beam element, equality [2.105] becomes a function of the elastic elongation (Δl) of the beam under tension T:

$$\sigma = \frac{T}{S} = E \times \frac{\Delta l}{l} \qquad [2.106a]$$

With all parameters *T, l, S and E* known, the calculation of the elastic elongation Δl from equality [2.106a], as a function of the axial stiffness *ES* of the beam, can be formulated as:

$$\Delta l = \frac{T}{ES} \times l \qquad [2.106b]$$

which allows us to calculate the stress σ and compare it to its counterpart σ_e.

COMMENT.– The deformation calculation has the advantage of determining the deformations of the material under load and deducing the stresses in it. Break calculation only allows us to think in terms of stresses, without being able to apprehend the deformations of the stressed material. This is the case for stability calculations (also called "break calculations" or "failure analysis"), presented and developed in section 2.5 of Volume 2, Chapter 2.

Numerical application

Consider a cylindrical mild steel beam with the following characteristics:

– section *S*: 0.01 m^2 (Hollow tube 198 mm/162 mm);

– modulus of steel: 2.1×10^5 MPa;

– elastic limit σ_e: 260 MPa;

– length *l*: 1.0 m.

The beam is subjected to a tension *T* of 0.5 MN.

From equation [2.105b], in break calculation, it comes to:

$$\sigma = \frac{T}{S} = \frac{0.5}{0.01} = 50 \text{ MPa} < \sigma_e = 260 \text{ MPa}$$

And, from equation [2.106b], in deformation calculation:

$$\Delta l = \frac{T}{ES} \times l = \frac{0.5}{2.1 \times 10^5 \times 0.01} \times 1 \cong 2.38 \times 10^{-4} \text{ m}$$

And, from equation [2.106a]:

$$\sigma = E \times \frac{\Delta l}{l} = 2.1 \times 10^5, \frac{2.38 \times 10^{-4}}{1} = 49.98\,\text{MPa} \cong 50\,\text{MPa} < 260 \rightleftarrows \text{MPa}$$

The two examples that we present below show how all of these "black boxes" of finite element calculation software (CESAR-LCPC, FLAC 2 and 3D, PLAXIS) work; software that is built following the same steps detailed in the two simple elastic beam examples that follow.

2.5.2.2. *Finite element solution of simple beam cases*

Example 1: One-dimensional beam under tension

Let us return to the example of the beam in Figure 2.38(a), and isolate section 10 (or "*finite element*" with a width of 10), for the purpose of successively analyzing:

1) the displacements, noted as $u(x)$ (Figure 2.39(a)); and

2) the equilibrium of the internal forces T on element dx (Figure 2.39(b)); an equilibrium based on the same principle as the developments carried out in internal stresses σ on element dx in section 2.1.3.5 and in Figure 2.8.

Based on the data in Figure 2.39, the equilibrium equation analogous to equation [2.13], in simplified (one-dimensional) form, is (Figure 2.39(b)):

$$\frac{\delta P}{\delta x} + T = 0 \qquad [2.107]$$

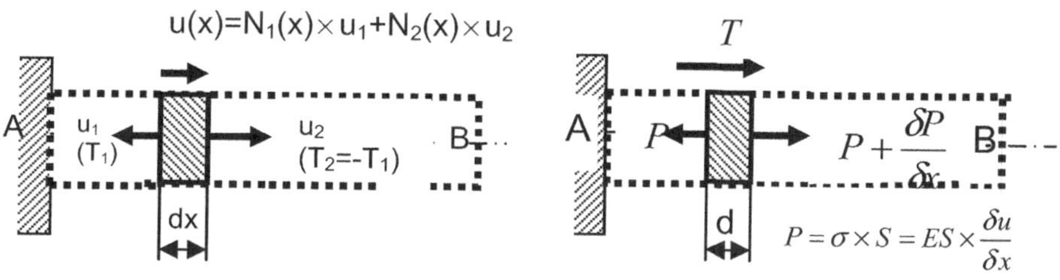

a) Displacements of finite element dx b) Tensions in finite element dx

Figure 2.39. *Finite element dx of a one-dimensional beam subjected to a tension T*

By introducing a one-dimensional behavioral law, analogous to that given by equation [2.106a], we get the following:

$$\sigma.S = ES \times \frac{du}{dx}(= P) \quad [2.108]$$

The equilibrium equation in equation [2.107] becomes:

$$ES \times \frac{\delta u^2}{\delta x^2} + T = 0 \quad [2.109]$$

Finite element calculations require choosing the "shape function" of the displacement $u(x)$. One of the simplest solutions is to consider a linear interpolation (other polynomial forms can be used, as in Example 2, which will follow this one).

The linear form consists of writing $u(x)$ in the following form:

$$u(x) = N_1(x) \times u_1 + N_2(x) \times u_2 \text{ (Linear)} \quad [2.110a]$$

Or in the form:

$$u(x) = [N_1 \qquad N_2] \times \begin{Bmatrix} u_1 \\ u_2 \end{Bmatrix} \quad [2.110b]$$

The linear interpolation functions $N_1(x)$ and $N_2(x)$ are easily determined from the boundary conditions given in Figure 2.40(a) and the linear form of $u(x)$:

$$u(x) = a.x + b \text{ (Linear)} \quad [2.111]$$

where, from Figure 2.40(a):

– for $x = 0 \rightarrow u(0) = \underline{u}_1$;

– for $x = l \rightarrow u(L) = u_2$.

This leads (Figure 2.40(b)) to:

$$\begin{cases} N_1(x) = (1 - \frac{x}{l}). \\ N_2(x) = \frac{x}{l} \end{cases} \quad [2.112]$$

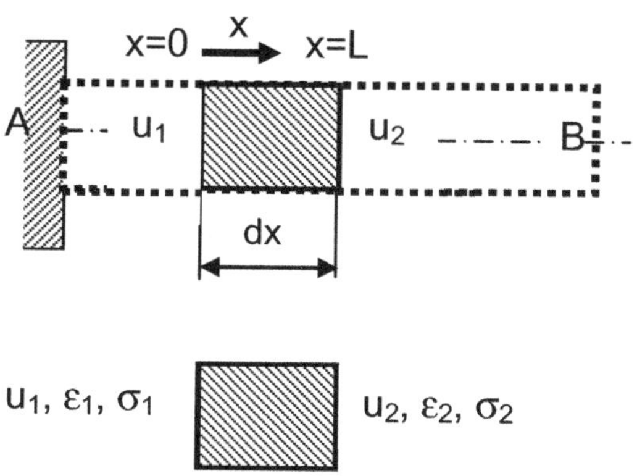

a) Boundary conditions

$$u(x) = a \times x + b \text{ (linear)}$$

$$= (1 - \frac{x}{l}) \times u_1 + \frac{x}{l} \times u_2$$

hence:

$$\begin{cases} N_1(x) = (1 - \frac{x}{l}). \\ N_2(x) = \frac{x}{l} \end{cases}$$

b) $N_1(x)$ and $N_2(x)$ interpolation functions

Figure 2.40. *Boundary conditions and forms of $N_1(x)$ and $N_2(x)$*

Equation [2.109] becomes:

$$ES \times \frac{\delta^2}{\delta x^2} \times [\mathrm{N}_1 \quad \mathrm{N}_2] \times \begin{Bmatrix} u_1 \\ u_2 \end{Bmatrix} + T = 0 \qquad [2.113a]$$

The Galleria method makes it possible (Finlayson 1972) to multiply the two terms of equality [2.113a] by the shape functions $N_1(x)$ and $N_2(x)$, and to integrate the form thus obtained over the whole element l, in order to express it by:

$$\int_0^l \begin{Bmatrix} N_1 \\ N_2 \end{Bmatrix} ES \times \frac{\delta^2}{\delta x^2} \times [\mathrm{N}_1 \quad \mathrm{N}_2] \times \begin{Bmatrix} u_1 \\ u_2 \end{Bmatrix} dx + \int_0^l \begin{Bmatrix} N_1 \\ N2 \end{Bmatrix} T dx = 0 \qquad [2.113b]$$

By direct application of Green's theorem, this is a relation that consists of writing:

$$\int N_1 \times \frac{\delta^2 N_2}{\delta x^2} dx = -\int \frac{\delta\, N_1}{\delta x} \times \frac{\delta\, N_2}{\delta x} dx \qquad [2.113c]$$

The following simple matrix form becomes:

$$\frac{ES}{l} \times \begin{bmatrix} 1 & -1 \\ -1 & 1 \end{bmatrix} \begin{Bmatrix} u_1 \\ u_2 \end{Bmatrix} = \frac{Tl}{2} \times \begin{Bmatrix} 1 \\ 1 \end{Bmatrix} = \begin{Bmatrix} T_1 \\ -T_2 \end{Bmatrix} \qquad [2.113d]$$

or:

$$\frac{ES}{2l}\times\begin{bmatrix}1 & -1\\-1 & 1\end{bmatrix}\begin{Bmatrix}u_1\\u_2\end{Bmatrix}=\begin{Bmatrix}-T\\T\end{Bmatrix} \quad [2.113e]$$

This is equivalent to the matrix equality in [2.44] or the general matrix formulation, the basis for solving finite element problems, as follows:

$$[K]\times\{U\}=\{P\} \quad [2.113f]$$

where (see Table 1.2 in Volume 4, Chapter 1) $[K],\{U\},\{P\}$ designate the stiffness matrix and the displacement and force vectors applied to the boundaries of the finite element model, respectively.

The matrix resolution of [2.113c] leads to the same displacements previously calculated (numerical application presented in section 2.5.2.1):

$$\begin{cases}u_1=-\dfrac{T}{ES}\times l\\ u_2=-u_1=\dfrac{T}{ES}\times l\end{cases} \quad [2.114]$$

Numerical application

With the numerical data of the application presented in section 2.5.5.1, it comes from the resolution of the matrix system [2.113b] at both ends of the finite element l (Figure 2.40(a)):

– displacement $u_1 = -2.38\times 10^{-4}$ m → deformation $|\varepsilon_1| = 0.238\,^0/_{00}$ and stress $|\sigma 1| = 50$ MPa (<260 MPa);

– displacement $u_2 = 2.38\times 10^{-4}$ m → deformation $\varepsilon_2 = 0.238\,^0/_{00}$ and stress $\sigma_2 = 50$ MPa (<260 MPa).

Example 2: One-dimensional beam under uniform pressure

Consider the example of a horizontal beam element AB, of inertia I, deformation modulus E (bending stiffness EI), length l and subjected to a uniformly distributed pressure $q(x) = q$ (Figure 2.41(a)). Based on the elements in Figure 2.41 and with the same approach as in Example 1 (equations [2.107]–[2.108]), the equilibrium equation, analogous to equation [2.109], is written as:

$$EI.\frac{\delta u^4}{\delta x^4}=q(x) \quad [2.115]$$

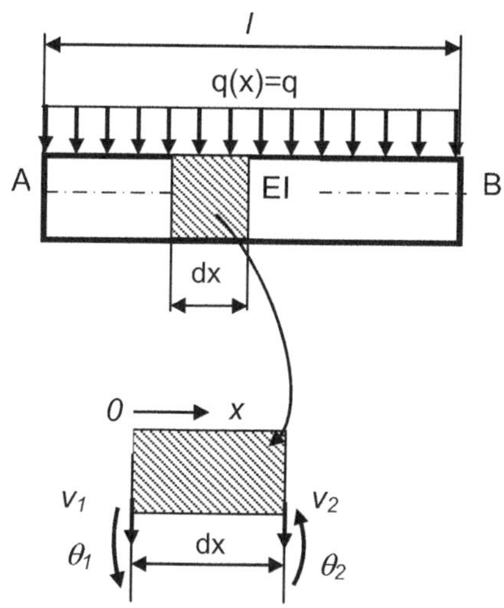

- At both ends of the finite element dx (Figure 2.41.a), the following are exerted:

- the vertical displacements v_1 and v_2,

- rotations θ_1 and θ_2, given by:

$$\theta_1 = \frac{\delta v_1}{\delta x_1}, \text{ and } \theta_2 = \frac{\delta v_2}{\delta x_2}$$

The displacement vector is given by:

$$u = \begin{Bmatrix} v_1 \\ \theta_1 \\ v_2 \\ \theta_2 \end{Bmatrix}$$

a) Beam element under pressure p(x)

b) Displacements at the ends of the element dx

Figure 2.41. *Beam element under uniform pressure q(x)*

Let us choose a cubic polynomial form (of order 3) of the displacement vector $u(x)$ (analogous form to equality [2.111]):

$$u(x) = a \times {}_0 + a_1 x + a_2 x^2 + a_3 x^3 \quad [2.116a]$$

or in matrix form:

$$u(x) = [N1 \quad N2 \quad N3 \quad N4] \times \begin{Bmatrix} v1 \\ \theta 1 \\ v2 \\ \theta 2 \end{Bmatrix} \quad [2.116b]$$

Now, according to the polynomial form in [2.116a] and the boundary conditions in Figure 2.41(a), the interpolation functions are expressed by the following polynomials:

$$\begin{cases} N_1(x) = 1 + \frac{2}{L^2}x^2 - \frac{3}{L^3}x^3. \\ N_2(x) = x - \frac{4}{L}x^2 + \frac{3}{L^2}x^3 \\ N_3(x) = -\frac{2}{L^2}x^2 + \frac{3}{L^3}x^3 \\ N_4(x) = \frac{6}{L}x^2 - \frac{6}{L^2}x^3 \end{cases} \quad [2.117]$$

With the same previous approaches (application of the Galleria method and Green's theorem and integration by parts), the introduction of the form functions in [2.117] into the equilibrium equation in [2.115] results in an equality that is analogous to the matrix form in [2.113d], that is to say:

$$\frac{EI}{L^3} \times \begin{bmatrix} 12 & 6L & -12 & 6L \\ & 4L^2 & -6L & 2L^2 \\ & & 12 & -6L \\ & Symmetry & & 4L^2 \end{bmatrix} \begin{Bmatrix} v_1 \\ \theta_1 \\ v_2 \\ \theta_2 \end{Bmatrix} = q \times \begin{Bmatrix} L/2 \\ L^2/12 \\ L/2 \\ -L^2/12 \end{Bmatrix} \quad [2.118]$$

Using the example of ideal beam conditions on simple retaining walls ($v_1 = v_2 = 0$), this makes it possible to easily find the traditional forms known in material resistance of θ_1 and θ_2, that is to say:

$$\theta_1 = -\theta_2 = -\frac{ql^3}{24EI}$$

COMMENT.– Examples 1 and 2 show that a finite element calculation (deformation calculation) goes through the following steps:

1) choice of the geometric model: spatial discretization (creation of the mesh);

2) choice of boundary conditions (also called limiting conditions);

3) choice of a behavioral model (behavioral law linking stresses to deformations);

4) construction of the stiffness and force matrix (forces at the nodes of the geometric model);

5) solving the system represented by equation [2.113] and equation [2.118];

6) calculation of displacements, deformations and stresses.

Indeed, finite element calculation leads to knowing everything about the "finite element" of the mesh and the requested material:

– the displacement state;

– the deformation state;

– the stress state;

– the plasticity points: when one of the stresses, σ_1 or σ_2, exceeds the elastic limit σ_e in the two previous examples or, in general, when the failure criterion (equation [2.60]) is verified.

2.5.2.3. *Generalization of finite element solving*

The two examples of an elastic beam under tension or bending, which we have presented to simplify the "finite element" concept, are very simple cases when compared to a soil problem that is numerically treated by finite element calculations.

The geometric model of various soils is the subject of a large mesh, which is sometimes very fine, especially in singular zones (at the base of the foundation, in contact with the wall, around a tunnel for example) comprising hundreds, even thousands, of meshes and thousands of nodes. In the nodes of each mesh, the $N_i(x)$ interpolation functions are generalized. Displacements are calculated at each mesh node, but stresses are generally assessed at the center of the mesh. For soils and other foundation elements, an appropriate behavioral law is introduced (section 2.2.2.3).

As an illustration, Figure 2.42(a) shows the selected mesh, displacement boundary conditions and loaded nodes of a PLAXIS-2D finite element numerical model, in order to study the interaction of deep excavations for the creation of three basement levels on an existing industrial building and a road traffic tunnel under a major boulevard, also in traffic.

The displacement boundary conditions are as follows:

– lateral faces of the model: zero horizontal displacements (nodes blocked in the horizontal direction and free in the vertical direction): $u = 0$;

– base of the model: zero vertical displacements (nodes blocked in the vertical direction and free in the horizontal direction): $v = 0$.

The loaded boundary conditions (loaded nodes on band AA in Figure 2.42(a)) represent the 20 kPa road overload of a major boulevard in traffic behind the retaining wall.

The constructed numerical model in Figure 2.42(a) includes around 1,200 three-node triangular elements and more than 3,000 nodes. The full stiffness matrix is normally at least 3,000 × 3,000 in size, which complicates and lengthens the numerical calculations. It is replaced, through numerical tricks, by a "band" matrix that is deduced from the fullwidth matrix.

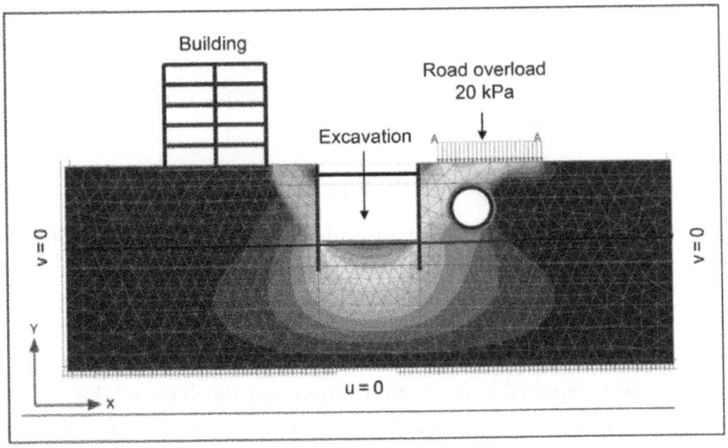

a) Two-dimensional numerical model: deep excavations in the vicinity of industrial buildings, tunnels and boulevards in traffic (PLAXIS-2D).

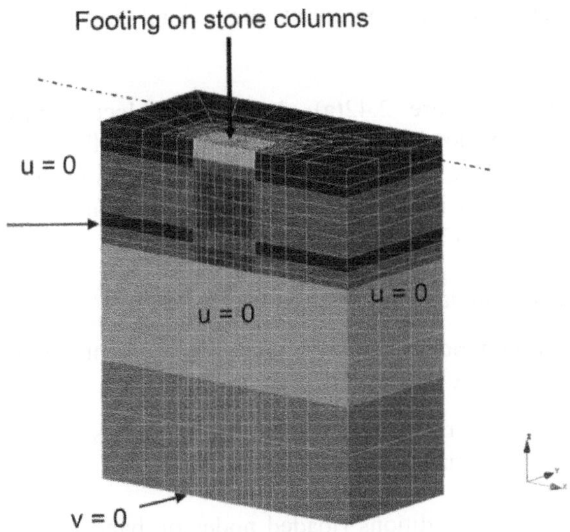

b) Three-dimensional numerical model: footing on stone columns.

Figure 2.42. *Examples of numerical modeling in finite elements (CESAR-3D). For a color version of this figure, see www.iste.co.uk/dhouib/geotechnics1.zip*

The second example presented in Figure 2.42(b) involves a three-dimensional model, developed on CESAR-3D, to understand the behavior of four stone columns under a shallow footing (Dhouib *et al.* 2004). The displacement boundary conditions are similar to the conditions of the 2D model in Figure 2.42(a): zero horizontal displacements ($u = 0$) on the lateral faces of the model and zero vertical displacements ($v = 0$) at the base of the model. The column heads and the surrounding soil are loaded by the service stress of the footing (125–200 kPa). Due to the symmetry of the model, only two stone columns are modeled under half of the foundation, which allows a significant gain in the size of the stiffness matrix and in the calculation time.

After this presentation "disseminating" the principle of finite element calculations, we study, analyze and comment on three distinct real-life projects by way of illustration, which are treated by finite element calculations in two dimensions (plane deformations) by PLAXIS-2D.

These three projects involve:

– Project 1: highway embankment;

– Project 2: retaining diaphragm wall with active tie-rods;

– Project 3: two sewage treatment plant rafts on vertical stiff inclusions.

In all three projects, theoretical finite element calculations are compared with displacement measurements at the project site.

2.5.3. *Project 1: Modeling and calculation of settlement of a road embankment*

2.5.3.1. *Project*

The project involves expanding the existing embankment from 2 × 2 to 2 × 3 lanes. As an alternative to the lightened polystyrene backfill solution to reduce the vertical stress on the sub-grade and settlements, a backfill solution on soils, improved by rigid inclusions, was selected.

In this application, the settlements of the highway platform without stiff inclusions are studied. The comparison of the settlements calculated by the "PLAXIS-2D" numerical modeling and the analytical methods with rigid inclusions will be compared in the application in section 2.8.2 of Volume 4, Chapter 2.

2.5.3.2. *Tolerance threshold of settlements*

The threshold settlements imposed by the project management are as follows:

– eligible absolute settlements of less than 30 cm;

– eligible differential settlements over 5 m of a cross-section less than or equal to 15 cm;

– eligible differential settlements between two cross-sections spaced at 10 m less than or equal to 10 cm.

2.5.3.3. *Geotechnical data*

Various cores and pressuremeter drillings (section 3.4.5 of Chapter 3) have been carried out and show the following lithology, from top to bottom:

– backfill of the existing highway;

– muddy clays that are more or less sandy (layer 1);

– clayey sands with a muddy passage (layer 2);

– dense sands with a sandstone passage;

– sandstone substratum.

Based on surveys and on-site tests, the geo-mechanical data selected for project justification are as follows:

Layers	Thickness (m)	p_L* (MPa)	E_M (MPa)	E_M/p_L*	α	E_s
Backfill in situ	3	0.65	7.7	12	1/2	15.4
Muddy clay (layer 1)	5	0.3	1.8	6	1/2	3.6
Clayey sand (layer 2)	4.5	0.41	4.8	12	1/3	14.4

p_l*: Limit pressure measured with a pressuremeter test (Figure 3.20);

E_M: Louis Ménard pressuremeter modulus (Figure 3.20);

α: Menard rheological coefficient (Table 3.7).

These three parameters are given in the references in Chapter 3 of this book.

E_s: Modulus of soil deformations.

Table 2.2. *Geo-mechanical data (mean section)*

Table 2.3 summarizes the compressibility and shear parameters measured in the laboratory, respectively (see sections 2.23 and 2.3 of Volume 2, Chapter 2 respectively).

Formations	γ (kN/m^3)	e_o	C_c	C_s	C_α	k (m/s)	c' (kPa)	φ' (degrees)
Backfill in situ	20	-	-	-	-	10^{-4}	5	30
Layer 1	18	1.16	0.40	0.040	0.006	10^{-9}	15	23
Layer 2	γ (kN/m^3)	-	-	-	-	10^{-8}	5	30

γ: Unit volume weight of the soil;

e_o : Void ratio at initial state;

C_c: Compression index;

C_s: Swelling index;

C_α: Secondary consolidation rate;

k: Permeability;

c': Effective cohesion;

φ': Effective friction angle.

Table 2.3. *Compressibility and shear parameters*

Figure 2.43 shows the typical profile of the existing and expanded embankment, and the underlying soil layers, as well as the water level that is moistening the soil layers underlying the backfills.

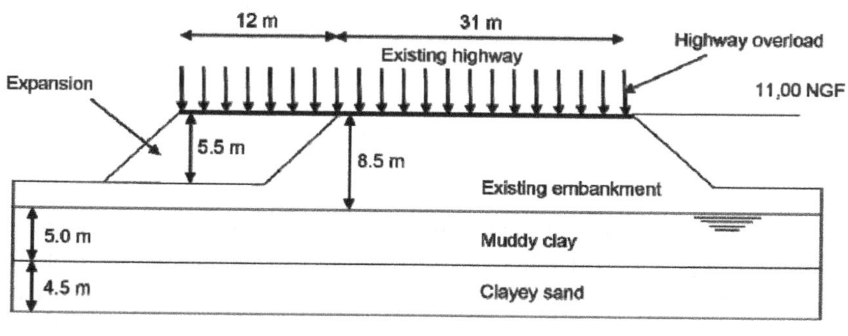

Figure 2.43. *Typical profile of the existing highway and the projected expansion*

Note that the water level is at an elevation of 2.50 NGF (at the base of the pre-existing embankment).

2.5.3.4. *PLAXIS-2D modeling*

Table 2.4 presents the different phases of work taken into account in the two-dimensional finite element modeling (PLAXIS-2D in plane deformations), as defined by the rheological model of soft soil behavior with creep ("Soft-Soil-Creep-Model" in the PLAXIS nomenclature) for the muddy clays and clayey sands that are assumed to be fine, soft and compressible soils, and by the Mohr-Coulomb model (see Figures 2.19 and 2.20) for the rest of the soils (that are considered rather powdery).

No.	Phase
0	Initialization of stresses in state K_0
1	Placement of the initial embankment and road overload
2	Consolidation over 30 years
3	Placement of the expanded embankment and road overload
4	Consolidation over 10 years
5	Consolidation over 30 years

Table 2.4. *Phases of the works that are taken into account in the PLAXIS-2D calculations*

The graphs in Figure 2.44 show the iso values of the soil settlements under the expanded embankment without stiff inclusions: maximum settlements (s_{max}) at time $t = 0$, at time $t = 10$ years and at time $t = 30$ years. Note that the settlements measured at the end of the embankment rise are 3.8 cm for 4 cm calculated (Figure 2.43(a)).

2.5.4. *Project 2: Study of the displacements of a retaining slurry trench wall*

As part of the extension of Line 1 of the Marseille metro between La Timone and La Fourragère, a perimeter diaphragm wall was built to support the excavations for the construction of the metro station, and the creation of 350 underground parking spaces under Boulevard Frangin and the tramway decks. The works were carried out in a delicate and complex geological and environmental context. Several panels were instrumented in order to follow the deformations of the wall, and to verify the compatibility of the measured displacements with the predictions of the calculations. In this application, the project criteria, the geological and geotechnical context, the data and assumptions of the modeling, and the displacements calculated by "PLAXIS-2D" using a rheological "Mohr-Coulomb" type of soil behavior model (see Figure 2.19 and equations [2.62]–[2.64]), are briefly presented and compared to on-site measurements on one of the instrumented panels, as well as to calculations based on the reaction coefficient method (see section 1.4 of Volume 4, Chapter 1) by a "RIDO" type calculation.

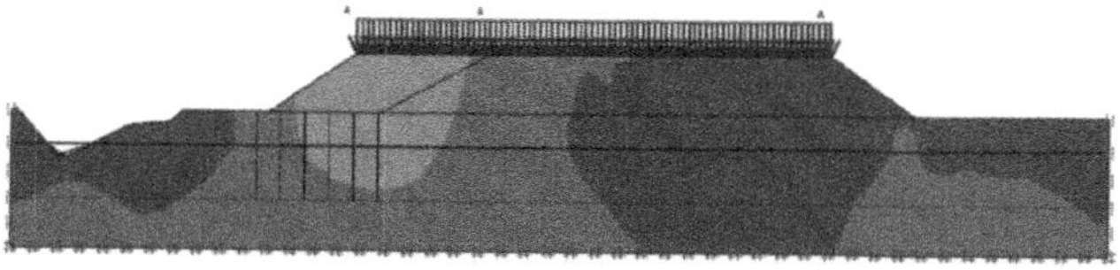

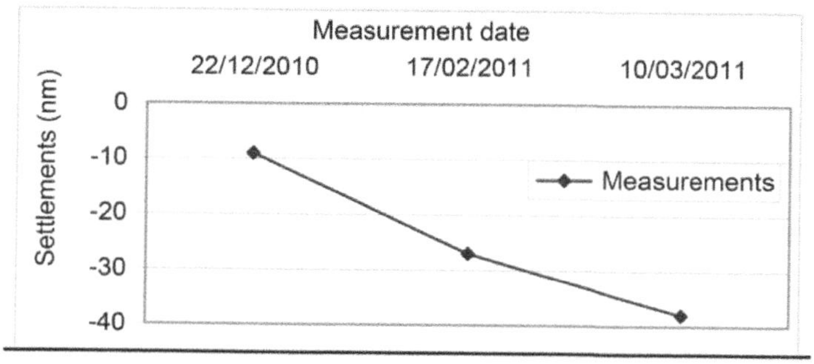

a) A t = 0, Calculated settlements s_{max} = 4.0 cm, measured settlements: 3.8 cm

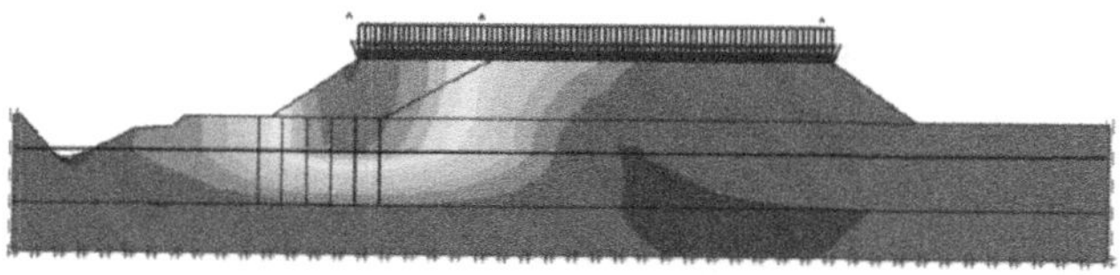

b) Calculated settlements at t = 10 years, s_{max} = 22.0 cm

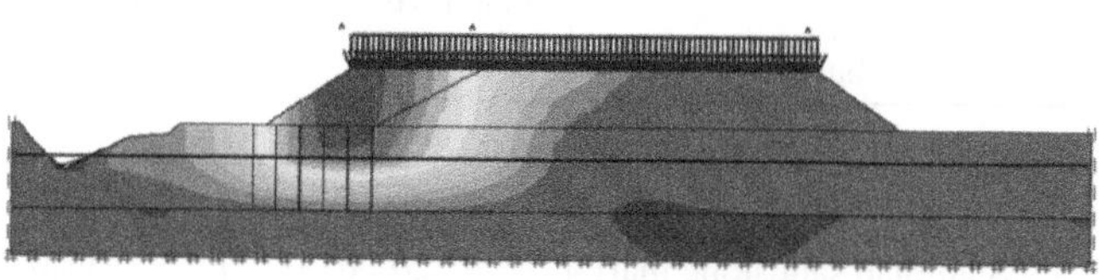

c) Calculated settlements at t = 30 years, s_{max} = 24.0 cm

Figure 2.44. *Iso values of consolidation settlements (PLAXIS-2D) and measurements at t = 0 (three months of embankment construction). For a color version of this figure, see www.iste.co.uk/dhouib/geotechnics1.zip*

2.5.4.1. *Project*

The project includes:

– a metro station;

– a parking lot with 350 spaces;

– the creation of decks under Boulevard Frangin and the Tramway.

The diaphragm wall is 100 m long and 45 m wide, with a height of 25 m. The wall has a total development of 10,450 m^2 and is anchored by 330 active tie-rods. The excavated volumes, under the shelter of eight pre-founded columns 28 m long, are 100,000 m^3. The concrete has a total volume of 24,000 m^3 for 24,000 kN of steel. The cost of the diaphragm wall and the tie-rods represents approximately 20% of the total amount of the contract. The overall period is 31 months, including 24 months of works.

2.5.4.2. *Geotechnical data*

Site geology

All of the on-site surveys and tests carried out have made it possible to establish several typical sections, which, from top to bottom, highlight the backfills and Stampian formations in the form of:

– silty-clay backfill;

– stampian sandstone and marl;

– stampian conglomerates;

– stampian sandstone and marl that is more or less sandy.

The dimensions of the base of the backfills and the various Stampian stratifications, as well as the average thicknesses, are presented in Table 2.5.

Soils	Base level (NGF)	Thickness (m)
Silty-clay backfill	46.00	2.8
Stampian sandstone/marl	41.00	5.0
Stampian conglomerate	40.00	1.0
± Sandy Stampian sandstone/marl	<29.00	>11.0

Table 2.5. *Different soil layers and their thickness*

Hydrological conditions

A water level has been retained at 40.00 NGF. This water level corresponds to the top of the more or less sandy marls.

Geo-mechanical site data

Geotechnical data from pressuremeter and mechanical laboratory tests are summarized in Table 2.6.

Formation	Volume weight γ (kN/m^3)	Shear characteristics				Mechanical characteristics		
		Short term		Long term		E_M module (MPa)	α coefficient	E_s module (MPa)
		c_u (kPa)	φ_{cu} (°)	c' (kPa)	φ' (°)			
Backfill	20	5	20	0	30	10	1/2	40
Sandstone/Marl	20	40	20	20	30	45	2/3	130
Conglomerate	22	80	25	40	30	45	2/3	130
Sandstone/Marl	22	100	25	50	30	150	2/3	450

E_M: Menard pressuremeter modulus;

α: Menard rheological coefficient;

E_s: *Soil* deformation modulus ($E_s=\beta.E_M/\alpha$).

Table 2.6. *Geo-mechanical data*

COMMENT.– The conventional soil deformation modulus (E_s), corresponding to the decompression of the soil, is multiplied by a factor β of 2 (β = 2), which is a usual value in this type of soil (clayey), and this can be higher in granular soils (sands, sands and gravels, gravels) where β can be set to 3–4, or even higher in gravel (β = 4–5).

2.5.4.3. *Selected field model and slurry trench wall*

The slurry trench wall was subject to instrumentation, including as a whole:

– seven inclinometer tubes with diameters of 102/114 mm;

– seven sonic auscultation profiles, each including six tubes with diameters of 50/60 mm.

The present application is limited to an instrumented panel, where the support includes two slurry trench walls:

– diaphragm wall molded into "trouser leg", exterior side;

– continuous principal retaining wall, which encloses the earthwork enclosure, and is the subject of this modeling.

On the typical profile ("soil model") deduced from the various tests and borings, the different soil layers are recalled, the geometry of the diaphragm wall is presented and the characteristics of the active tie-rods are described. Figure 2.45 illustrates the typical profile studied.

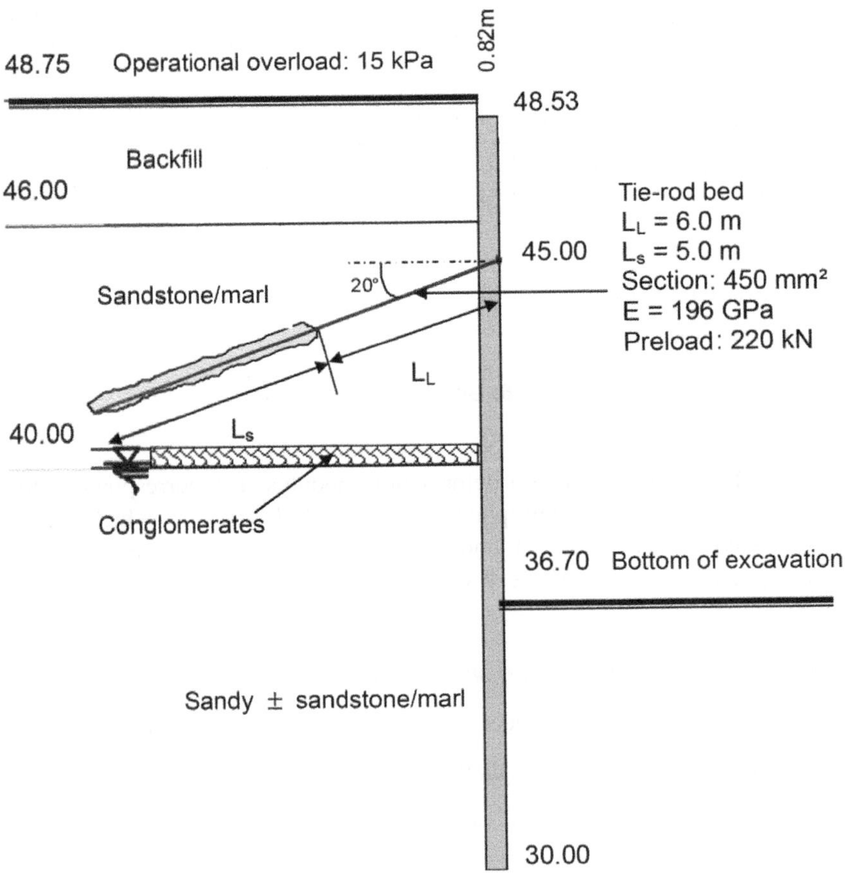

Figure 2.45. *Profile of the studied panel ("field" model and slurry trench wall). For a color version of this figure, see www.iste.co.uk/dhouib/geotechnics1.zip*

Figure 2.46 shows two pictured views of the excavations during the earthwork.

a) General view of the excavation

b) South-western profile of the slurry trench wall

Figure 2.46. *Views of the excavation during the construction of the slurry trench wall and the excavation works. For a color version of this figure, see www.iste.co.uk/dhouib/geotechnics1.zip*

2.5.4.4. *Finite element modeling (PLAXIS-2D)*

Deformed model and comparative displacements

A plane deformation model is developed using:

– an "elasto-plastic" behavioral model, the "Mohr-Coulomb" type, for all the soils in the application;

– a purely elastic behavior for the retaining diaphragm wall.

This is all done while simulating the various phases of excavation that are taken into account in the work, in the modeling.

Figure 2.47(a) shows the deformed model in the final excavation phase of the studied panel.

Figure 2.47(b) shows the evolution of horizontal displacements as a function of depth (NGF), measured by the inclinometer tubes and calculated by PLAXIS-2D and RIDO (reaction coefficient method presented in section 1.4 of Volume 4, Chapter 1).

The comparison of horizontal displacements shows that:

– for low levels of wall deformation, the two types of approaches are very close (finite elements: 10 mm; reaction coefficient method: 6 mm);

– these displacements are more or less in line with the measured displacements of 6 mm.

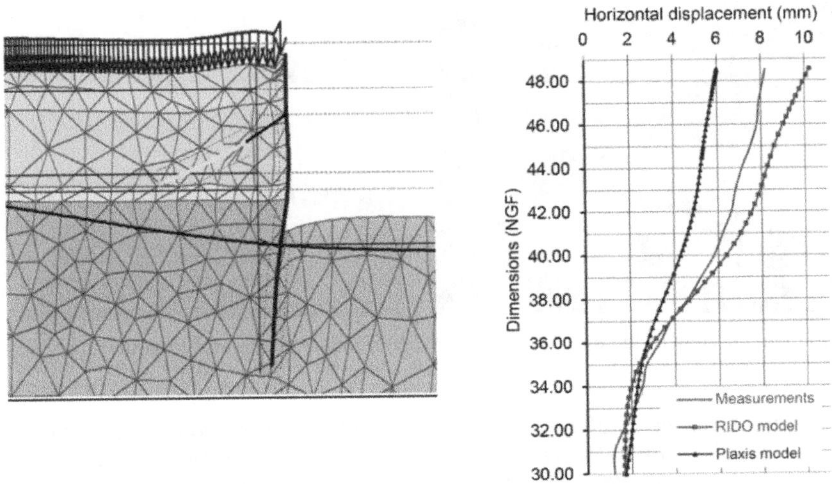

a) Deformed mesh

b) Measured and calculated displacements (result adjustment)

Figure 2.47. *Finite element model of the studied panel (Dhouib* et al. *2007). For a color version of this figure, see www.iste.co.uk/dhouib/geotechnics1.zip*

2.5.5. *Project 3: Modeling of rafts on vertical stiff inclusions*

Figure 2.48 illustrates two cases of justification, in plane deformations with the Mohr-Coulomb rheological model by "PLAXIS-2D" finite element calculations, of two rafts of two different sewage treatment plants (STEP). Both rafts are based on vertical stiff inclusions (see section 2.6.5 of Volume 4, Chapter 2).

Figure 2.48(a) and Figure 2.48(b) show the view of the raft and its settlement shape. These settlements, which are 15 mm at the center, are accentuated at the edge; in other words, they are 28 mm, due to the presence of heavily loaded peripheral walls. The settlements measured at the periphery of the wall, by four distinct reference marks, vary between 12.7 and 23.7 mm, with an average of 16.5 mm, remaining lower than the settlements of 28 mm calculated at the edge of the raft. In this particular case, theoretical calculations based on the Combarieu method (Combarieu 1988) lead to total settlements of 16 mm under the raft.

Figure 2.48(c) and Figure 2.48(d) illustrate the theoretical results of a study of the "sinkhole condition" with the modeling of the raft and the sinkhole in the center

(Figure 2.48(c)), and the settlements ("base") of the raft (Figure 2.48(d)). This project, treated in judicial expertise for companies' claims, was validated by the justice expert on the basis of a 4 m-diameter sinkhole in service (verification at the Serviceability Limit States: SLS), and a 6 m one at the Ultimate Limit States (ULS). For sinkholes larger than 6 m (8 and 10 m in this case), the client accepted the repair of the raft in the case of its ruin, following the future activity of a sinkhole of 8 and 10 m in diameter. The given reason for the justice expert imposing this raft verification for diameters of 8 and 10 m was the formation of an old sinkhole, of 10 m in diameter, a few kilometers away from this station extension project around 20 years ago.

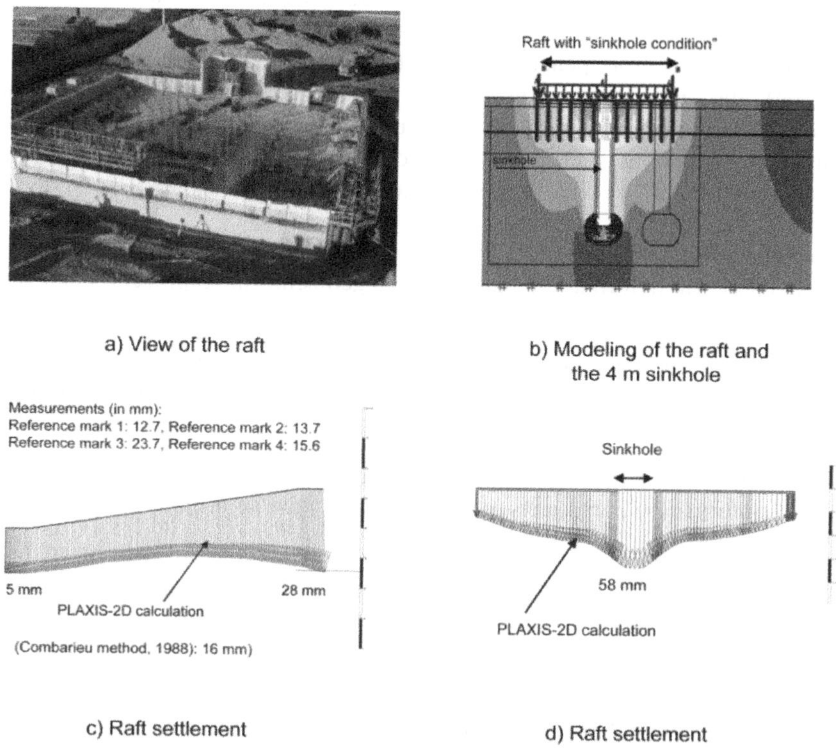

Figure 2.48. *Finite element calculation by "PLAXIS-2D" of two sewage treatment plant rafts. For a color version of this figure, see www.iste.co.uk/dhouib/geotechnics1.zip*

2.5.6. *Some lessons to be learned from deformation calculations*

The reminder in section 2.5.2.1, which was presented in order to distinguish the difference between "analysis (calculation) at failure" and "deformation analysis", was intended to show that deformation calculation has the advantage of determining the deformations of the material under load and deducing the stresses in it, whereas analysis (calculation) at failure only allows reasoning in terms of stresses, without being able to understand the deformations of the material under load. This advantage, which does exist, is deceptive in geotechnical engineering.

Indeed, nowadays, soil mechanics does not offer deformation parameters that can be directly measured in the soil and, moreover, that are reliable. The famous "conventional deformation modulus" of soil injected in deformation calculations, which some geotechnical enthusiasts call "elastic modulus" in finite element codes[3], or worse yet "Young's modulus", is generally deduced in France and, fortunately, from on-site tests. Whether the on-site measurements come from Louis Menard's pressuremeter tests or from penetration tests (if the soil allows it in terms of refusal penetration), the engineer must be "ingenious" and look for a good correlation between this conventional deformation modulus of the soil to be injected into the deformation calculations and the measured pressuremeter modulus, and/or the measured point resistances with the penetrometer (see Chapter 3).

Faced with this new interest, from young engineers and the seniors who supervise them, in deformation calculations, geotechnical engineers must keep certain rules of wisdom and common sense:

– any deformation calculation can only serve to guide the designer to have a "vague" idea of the order of magnitude of the deformation phenomena of the structure;

– before proceeding, it is necessary to analyze the soil data, check it and choose the "good" deformation parameters to be injected into the deformation calculations;

– then, take a pencil and paper and turn on the regular calculator so as to apprehend the order of the phenomena on the basis of the chosen parameters, before entering the "black box" that is calculation software in finite elements, finite differences and boundary method;

– after having performed the deformation calculations, take the time to compare the results with the "regular" calculation;

– prepare the monitoring program of the deformations of the structure, keeping in mind that, in geotechnics, "*to draw a line, two measurement points are better than a hundred theoretical points*".

3 From CESAR-LCP, FLAC 2 to 3D, to PLAXIS (the most used in France, as there are many others in the world). For the same project with identical data, all of these software must give the same calculation results, with a level of precision like that of the convergence of the iterations (in the estimation of the deformations).

This is a wise "conclusion", complemented by the other conclusions at the end of Chapter 3 and Volume 3, Chapter 3.

2.6. References

Barbas, A. and Franck, R (1982). Utilisation de la méthode des éléments finis en mécanique des sols dans le domaine de l'élasto-plasticité. L.P.C. research report n° 116, Paris.

Belkeziz, A. and Magnan, J-P (1982). Analyse numérique de la consolidation bidimensionnelle – méthodes de résolution numérique. L.P.C. research report n° 115, Paris.

Bernard, F. (1983). La méthode des éléments finis en viscoélasticité et en viscoplasticité : application à la mécanique des sols. L.P.C. research report n° 123, Paris.

Brinkgreve, R.B.J. (2016). *PLAXIS-2D – Finite Element Code for Soil and Rock Analysis.* A.A. Balkema Publishers, Delft, The Netherlands.

Combarieu, O. (1988). Amélioration des sols par inclusions rigides verticales : application à l'édification de remblais sur sols médiocres. *Revue française de géotechnique*, 44, 57–70.

Coulomb, C.A. (1776). Essai sur une application des règles des maximis et minimis à quelques problèmes de la statique, relatifs à l'architecture. *Mémoires de mathématiques de l'Académie Royale des Sciences*, 7, 343–382, Paris.

Cundall, P.A. (1976). Explicit finite difference methods in geomechanics. *International Journal for Numerical and Analytical Methods in Geomechanics*, 1, 132–150.

Darve, F., Boulon, M., Chambon, R. (1978). Loi rhéologique incrémentale des sols. *Journal de mécanique*, 17(15), 679–716.

Dhat, G. and Touzot, G. (1987). *Une présentation de la méthode des éléments finis*, 2nd edition. Maloine S.A. Editeur, Paris.

Dhouib, A. and Robinet, J-C. (1987). Domains of behavior and behavior of reinforced soils. *International Conference Series on Advances in Numerical Methods in Engineering: Theory and Application, NUMETA*, Swansea.

Dhouib, A., Bougeois, E., El Jouhari, K., Leroi, B. (2004). Comportement des semelles sur colonnes ballastées. *ASEP-GI 2004*, Dhouib, A., Magnan, J.P., Mestat, P. (eds). Presses de l'Ecole Nationale des Ponts et Chaussées, Paris.

Dhouib, A., Lelièvre, S., Idrac, P., Betoux, O. (2007). Station "La Blancarde" à Marseille : comportement d'une paroi moulée à ancrages actifs. *Revue Travaux*, 844, September.

Di Maggio, F.L. and Sandler, I.S. (1971). Material model for granular soil. *ASCE*, 97(EM3), 935–950.

Di Maggio, F.L., Sandler, I.S., Baladi, O.Y. (1976). Generalized cap model for geotechnical materials. *ASCE*, GT7.

Durand, E. (1961). *Solutions numériques des équations algébriques*, Volume II. Edition Masson et Cie, Paris.

Finlayson, B.A. (1972). *The Method of Weighted Risiduals and Variational Principles*. Academic Press, New York.

Galerkin, B.G. (1915). Road and plates. Series solutions of some problems of elastic equilibrium of rods and plates. *Vestnik inzhenerov i tekhnikov*, 19, 897–908.

Hardin, B.O. (1968). Vibrations modulus of normally consolidated clay. *ASCE*, 95(SM6), 1531–1537.

Hardin, B.O. and Drnevich, V.P. (1972). Shear modulus and damping in soils: Measurement and parameter effects. *Journal of Soil Mechanics and Foundation Division, ASCE*, 98(7), 667–692.

Hill, R. (1950). *The Mathematical Theory of Plasticity*. Oxford University Press, London.

Hujeux, J.C. (1985). Une loi de comportement pour le chargement cyclique des sols. E.R.A. report, C.N.R.S., 1029, Paris.

Idriss, I.M. (1979). Nonlinear behaviour of soft clay during cyclic loading. *ASCE*.

Imbert, J-F. (1979). *Analyse des structures par éléments finis*. Editions Cépaduès, Toulouse.

Kondner, R.L. (1963). Hyperbolic stress-strain response: Cohesive soils. *Journal of the Soil Mechanics and Foundations Division, ASCE*, 89(1), 115–143.

Lemaître, J. (1984). Loi de comportement viscoélastique. MMB Course, Mechanical Engineering, Paris VI.

Leroueil, S., Magnan, J.-P., Tavenas, F. (1985). *Remblais sur argiles molles*. Technique et Documentation, Lavoisier, Paris.

Luong, M.P. (1979). Les phénomènes cycliques dans les sables. *Journée de rhéologie E.N.T.P.E*, April.

Magnan, J.P. and Mestat, P. (1992). Utilisation des éléments finis dans les projets géotechniques. *Annales du BTP*, 509, 83–107.

Magnan, J.P. and Mestat, P. (1997). Loi de comportement et modélisation des sols. C218. Traité Construction, Volume C21.

Mestat, P. (2002). NUMGE 2002: Numerical Methods in Geotechnical Engineering. *5th European Conference, Paris*, 4–6 September.

Mroz, Z., Norris, V.A., Zienkiewicz, O.C. (1978). An anisotropic hardening model for soils and its application to cyclic loading. *IJNAM in Geomechanics*, 2, 203–221.

Nova, R. and Wood, D.M. (1979). A constitutive model for sans in triaxial compression. *International Journal for Numerical and Analysis Methods in Geomechanics*, 3(3), 255–278.

Pastor, M., Zienkiewicz, O.C., Leung, K.H. (1985). A simple model for transient soil loading in earthquake analysis II: Basic model and its applications. *IJNAM in Geomechanics*, 9, 477–498.

Pecker, A. (1984). *Dynamique des sols*. Edition des Presses de l'Ecole Nationale des Ponts et Chaussées, Paris.

Prager, W. (1955). The theory of plasticity: A survey of recent achievements. *Proceedings of the Institution of Mechanical Engineers*, 169.

Prager, W. (1958). *Problèmes de plasticité théorique*. Dunod, Paris.

Prat, M. (1995). *La modélisation des ouvrages*. Edition HERMES, Paris.

Roscoe, K.H. and Burland, J.B. (1968). On the generalized stress-strain behaviour of wet clay. *Engineering Plasticity*. Cambridge University Press, UK.

Roscoe, K.H. and Poorooshasb, H.B. (1963). A theoretical and experimental study of strains in triaxial compression tests on normally consolidated clay. *Géotechnique*, 13(1).

Roscoe, K.H., Schofield, A.N., Wroth, C.P. (1958). On the yielding of soils. *Géotechnique*, 8(1).

Roscoe, K.H., Schofield, A.N., Thurairajah, A. (1963). Yielding of clays in states wetter than critical. *Géotechnique*, 13(3), 211–240.

Rowe, P.W. (1962). The stress dilatancy relation for static equilibrium of an assembly of particles in contact. *Proceedings of the Royal Society, A*, 269, 500–527.

Salençon, J. (1980). *Calcul à la rupture et analyse limite*. Edition des Presses de l'Ecole Nationale des Ponts et Chaussées, Paris.

Salençon, J. (1983). *La Visco-élasticité*. Edition des Presses de l'Ecole Nationale des Ponts et Chaussées, Paris.

Salençon, J. and Halphen, B. (1984). Cours d'élasto-plasticité. DEA Mechanical Engineering Course, Ecole Nationale des Ponts et Chaussées, Paris.

Schlosser, F. (2006). *Eléments de mécanique des sols*. Edition des Presses de l'Ecole Nationale des Ponts et Chaussées, Paris.

Schmertmann, J.H. (1970). Stabtic cone to compute static settlement over sand. *Journal of Soil Mechanics & Foundations Engineering Division*, 96, 1011–1043.

Schmertmann, J.H. (1978). Guidelines for cone penetration test performance and design – US Department of Transportation. Federal Highway Administration (FHA), Washington, DC.

Schmertmann, J.H., Hartman, J.P., Brown, P.R. (1978). Improved strain factor influence diagrams. *Journal of Soil Mechanics & Foundations Engineering Division*, 104, 1031–1035.

Schultze, E. and Menzenbach, E. (1961). *L'essai de pénétration standard et la compressibilité des sols*. Congrès International de Mécanique des sols, Paris.

Seed, H.B. (1976). Evaluation of soil liquefaction effects on level ground during earthquakes. In *Liquefaction Problems in Geotechnical Engineering, ASCE National Convention*. ASCE, Philadelphia.

Seed, H.B. and Idriss, I.M. (1970). Soil moduli and damping factors for dynamic response analysis. Report 70-10, Earth Engineering Research Center, University of California, Berkeley.

Skempton, A.W. (1948). Vane test in the alluvial plains of river firth near Grangemouth. *Géotechnique*, 1(2), 111–125.

Smith, I.A.A. (1972). Stress and strain in a sand mass adjacent to a model wall. PhD Thesis, University of Cambridge, UK.

Smith, I.M. (1982). *Programming the Finite Element Method with Applications to Geomechanics*. John Wiley and Sons Inc., New York.

Stroud, M.A. (1971). The behaviour of sand at low stress levels in the simple apparatus. PhD Thesis, University of Cambridge, UK.

Tancelin, M.E. (1952). Mesure des pressions de terrain dans les mines de fer de l'Est. *Annales de l'ITBTP*, n°58, October.

Taylor, G.I. and Quinney, H. (1931). The plastic distortion of metals. *Philosophical Transactions of the Royal Society*, 230, 323–362.

Teng, W.C. (1966). *Calcul des fondations et mur de soutènement*. Eyrolles, Paris.

Terzaghi, K. (1951). *Mécanique des sols appliquée*. Dunod, Paris.

Terzaghi, K. and Peck, R. (1957). *Mécanique des sols appliquée*. Dunod, Paris.

Tresca H. (1864). Mémoire sur l'écoulement des corps solides soumis à de fortes pressions. *CR Académie des Sciences Paris*, 59, 754.

Vermer, P.A. (1978). A double hardening model for sand. *Géotechnique*, 28.

Waschkowski, E. (1983). Le pénétromètre dynamique. *Bulletin de Liaison des Laboratoires des Ponts et Chaussées*, n°125.

Wild, P. (1977a). Simple model of granular material in finite element solutions. *Arch. Hydro.*, Tom XXIV. Zeszyt, Poland.

Wild, P. (1977b). Two invariants depend model of granular media. *Arch. of Mechanics*, 29(6), 799–809.

Wild, P. (1979). Mathematical and physical foundations of elasto-plastic models for granular media. French–Polish Symposium, 25–30 March, Paris.

Zienkiewicz, O.C. (1977). *The Finite Element Method*. McGraw-Hill, London.

Zienkiewicz, O.C. (1982). Basic formulations of static and dynamic behavior of soils and other poros media. *Applied Mathematics and Mechanics*, 3, 457–468.

Zienkiewicz, O.C., Leung, K.H., Pastor, M. (1985). A simple model for transient soil loading in earthquake analysis I: Basic model and its applications. *IJNAM in Geomechanics*, 9, 453–476.

3

Determination of the In-situ Geotechnical Parameters of Soils[1]

What a poor, orphaned science, with no experience!

Ammar DHOUIB

3.1. Introduction

3.1.1. *The purpose of testing in geotechnical projects*

The problems presented by construction projects require a large number of parameters to be determined. These include the identification and classification of soils, the characterization of their physical, hydraulic and mechanical properties, as well as the application of soil behavior models across a whole range of static, monotonic or dynamic loads, applied to soil throughout its history by various structures. These range from simple ones (houses, low-rise buildings, small bridges and embankments) to complicated ones (large bridges and viaducts, high-rise buildings (HRB), large embankments and dams, etc.).

In Chapter 1 of this book, we reiterated and summarized the main parameters of soil identification and classification. In this chapter, we look at the tests that make it possible to determine these parameters on site, as well as actual applications of soil identification and classification parameters with respect to working platforms

1 In memory of Louis Ménard.

for roadways and highways, as well as rafts and paving for civil and industrial buildings.

Chapter 2 defines the relations between the stresses and deformations of soils, and the behavioral laws of the latter under various loads: static monotonic and cyclic or dynamic. It also gives the essential mechanical parameters of soils: deformation modulus (E), Poisson's ratio (ν), shear modulus (G) and percentage of damping (β). In this chapter, we present, detail and comment on all of the tests used to determine the geotechnical soil parameters used for the design and justification of structures, in terms of bearing capacity and deformability. These parameters are generally determined on site or in a laboratory. For major projects involving engineering structures, buildings with several basement levels, road and highway embankments, and underground structures, justification of the projects is essentially based on on-site tests, with the aim of extracting so-called "remolded" or "undisturbed" samples and taking on-site measurements of the soil parameters (physical, mechanical and hydraulic), as well as laboratory tests on taken samples. It is a campaign of *investigation*, *survey* and *study of soil*.

3.1.2. *On-site and laboratory tests: which to choose*

The current tendency in France is to increase the number of on-site and laboratory tests, while favoring on-site tests because laboratory tests run up against two essential problems. The first is that these tests are carried out on "small" samples, compared to the "large" size of the soil and its heterogeneities, so there is a *representativeness* problem.

The second is related to the *remolding* of the samples, referred to as "undisturbed" samples, during their collection and their transport to the laboratory for mechanical compression or shear tests. Another reason in favor of on-site tests is the development of new equipment and new means for "large-scale" on-site measurement of soil characteristics, so as to ensure a better assessment of the measurements. Here again, there are still obvious imperfections in certain tests that carry out local measurements, whether it be for the determination of physical properties (density meter measurements), or mechanical properties (pressuremeter or phicometer tests), or for the hydraulic properties of the soils (small-scale permeability measurements using the Lefranc test or the Lugeon test).

3.2. General methodology of geotechnical studies

3.2.1. *Phases and types of standard geotechnical studies*

The type and extent of a soil survey depends on the geotechnical project. The construction of a bridge or the digging of a tunnel is totally different from the building of a house or the raising of a small embankment to establish a road. This is why precise knowledge of the project (its footprint, the lowering service loads, the tolerances in terms of absolute and differential settlements, the destination of the structure and its environment) is essential for the geotechnical engineer, in order to define the soil survey campaign that is best adapted to the project, and to be able to optimize the foundation, support, earthworks and soil treatment solutions. In general, the study of a geotechnical project can be done in one, two or more phases according to the schematic in Figure 3.1:

– implementation phase of the *preliminary soil assessment campaign*, sampling and testing on site and in a laboratory. In this phase, the assessment is preliminary when the project is not yet defined or it is to be adapted according to the soils encountered (e.g. the creation of a subdivision in a zone of "bad" soil, the number of basements under a building, the length of a viaduct crossing a zone of compressible soil, which can be partially improved and backfilled so as to reduce the length and the cost of a viaduct, the choice between an open or a covered trench or an underground structure in order to avoid the digging of a tunnel or to reduce its length);

– execution phase of the *definitive soil assessment campaign*, the collection of samples and testing on site and in a laboratory, when the project is fully defined (floor space, adjoining structures, number of basements and floors for a building; height and footprint of a road, motorway or railway embankment; floor space, cover and immediate environment of an underground structure) and all its criteria are fixed (load-bearing capacity, wind forces, dynamic actions: machine vibrations, earthquakes, water table fluctuations and flooding);

– implementation phase of the *complementary soil assessment campaign* to understand and recognize particular aspects of the project (assessment of the existing foundations of adjoining buildings, determination of the position of the water table or of random circulations, or perched water tables, water table lowering tests prior to the construction of basements or the passage of a tunnel drilling machine, the search for natural cavities: gypsum or chalk dissolution pockets, or man-made cavities such as underground or open-cast quarries, war trenches and chalk pits).

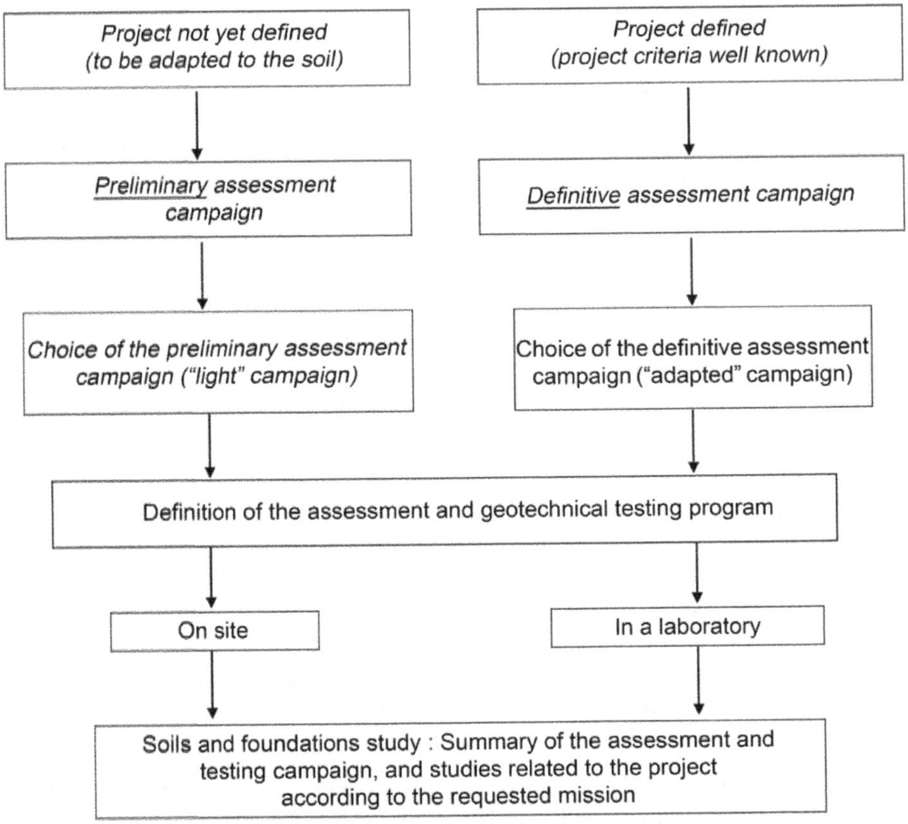

Figure 3.1. *General schematic of the phases of an assessment and study campaign for geotechnical projects*

3.2.2. *Standard geotechnical missions*

In France, since 2000, the studies of geotechnical projects have been standardized and conform to different geotechnical missions, detailed in Standard NF P94-500 of June 2000, which was revised in December 2006 and November 2013. This standard defines five main missions:

– survey and test mission: (formerly *G0*) or *IG* (Geotechnical Investigations);

– geotechnical feasibility mission: pre-project study *G1* (*G11* and *G12*), which, in the 2013 version, became *ES* (Site Study) and *PGC* (General Construction Principles);

– geotechnical project study mission: *G2* (Phases 1 and 2), changed to *G2 AVP* (Preliminary project), *G2 PRO* (Project) and *G2 DCE/ACT* in 2013 (Enterprise Consultation File/Assistance to the Works Contract);

– mission of geotechnical study of execution: *G3*;

– mission of geotechnical supervision of execution: *G4*;

– mission of geotechnical diagnosis: *G5* (on part or all of the work, with or without damage).

3.2.2.1. *Mission (G0-ING): Execution of surveys and tests*

According to the initial 2000 version of Standard NF P94-500, the G0 – ING (Geotechnical Investigations) mission is intended for "the execution of surveys, tests and geotechnical measurements".

This campaign, of on-site soil assessment using surveys, testing and mechanical measurements, as well as taking samples for identification tests and/or mechanical characterization in the laboratory, gives rise to a factual report summarizing the results of the investigations, without any implication with regard to the sizing of part or the whole of the project. According to the terms of the standard, this mission "excludes any study or consulting activity, as well as any form of interpretation". In this mission, the assessment can be preliminary or definitive but it cannot, in any case, be applied to geotechnical construction projects without the sequence of other standard geotechnical missions that follow.

3.2.2.2. *Mission G1: Preliminary geotechnical study*

This mission is devoted to the "geotechnical feasibility study", in other words, the study of the preliminary project which must decide the general principles and conditions of project implementation. It includes two parts.

Phase *ES*: site study (Mission G11: Preliminary study of the preliminary project)

This is a mission that consists of giving the general principles of the adaptation of the works to the soils encountered (in terms of foundations, support structures and earthworks), without any implication on the pre-designing of part or the whole of the works.

This study can be based on documentary research or a survey and test study, carried out within the framework of a "mission G0" (based on a soil assessment campaign, sample collection and tests on site and in a laboratory), making it possible to specify the geological, hydrogeological and environmental conditions of the project site.

COMMENT.– Any campaign carried out within the framework of mission G1 Phase ES (G11) can only be preliminary; it does not give rise to any pre-designing of the works and must be supplemented by a study of the General Principles of Construction (PGC) and of the preliminary project G2AVP (G12).

Phase: Preliminary study design: APS (or formerly Mission G12)

After a survey and test mission, and a preliminary geotechnical feasibility study *ES* (G11), the *PGC Phase* mission requires the pre-sizing of part or the whole of the project by creating a geotechnical study report, including:

– justification of the project, by detailing data relating to the soils and the presence of water, the type of foundations, the bearing capacity of the soils and the resulting absolute and differential settlements, the data relating to the support structures and the water, as well as all of the recommendations inherent in the execution of the ancillary works: paving on the ground, earthworks, partitions, etc.;

– presentation of examples of pre-designing of foundations, parameters relating to the calculation of supports, slabs, embankments and possibly alternative foundation solutions: improvements and treatment of soils.

COMMENTS.– The campaign carried out within the framework of mission ES (G12) can be definitive and, according to the requirements of the project, can also be followed by a complementary campaign (within the same framework of this mission or within the framework of a specific G5-type mission, treating a specific part of the work).

Any DCE (*dossier de consultation des entreprises* – enterprise consultation file) requiring the preliminary design of the project (contract of construction without design) must be accompanied, at the very least, by a geotechnical study G12 (G2 AVP). It is a minimal mission, followed by a mission of geotechnical study of project G2. The company then has the task of diligently building mission G3, which is entrusted to it in accordance with Standard NF P94-590.

3.2.2.3. *Mission G2: Geotechnical study of the project*

After the G2 AVP mission (improved G12), mission G2PRO is intended for the design of geotechnical projects and is generally associated with a project management mission. After the G0 and G1 missions (ES and PGC), this mission gives rise to a design study, in-depth calculations and justification of the geotechnical project. It includes two phases.

1) *Phase 1*, in which the geotechnical engineer:

– does or does not validate the geotechnical assumptions set by mission G1;

– establishes the notes on designing and develops the methods of execution for the geotechnical works in all their dimensions (definitive foundations with bearing capacities and absolute and differential settlements, types of supports and their definitive design, earthworks, embankments and excavation difficulties, paving and implementation solutions, rafts and conditions of implementation and stability, embankments and construction phases, as well as conditions of stability in phases of loading and exploitation, underground works and conditions of excavation, support and stability according to the environment: cover, geological disturbances, partitions, and provisions with respect to water, run-off or free, loaded or perched water tables;

– provides an approach regarding the quantities useful for the works (volume of concreting for foundations, density, length and reinforcement of shafts for a deep foundation solution on piles or meshes, and quantity of nails for a nailed wall, backfill cubature for road, motorway or railway embankments) and approximations of costs of the works and execution times of the geotechnical works.

2) *Phase 2*, where the geotechnician:

– details the calculation notes, draws up the plans, estimates the price lists and sketches the provisional work schedules;

– participates in the analysis of the companies' financial offers and assists the client in the selection of the company to be awarded the work.

COMMENT.– After the G1 mission, the G2 mission is the most important step in the development of the geotechnical projects because it makes it possible to analyze all the parts of the project in detail and to optimize the main elements: justification and validation of the solution or the variant of foundations, calculations, designing and justification of the supports, by seeking to optimize the quantities of anchoring, limitation of the stages of raising an embankment so as to reduce time and cost of the earthworks, finer study of lowering of the water table, etc.

3.2.2.4. *Mission G3: Geotechnical study of execution: EXE*

In a way, this mission is an extension of the G2 design mission but in the execution phase (EXE phase), given that it aims to do the following.

In the study phase

– verify and validate the geotechnical data and assumptions of the G2 mission;

– study all the components of the geotechnical works in detail and check the supporting calculations (for foundations, supports, pavements, rafts, water table drawdown, backfills and underground works);

– define the methods and the conditions of the realization of the works: phasing of implementation, follow-up and control (during execution).

In EXE phase (execution)

– auscultate and monitor geotechnical structures;

– verifiy the G2 geotechnical data and assumptions during construction;

– document the end of works and maintenance of the works in the form of a record.

3.2.2.5. *Mission G4: Geotechnical supervision of execution*

The aim of this mission is to "control" the uncertainties and geotechnical hazards in the field, following the studies carried out in the framework of the G2 and G3 missions. It constitutes the adaptation stage of the geotechnical works on soils actually encountered. It therefore consists of:

– monitoring and adapting the conditions of execution of the geotechnical works, if necessary (e.g. modification of the backfill following poor soil performance, increase of the stresses at the SLS under foundations following the discovery of a good soil, or deepening of the anchoring of the foundations in backfill pockets, or decompressed soil, reduction of pumping wells due to low dewatering flow, injection and treatment of soils following the unexpected encounter of gypsum dissolution cavities or of karstic origin in the environment where the piles are made, etc.);

– defining the monitoring programs of the structures according to the behavior of the soils encountered and the "response" of the geotechnical structures (installation of complementary piezometers around the pumping wells, installation of inclinometers to follow the deformations of a wall, increase of impedance tests to control the proper execution and the continuity of the piles, bringing the support arches closer and multiplying the measurements of converging points in a tunnel, etc.;

– establishing the tolerance threshold values according to the nature of the structure of the works, which are compatible with the behavior of the geotechnical works (permissible deformations of a wall or convergence of an underground work, permissible tension force in a tie-rod or in a nail, acceptable residual settlement in order to take up a new stage in the raising of an engineered backfill, etc.);

– drawing up regular reports and summary notes on the results of monitoring and measurements (embankment settlements and changes in pore pressures in the soil under the embankment, wall displacements and tension of tie-rods);

– participating in the closure of the works (processing the end-of-works file, correspondence with the client and the other intervening parties of the geotechnical works, and signing the formal record of receipt);

– establishing or participating in the implementation of recommendations and a maintenance program for geotechnical structures: implementation of procedures for regular inspection of drainage structures (drainage pits of road or rail platforms, pits under drainage rafts and weepholes of retaining walls), tension control in permanent anchors, etc.

Comments common to all G2 to G4 missions

In the geotechnical projects of these decades of evolution and progress, safety factors – typically three – are imposed on the stability of works, which are sensible and legitimate but it is left "up to" the geotechnicians to multiply these factors by two, or even by three, often leading to very pessimistic conclusions. We will give some typical cases later in the practical examples.

If the conditions of part or the whole of the project require it, the three missions (G2, G3 and G4) can call on a mission of surveys, tests and geotechnical measurements (G0), and for a factual report of all the services. This mission can be specific (e.g. measurement of soil permeability in "large volume" by a pumping test, so as to better define the conditions of drawdown, shear tests for the determination of the phicometer shear characteristics on site or in a laboratory: with the triaxial apparatus or the Casagrande box) or more detailed (implementation of a set of complementary surveys and measurements for the validation of all the parameters intervening in the justification of all the components of the geotechnical project).

The G2 mission is mandatory in order to get a signature for a lump sum contract for foundations.

3.2.2.6. *Mission G5: Geotechnical diagnosis*

This mission is always specific, in other words, limited to only part of the work without any implication on all of the works of the geotechnical project.

It involves either a "*Geotechnical diagnosis before, during or after the construction of a work without damage*", which includes:

– if necessary, defining a mission of surveys and geotechnical tests (e.g. to check the bearing capacity of the soils or the shear parameters of the soils, or the permeability of the soils before lowering the water table);

– validating the data and hypotheses of the G0 mission which has been purposely implemented, or fixed beforehand within the framework of the other missions (G1, G2, G3 or G4);

– deepening the study relating to a well-defined part of the project (e.g. the estimation of the absolute and differential settlements of several foundations in order to check the compatibility of these settlements with the permissible settlements of the projected or already-implemented structure, or resumption of the support calculations on the basis of shear parameters, measured within the framework of the G0 mission or fixed beforehand by the other missions, which relates to the detailed study of the system of dewatering: identification of the flow rates, number of pumping wells and depth of drawdown after the permeability of the soils encountered during a G0 mission has or has not been measured).

Or a "*Geotechnical diagnosis on a work with damage*". This mission applies to claims on existing works or those under construction. In the existing works, the G5 mission particularly concerns drought problems, where pavilions founded on clays undergo cycles of swelling-shrinkage that break up the structure, as well as land upheaval under slabs laid on swelling soils (clay or cinder). These finally mentioned damages on large-surface buildings (supermarkets in particular) are very expensive because of the operating losses and repairs.

As part of this assignment, the geotechnician typically:

– defines an investigation campaign for the assessment of the soils and the foundations of the works (a series of surveys, tests and geotechnical measurements, accompanied by reconnaissance excavations of the existing foundations within the framework of the G0 mission: for example, to check the bearing capacity of the soils or the level of the foundations and to quantify the swelling potential of the clays, or a few drillings through a slab to assess the nature of the sub-base of its foundation, with sampling and analysis in a laboratory);

– analyzes the results of the geotechnical investigations in detail, in an attempt to understand and find the probable causes of the disaster;

– gives a first approach for the possible remedies: for example, an underpinning solution by micropiles of a pavilion, with an example of pre-designing.

COMMENT.– Any repair solution proposed under the G5 mission must be subject to a subsequent geotechnical project study (Mission G2).

For a better understanding of the geotechnical missions of Standard NF P94-500, we draw the reader's attention to the diagram in Figure 3.2, which summarizes the different missions, their fields of application and the conditions of their use in a large and wide range of construction works.

3.2.3. *Blueprint of a geotechnical study*

It is difficult to establish a blueprint for a good geotechnical study because the project is generally poorly known or poorly defined, and there are many parties involved, often without prior synergy to properly define the project criteria. This results in the geotechnician not getting all the elements necessary to submit a proposal for a survey campaign (be it preliminary or final), nor a geotechnical study adapted to the project and its precise criteria. In general, geotechnicians have little or no knowledge of geotechnical projects when they receive mail requests for quotations for geotechnical studies prior to the construction of the project. Often, with private building projects, the requests come from promoters, contractors, architects or project managers appointed by the project owner or a delegated project owner. The municipalities often delegate project management to the DDE (*Directions Départementales de l'Equipement*: Departmental Directorates of Equipment).

In most cases, the geotechnician receives a site plan, a ground plan (sometimes without a scale) and a letter asking them to send a "price offer" for a geotechnical study within a determined time (one to two/three weeks). The project criteria (floor space, load-bearing capacity and environmental conditions: adjoining buildings) are often not disclosed.

For more "important" projects such as specific buildings (like high-rise buildings, engineering structures, highway or railway embankments and underground structures), the clients (promoters and wise contractors, general councils and DDE, SNCF) are less "frugal" and provide more elaborate consultation files with site plans, ground plans of the existing and future structures, as well as some standard sections indicating some aspects of the project (number of basements and levels, horizontal and vertical forces, road or railway stresses, detailed geological sections of the route and tunnel covering).

Another disappointing aspect of geotechnical studies is that clients are increasingly reducing geotechnical services to the bare minimum under the pretext of unbridled financial logic. With poor or little knowledge of the project by the geotechnician and the budgetary restrictions inflicted on geotechnical studies, as well as the lowest bidder "policy" (the cheapest), geotechnicians are creating more and more pessimistic reports, which in turn negatively affect the projects even further.

Request for quotation file, including at least:

- Site plan, and site access requirements;
- History of the site (occupations, modifications of the ground);
- Topographical plan at the time of the "Surveys and tests" mission;
- Network map (EDF, GAS, Water, Telecommunications, various for DICT*);
- Ground plan: existing works and future project;
- Project criteria: floor space, typical sections and load descent;
- Environmental conditions: partitions and their states.

↓

Development of the geotechnical offer

- Site visit;
- Study of all the documents submitted with the request for quotation.

CONTENT OF THE OFFER:

- Reminder of the stakeholders;
- Definition of the requested mission;
- Project criteria;
- Consistency of the assessment and testing campaign;
- Content of the summary report (according to the requested mission);
- Conditions of intervention: Planning, DICT, Report delivery, Regulations

↓

Presentation of the offer to the client and negotiation

- Presentation and support of the presented offer;
- Knowledge of the real needs of the client;
- Possible adaptation to the client's needs

↓

Preparation of the on-site intervention after the client's written agreement

- DICT;
- Preparation of the intervention: Contact with the client and any concerned party (Access conditions, networks, neighbors, water supply source, working hours according to environmental constraints)

↓

On-site intervention

- Implementation of services in permanent contact with the client;
- Adaptation, with the client's written agreement, of services according to the soils encountered and the criteria of the project (in-depth surveys, research of unexpected anomalies).

↓

Preparation of the summary report of the assessment, tests and studies related to the project in accordance with the requested mission:

Presentation of the conclusions of the report to the client and all parties concerned (Project Manager, Architect, Control Office, Structural Engineering Office)

*DICT: *déclaration d'intention de commencement de travaux*
(declaration of the intention to begin work)

Figure 3.2. *Good progress of the geotechnical study*

Today, geotechnical studies are summarized by the following "simple" sentence: "The client counts his money in order to do the bare minimum and at the lowest price, so the geotechnician makes his report for his insurer without any concern for the optimization of the work". The CCTP (*Cahier des Clauses Techniques Particulières*: Special Technical Clauses Handbook) completes the sad loop: "It is up to the contractor to carry out the surveys that they consider useful ...". However, the rule is simple: a good study must be the subject of a good technical offer because the "good" soil report begins at the submission of the offer, so that at the end of the project, the project owner and the company are winners in terms of the famous GMP, the "Guaranteed Maximum Price". Let us hope that this is the future in a world in structural crisis!

One last word on the "observational method" which, in an "interactive" approach, consists of: (1) calculating (sensitivity calculation with a cursor between minimum and maximum parameters); (2) instrumenting and monitoring; and (3) reacting if the set thresholds are not respected. It can be summarized as CFA: Calculate-Follow-Act. Although it is very useful in large construction projects, it is not valued by project owners because it sows the seeds of doubt.

The diagram in Figure 3.2 summarizes the basic documents and bidding steps for a geotechnical study.

3.3. Inventory of field tests in France

In France, the following on-site and laboratory tests are carried out today.

3.3.1. *Soil identification and classification tests*

Soil identification and classification tests are the foundations of geotechnical engineering (Chapter 1 of this book) because they make it possible to:

– define the precise nature of the soil (sand, gravel, clay, silt, peat, marl, etc.). This is the soil *identification* phase;

– classify the identified soils in a conventional way. This is the soil *classification* phase.

Identification and classification tests are very important and instructive for all geotechnical projects. They are used particularly for the study of road platforms and slab and raft foundations and are very useful.

The main tests are as follows:

– grading analysis test (Standard NF P94-056 and 94-057, 1996b and 1992);

– Atterberg limit test (Standard NF P94-051 and 94-052-1, 1993 and 1995);

– blue value test: VBS (Standard NF P94-040, 1993), equivalent to sand ES (Standards XP P18-597 and 598);

– measurement of the % of organic material (Standard NF P94-055, 1993).

3.3.2. *Laboratory compression tests*

– uni-axial compression test (Standard NF P94-077, 1997a);

– oedometer compressibility and creep tests (Standard NF P94-090-1, 1997b and NF P 94-091, 1995a).

3.3.3. *Laboratory shear tests*

– Laboratory vane test (Standard NF P94-072, 1995a);

– shear tests with the Casagrande box (Standard NF P94-0711 and 2, 1994) and with triaxial apparatus (Standard NF P94-070, 1994 and NF P94-074, 1994);

– shear tests with triaxial apparatus (Standard NF P94-070, 1994 and NF P94-074, 1994).

3.3.4. *Legend for parameters measured in the laboratory*

– c', φ': effective cohesion and internal friction angle determined with the triaxial apparatus (Standard NF P94-074, 1994) or with the direct shear box (Standard NF P94-071, 1994);

– c_u: undrained cohesion determined with the triaxial apparatus (Standard NF P94-074, 1994);

– λ_{cu}: rate of change of undrained cohesion as a function of consolidation stress $\lambda_{cu} = \Delta c_u / \Delta \sigma'_3$ (Standard NF P94-074, 1994);

– C_s: swelling index determined with an oedometer (Standard NF P94-090-1, 1997b);

– C_c: compression index determined with an oedometer (Standard NF P94-090-1, 1997b);

– e_0: void index determined with an oedometer (Standard NF P94-090-1, 1997b);

– C_α: secondary compression index determined with an oedometer (Standard NF P94-091, 1995b);

– c_v: consolidation coefficient determined with an oedometer (Standard NF P94-090-1, 1997b);

– σ'_p: pre-consolidation pressure determined with an oedometer (Standard NF P94-090-1, 1997b);

– k: permeability coefficient determined in a laboratory (Standard NF P94-090-1, 1997b) or on site by pumping tests (Standard NF P94-130, 1992), Lugeon test (Standard NF P94-131, 1994) and Lefranc test (Standard NF P94-132, 1992).

3.3.5. *Tests to determine the mechanical properties of soils on site*

– On-site vane test (Standard NF P94-112, AFNOR 1991a);

– cone penetration test: CPT (Standard NF P94-113, AFNOR 1996a);

– cone penetration test with pore pressure measurement: Piezocone CPU (Standard NF P94-119, 1996a);

– dynamic penetrometer tests (Type A, Standard NF P94-114, AFNOR 1990a, and Type B, Standard NF P94-115 AFNOR 1990b);

– standard penetration test: SPT (Standard NF P94-116, AFNOR 1991b);

– Louis Ménard pressuremeter test: MPT (Standard NF P94-110-1, AFNOR 2000 and NF P94-110-2, AFNOR 1999);

– phicometer test (Standard NF P04-120, AFNOR 1997c);

– surface loading test with a plate (LCPC operating mode, 1973), and Standard NF P94-117-1 and Standard NF P94-117-3 from 2000 to 2008);

– surface loading test with the Dynaplate test (LCPC newsletter no. 122).

3.3.6. *Legend for parameters measured on site*

– p_l: limit pressure measured with a Ménard pressuremeter (Standard NF P94-110);

– p_f: creep pressure measured with a Ménard pressuremeter;

– E_M: pressure modulus measured with a Ménard pressuremeter (Standard NF P94-110);

– q_c: point resistance measured with a static penetrometer (Standard NF P94-113, 1996a) or with a piezocone (Standard NF P94-119, 1995);

– f_s: lateral friction measured with the static penetrometer (Standard NF P94-113, 1996a) or with a piezocone (Standard NF P94-119, 1995);

– N_{SPT}: number of blows determined with a drill penetrometer (Standard NF P94-116, 1991);

– c_u: undrained cohesion (shear strength) of the "intact" soil, measured with a field vane test (Standard NF P94-112, 1991);

– c_R: undrained cohesion (shear strength) of the remolded soil, measured with a field vane test (Standard NF P94-112, 1991);

– s_r: sensitivity coefficient (c_u/c_r).

3.3.7. *Tests for the measurement of soil permeability*

– Pumping test (Standard NF P94-130);

– Lefranc test (Standard NF P94-131);

– Lugeon test (Standard NF P94-132).

3.3.8. *Tests to measure the dynamic soil parameters*

– Laboratory:

- free vibration test;

- resonance test;

- forced vibration test.

– On site:

- tests carried out from the surface: seismic refraction, sustained vibration of a mass;

- tests performed in a boring: "down-hole", "up-hole" and "cross-hole" tests.

3.3.9. *Specific tests*

These are specific tests for the determination of soil and rock parameters, either in a laboratory (AFNOR, Volume 1: Laboratory tests, 1999) or on site (AFNOR, Volume 2: On-site tests, 1999).

The focus here is on the on-site determination of soil parameters required for the calculation, sizing, optimization and justification of geotechnical works. The most useful laboratory tests are presented, analyzed and commented on in Volume 2, Chapter 2 of this book.

3.4. On-site parameter determination tests

In France, the performances of on-site tests for the determination of geotechnical parameters are currently as follows.

3.4.1. *Vane test*

3.4.1.1. *Principle and test device*

The field vane test was invented in Sweden at the beginning of the last century. Dubbed the "vane test", its purpose is to measure the undrained cohesion (noted as c_u) of soft and compressible fine soils. In France, the French Standard NF P94-112 and the LCPC operating mode (Dunod 1973) describe the test and the parameters measured.

The test setup includes:

– a drill stem;

– pallets called "reels";

– a rotation and measurement system of the angular variation with a manometer.

Figure 3.4 shows a cohesive, soft, compressible fine soil mass and the schematic cross-section of the test setup (Figure 3.3(a)), and a field photo (Figure 3.3(b)).

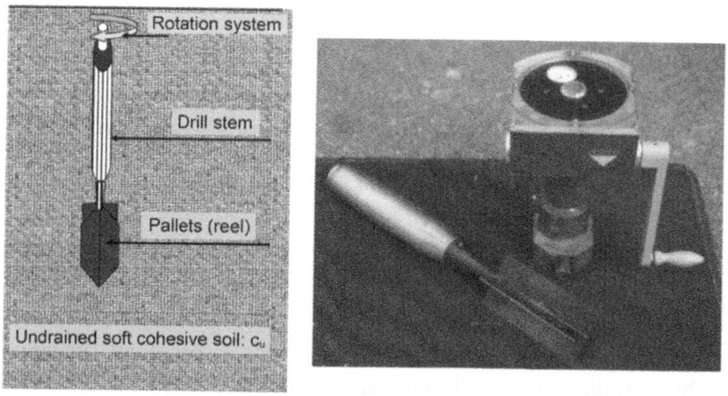

a) Fine soft soil mass and vane test

b) Test device (Photo from FONDASOL site)

Figure 3.3. *Field vane test. For a color version of this figure, see www.iste.co.uk/dhouib/geotechnics1.zip*

3.4.1.2. *Measurement of undrained cohesion (c_u): experimental curve*

The execution of the test, defined by the LCPC operating mode (Dunod 1987) and the French Standard NF P94-112 (1991), is as follows:

– the device (drill stem and reel) is driven into the soft soil at a sinking speed:

- from 0.8 to 1.3 cm/s for *slow tests*,

- from 8.5 to 13.0 cm/s for *fast tests*;

– the whole device is subjected to a leading rotation speed of 18 degrees/min;

– the leading angle of rotation (θ) is noted as a function of the torque (M_0), read directly on a manometer.

Figure 3.4 shows the typical curve of the variation of the leading angle of rotation (θ) as a function of the manometer reading.

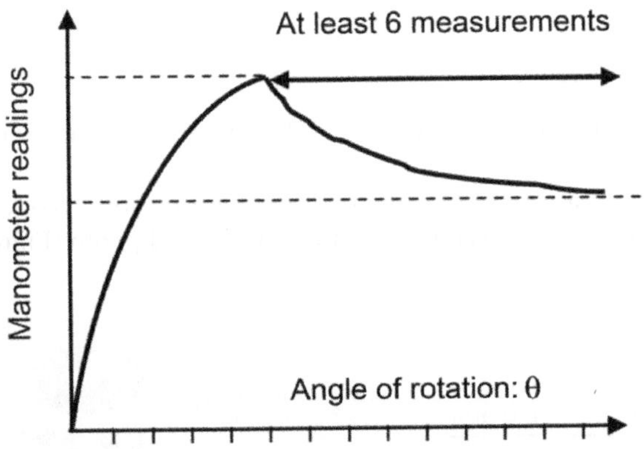

Figure 3.4. *Experimental curve of undrained soil cohesion measurement using the field vane test*

3.4.1.3. *Practical determination of undrained soil cohesion (c_u)*

Undrained soil cohesion is determined from a practical interpretation in "plasticity" by introducing the following assumptions.

1) The soil is fine anisotropic with undrained horizontal c_{uh} and vertical c_{uv} cohesion.

2) The total moment is the torque M_0 applied at the head.

3) The shear stress (τ) is identical to the undrained cohesion (c_u) in both horizontal and vertical directions.

In the vane test, the pallets delineate a shear surface, where the soil is sheared on the two horizontal flat surfaces Γ_R of radius R, and the lateral surface Γ_h of height h. Figure 3.5 shows the three shear surfaces delineated by the reel. The balance of the moments (M_R on Γ_R and M_h on Γ_h) leads to the following equalities:

– On the lateral surface Γ_h:

$$dM_h = \int_0^h 2\pi R \times c_{uv} \times R\,dz \qquad [3.1]$$

After integration, this leads to the following expression:

$$M_h = 2\pi \times R^2 \times c_{uv} \times h \qquad [3.2]$$

– On the flat horizontal surface Γ_R:

$$dM_R = \int_{\Gamma_h} c_{uh} \times r\,ds = \int_{\Gamma_h} c_{uh} \times r \times 2\pi r\,dr \qquad [3.3]$$

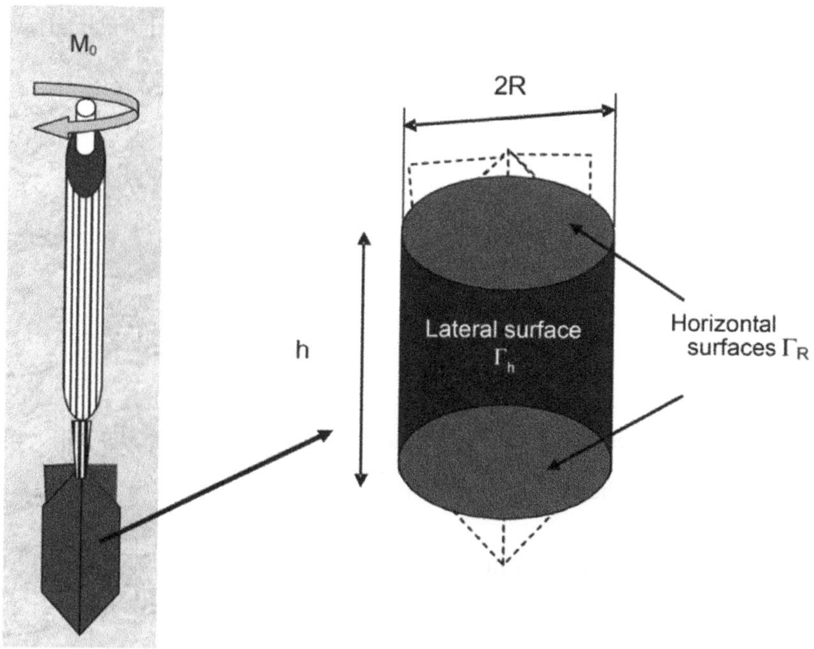

Figure 3.5. *Shear surfaces in the vane test. For a color version of this figure, see www.iste.co.uk/dhouib/geotechnics1.zip*

After integration, this leads to the following expression:

$$M_R = 2\pi \frac{R^3}{3} \times c_{uh} \qquad [3.4]$$

Now, the total moment M is given by:

$$M = M_h + 2M_R = M_0 \qquad [3.5]$$

By introducing equality [3.2] and equality [3.4], we obtain the relation linking the torque M_0, applied at the head, to the two undrained cohesions mobilized in the two horizontal (c_{uh} on Γ_R) and lateral (c_{uv} on Γ_h) surfaces of an *anisotropic fine soil*, in other words:

$$M_0 = 4\pi \frac{R^3}{3} \times (c_{uh} + \frac{3}{2} \times c_{uv}) \qquad [3.6]$$

In geotechnical projects of today, the vane test is interpreted under the assumption that the *soil is fine and isotropic*, which means that the undrained cohesion (c_u) of the soil is the same in both surfaces (vertical and horizontal), which makes it possible to write the following:

$$c_{uh} = c_{uv} = c_u \qquad [3.7]$$

As well as express the latter, from equation [3.6] and equation [3.7]:

$$c_u = \frac{M_0}{2\pi R^2 \times (H + \frac{2}{3} R)} \qquad [3.8]$$

COMMENTS.–

1) According to tests carried out by Aas (1965) on clays, the ratio of the horizontal cohesion to the vertical cohesion $\frac{c_{uh}}{c_{uv}}$ is in the order of:

– 1.1 for slightly over-consolidated clays;

– 1.5–2 for normally consolidated clays.

2) Tests carried out by the author have also shown that this ratio is in the order of:

$\frac{c_{uh}}{c_{uv}} = \frac{1}{K_0}$ (K_0: resting earth coefficient, Volume 4, Chapter 1)

The vane tests currently practiced in France use pallets, whose dimensions (Figure 3.3 and Figure 3.5) are approximately:

$$h \approx 4R \qquad [3.9]$$

This makes it possible to simplify equation [3.8] and to express the undrained cohesion (c_u) of a fine soil, assumed to be isotropic, by the relation used in practice:

$$c_u \approx \frac{M_0}{7.2\, R^3} \qquad [3.10]$$

3.4.1.4. *Measured parameters and relations between parameters*

The vane test measures:

– the undrained cohesion of “intact” soil, c_u;

– the undrained cohesion of “remolded” soil, c_u^r.

Figure 3.6 shows the graphical representation of undrained cohesion (c_u) in the normal σ and tangential τ stress plane (Figure 3.6(a)), and the typical curve of the vane test (Figure 3.6(b)).

Undrained cohesion of “intact” soil: c_u

The undrained cohesion of “intact” soil, noted as c_u or s_u (especially in English-speaking countries), corresponds to the envelope of Mohr’s circle (Figure 2.6). The undrained cohesion of “intact” soil corresponds to the maximum shear strength (τ_{max}) of the undrained fine soil and allows us to write (Figure 3.6(a) and Figure 3.6(b)):

$$\tau_{\max} = c_u \qquad [3.11]$$

and the simple compressive strength (R_c) of the soil is related to the undrained cohesion (c_u) by the equality:

$$R_c = 2c_u \qquad [3.12]$$

Undrained cohesion of “remolded” soil: c_u^r

The undrained cohesion of the remolded soil (c_u^r) is a minimum value of undrained soil cohesion and corresponds to the residual shear strength of the soil (Figure 2.18(a)), so we can write:

$$\tau_r = c_u^r \qquad [3.13]$$

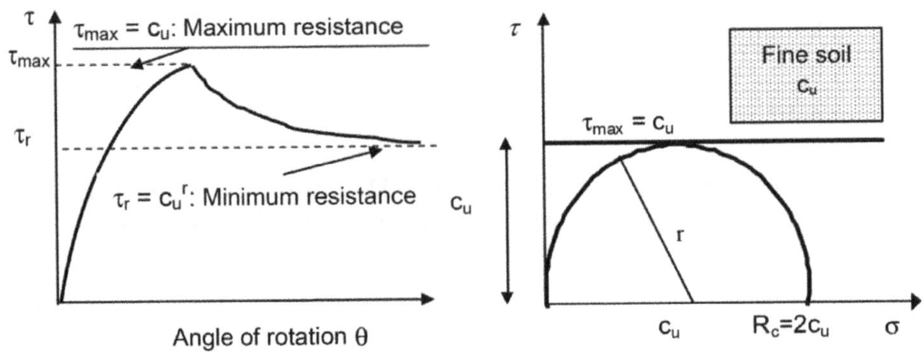

a) Vane curve b) Undrained cohesion of "intact" soil

Figure 3.6. *Vane test parameters*

The ratio of the two measured cohesions makes it possible to define the sensitivity coefficient, noted as S_t, expressed by the ratio of the maximum undrained cohesion to the residual undrained cohesion, that is to say:

$$S_t = \frac{c_u}{c_u^r} = \frac{\tau_{max}}{\tau_r} \quad [3.14]$$

COMMENTS.–

1) The theoretical interpretation in linear elasticity leads to a so-called "tangential" equation (Figure 3.7(a)), which has a form that is analogous to equality [2.47]:

$$\frac{\partial \tau_\theta}{\partial r} + 2\frac{\tau_\theta}{r} = 0 \quad [3.15]$$

2) The tangential shear stress (τ_θ) is related to the torque (M_0) by the following equality:

$$M_0 = 2\pi \times hR^2 \times \tau_\theta \quad [3.16]$$

3) The tangential stress (τ_θ) is related to the distortion (γ_θ) through the shear modulus (G) by the following linear relation:

$$\tau_\theta = G \times \gamma_\theta \quad [3.17]$$

4) When the deformations of the soil are negligible and the reel is slimmed down (stiff and infinitely long solid cylinder), the linear elastic distortion (γ_0) corresponding to the tangential stress (τ_0), mobilized under the assumption of a linear elastic behavior and isotropic soil, is related to the shear modulus (G_0) at the origin (Figure 3.7(b)) by:

$$\tau_0 = G_0 . \gamma_0 \quad [3.18]$$

where the distortion (γ_0) is twice the effective rotation (ω_0) of the reel (Figure 3.7(b)), or:

$$\omega_0 = 2\gamma_0 \quad [3.19]$$

This allows the shear modulus (G_0) to be determined from the vane test (Figure 3.7(b)) by:

$$G_0 = \frac{\tau_0}{2\omega_0} \quad [3.20]$$

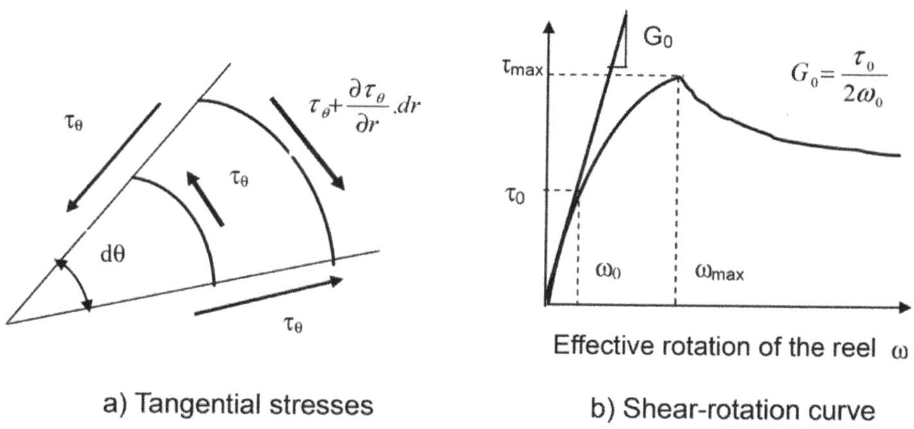

Figure 3.7. *Analysis of the vane test in linear elasticity*

3.4.1.5. *Soils involved*

The vane test can only be performed in soft soils where it is possible to prepare the test hole by direct drilling (sufficiently soft soil) or by pre-drilling with an auger (less soft soil) and to rotate the test device (shafts and reel) to shear the soil. The test is performed relatively well in the soils shown in Table 3.1.

Soil	State	Non-drained cohesion c_u (kPa)	Notes
Clay	Very soft	10–20	Fast or slow operating
	Soft	20–50	
	Not very stiff	50–80	
Sand and silt	Clean and permeable	Low (5–10)	Difficult to interpret
Silt	Organic	10–30	Fast or slow operating
Peat	Amorphous	20–50	Same as clay
	Decomposed	Low (10–15)	Same as clay
	Fibrous	-	Several peaks

Table 3.1. *Field of application of the vane test and characteristic values of undrained cohesion of typical soils*

3.4.1.6. *Influential factors*

The influencing factors affecting the measurement of undrained soil cohesion by the field vane test are:

– the speed of rotation;

– the degree of remolding.

Speed of rotation

Increasing the speed of rotation may harden the material and thereby increase the undrained cohesion of the sheared soil. Aas (1965) compares the results of slow tests (duration of 2–4 days) and rapid tests, and notes that the latter give 20%–40% higher undrained cohesions than those obtained by the former. This result is compatible with the conclusions of Caquot and Kérisel (1956), who note that the increase in deformation rate results in a "hardening" of the clays.

These results can be explained (Roy and Leblanc 1986) by the fact that when deformation rates are very low, pore pressures have time to dissipate so as to improve material consolidation. In rapid tests, there is an increase in pore pressure and therefore a slight decrease in undrained soil cohesion which, according to Aas (1965), does not exceed 5%.

Level of remolding

Cadling and Odenstad (1950) define a level of remolding (d_r), which is a function of the total vane thickness (e) and radius R (Figure 3.6):

$$d_r = \frac{4e}{2\pi R} \qquad [3.21]$$

Based on slow tests on clays, Roy and Leblanc (1986) conclude that undrained cohesion decreases with the increase in the level of remolding (d_r), and that this decrease is approximately linear. To quantify the impact of the remolding on undrained soil cohesion, Aas (1965) found that cohesion increased from 28% to 52% 24 hours after drilling the device into the clays.

Failure cylinder thickness

During the test, the pallets delineate a shear surface where the soil is sheared off the lateral surface (Figure 3.6) with a diameter of $2R$. Aas (1965) and Roy and Leblanc (1986) highlight the phenomenon of progressive failure and observe the formation of a zone around the pallets. These observations are consistent with the findings of Skempton (1948) and Arman *et al.* (1975). To account for the failure zone around the pallets, Skempton (1948) suggested adopting a shear "cylinder" with a diameter of $2R'$, and not $2R$, where he estimated that:

$$2R' = (1 + 0.5) \times 2R \qquad [3.22]$$

This would lead to a decrease in undrained cohesion c_u (equation [3.10]) of about 15%.

Reel size

The impact of reel size has not been specifically studied in the literature. Generally speaking, the lowest cohesion corresponds to reels with the highest $h/2R$ ratios. Aas (1965) attributes this result to progressive failure around the shear cylinder delineated by the reel (equation [3.22]).

3.4.1.7. *Typical vane test profile and undrained cohesion corrections*

Figure 3.8(a) illustrates a typical vane test profile where the following are successively carried out as a function of depth:

– the undrained cohesion of "intact" soil, c_u;

– the undrained cohesion of "remolded" soil, c_u^r.

Figure 3.8(b) shows the sensitivity coefficient S_t (ratio of the two parameters) as a function of depth.

Correction of undrained cohesion according to the soil plasticity index

Based on studies of landslides of embankments built on plastic clays, Bjerrum (1972) states that the undrained cohesion actually involved in the soil in situ is lower than that measured with a vane test on the same soil. The author links the decrease in cohesion to the plasticity of the clays and introduces a correction coefficient μ_r, expressing the corrected cohesion (c_u^c) to the cohesion measured by the vane test (c_u), in other words:

$$c_u^c = \mu_r \times c_u \qquad [3.23]$$

where μ_r is a function of the plasticity index (I_P) of clays and whose empirical evolution as a function of I_P is given by the author in Figure 3.8(c).

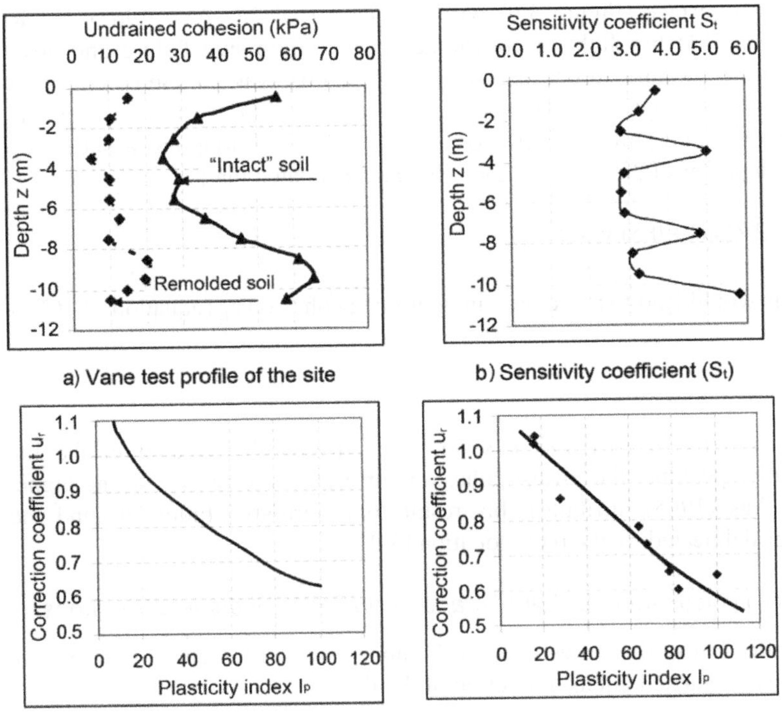

c) According to Bjerrum (1972)

d) According to Pilot (1972)

Figure 3.8. *Typical vane test profile and evolution of the undrained cohesion correction coefficient c_u as a function of the plasticity index I_P*

Based on the analysis of the failure of several embankments that are founded on very soft clays, Pilot (1972) proposes an undrained cohesion correction similar to that of Bjerrum (equation [3.23]), as well as a correction curve also quite similar to that of Bjerrum (Figure 3.8(d)).

The correction of the undrained cohesion (c_u) according to the soil plasticity index (I_P) is essential in geotechnical projects: the case of embankments for example, which is founded particularly on plastic to very plastic clays. This requires that on-site vane tests be accompanied by laboratory measurements of the plasticity index over the entire height of the vane test profile.

3.4.1.8. *Role of the field vane test in geotechnical project studies*

The vane test is not widely used in France. It is a discontinuous measurement because the tests can be carried out every meter, or even every 0.5 m. As mentioned in section 3.3.1.5, the test is technically feasible in soft soils. It is generally used to assess the bearing capacity of soft soils under embankments, and the general or local short-term stability of earthworks. However, it is unable to estimate the settlements of geotechnical structures.

3.4.2. *Cone penetration test (CPT)*

Much like the vane test, the cone penetration test is very old, with its first uses dating back to 1929 in the United States of America. From 1932, the test was quickly developed at the Delft Soil Mechanics Laboratory in the Netherlands. The test has been widely used throughout the world, and also sometimes locally in France (Sanglera 1965), although it is well ahead of the pressuremeter for historical reasons inherent to French geotechnical culture. The French Standard NF P94-113 describes the test.

3.4.2.1. *Principle of the test*

The test is simple in its principle, given that it consists of:

– applying a vertical force on a drill stem equipped with a tip (the cone itself);

– drilling in a static way (without vibration) and without rotation of the assembly, continuously;

– measuring the active forces by the tip:

- the reaction at the tip of the device, called *point resistance* and noted as q_c,

- the *resistance to the lateral friction*, noted as f_s, when the tip, equipped with a friction sleeve at its base, is fixed in relation to the drill stem and is of the same diameter as the latter;

– deducting a "*friction ratio*" from q_c and f_s, noted as F_R, and a "*friction index*", noted as I_f.

The drilling is done at a constant speed of 20 mm/s, with a tolerance of ± 5 mm. The measurement is done continuously or at least every 200 mm. The diameter of the point is 36 mm (or more than 36 mm for the piezocone).

3.4.2.2. *Relation between parameters measured with traditional penetrometers*

Generally speaking, there are a large number of cone penetrometers in France and abroad, as well as cone penetrometers with pore pressure measurement known as "*Piezocône*" in France and "*Mémocône*" in Scandinavian countries.

The cone penetrometer is analogously comparable to a foundation pile. It is a mini-pile where, under a given load (Q), the point resistance force (Q_p) of the shaft and the lateral friction resistance force (Q_s) involved on the lateral surface of the pile, are mobilized.

Figure 3.9 shows an analogous diagram between the vertically loaded pile (Figure 3.9(a)) and the cone penetrometer, with a friction sleeve for the possible measurement of lateral friction during the drilling of the device in the soil (Figure 3.9(b)). Traditional cone penetrometers therefore allow the following to be measured (Figure 3.9(b)):

– the total force Q_t, which is the total force necessary to drill in the whole device: drill stem and tip;

– the total apparent tip force Q_c, which designates the force necessary to just take up the cone of the penetrometer tip. If the total solid section of the point is A_c, then the apparent tip resistance q_c is given by:

$$q_c = \frac{Q_c}{A_c} \qquad [3.24]$$

– the total frictional force Q_{st}, which represents the force obtained by the difference between the total force Q_t and the apparent total force Q_c mobilized at the tip, in other words:

$$Q_{st} = Q_t - Q_c \qquad [3.25]$$

– the lateral friction force Q_s mobilized locally on the sleeve, which is the total force required to take up the friction sleeve onto the lateral surface of the sleeve A_s. The local unit lateral friction (on the sleeve) is given by:

$$q_s = \frac{Q_s}{A_s} \qquad [3.26]$$

From these main parameters, the following dimensionless parameters are deduced:

– the friction ratio F_R (or R_f), which is the ratio of the unit lateral friction q_s to the point resistance q_c, both measured at the same depth, that is to say:

$$F_R = \frac{q_s}{q_c} \qquad [3.27]$$

– the friction index I_f, which is the quotient of the apparent point resistance q_c by the local unit lateral friction q_s, also measured at the same penetration depth, in other words:

$$I_f = \frac{q_c}{q_s} \qquad [3.28]$$

COMMENT.– The two ratios F_R (equation [3.27]) and I_f (equation [3.28]) are expressed in percent.

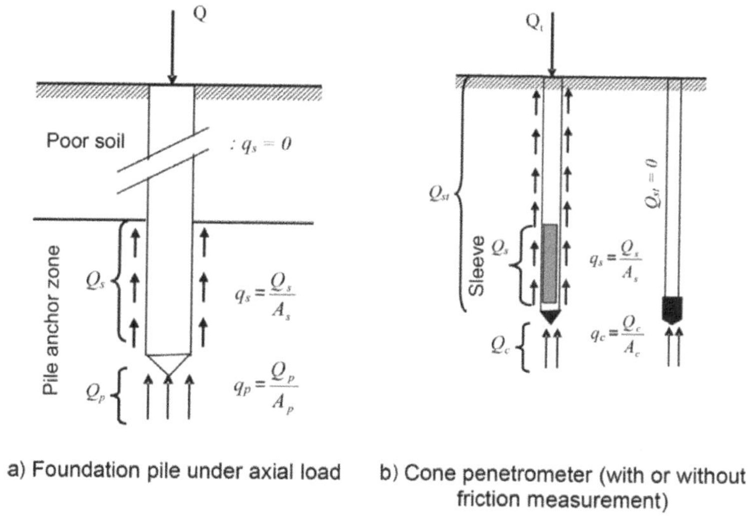

Figure 3.9. *Analogy between a foundation pile and a cone penetrometer*

3.4.2.3. *Piezocone test (CPTU)*

Principle of the test

Let us first note that the "piezocone" is a "tip piezometer" used to measure the water level (in the form of pressure) by a specific device and the resistance of the tip of the cone.

The piezocone is thus a cone penetrometer which, in addition to the parameters measured by traditional devices (equalities [3.24]–[3.28]), makes it possible to measure the pore pressures developed in the soil during the drilling of the device. These represent, in relation to the atmospheric pressure, the pressure of the water mobilized at the depth reached by the tip and measured by a pressure filter, located just above the conical tip.

The test also allows the measurement of the total head force (Q_t) and the lateral friction force (Q_s) on the sleeve located above the penetration cone. When at depth for the pressure test, the drilling is stopped, and the pore pressure (u) is immediately measured as a function of time in order to observe its evolution.

Let us use Δu to note the pore pressure deviation (u, measured at a time t after the drilling has been stopped) in relation to the pore pressure (u_0, prevailing in the soil before the test), and use Δu_h for the pore pressure deviation (u_h, measured when the drilling has been stopped) in relation to u_0 too.

According to Standard NF P94-119, the pore pressure dissipation measurement is stopped when the deviation Δu has decreased by more than half, when compared to the deviation Δu_h, that is to say:

$$\Delta u = u - u_0 \leq \frac{1}{2}\Delta u_h = \frac{1}{2}(u_h - u_0) \qquad [3.29]$$

The time required for equality [3.29] to be verified:

$$u - u_0 = \frac{1}{2}(u_h - u_0) \qquad [3.30]$$

is conventionally noted as: t_{50}.

Relations between parameters measured at the piezocone

Let us note u_0 as the initial pore pressure existing in the soil before the drilling of the device (this pressure is the hydrostatic pressure that is undisturbed by the drilling or by the rapid loading of the soil, and known beforehand by an appropriate site survey). Let us note u as the pore pressure which prevails in the soil, during drilling, at the base of the tip (same depth as u_0). The variation (Δu) of the pore pressure at this depth, at a given time t, is then expressed by:

$$\Delta u = u - u_0 \qquad [3.31]$$

This is an algebraic value because the value of the pore pressure (u), mobilized at the base of the tip, can be negative (such is the case of clays, for example).

The mobilization of the pore pressure (u) in terms of the filter element results in a total force Q_u being developed in the annular space (A_c – A_u), located above the penetration cone. The total drilling load (Q_t) applied is balanced by the water action (Q_u) and the mobilized load at the tip (Q_c). Figure 3.10 shows a diagram of the tip with the friction sleeve and the pore pressure and sealing device overlaying the cone.

The total resistance mobilized at the tip of the cone (noted q_t) is then:

$$q_t = \frac{Q_t}{A_c} \qquad [3.32]$$

The total charge Q_t is the sum of the two terms Q_u and Q_c; hence:

$$q_t = \frac{(Q_c + Q_u)}{A_c} \qquad [3.33]$$

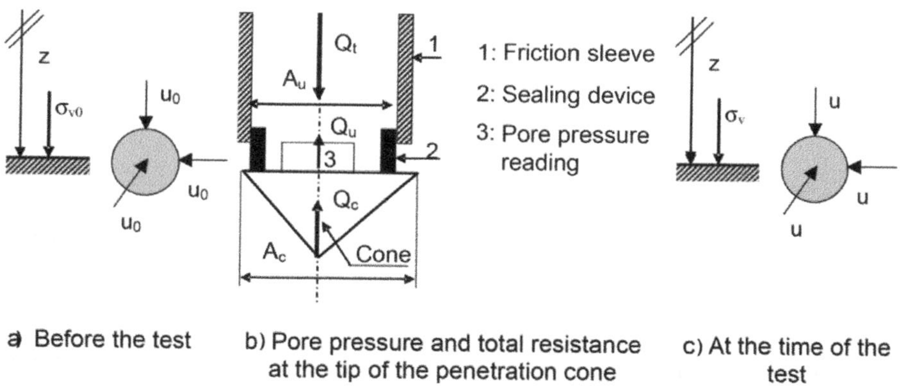

Figure 3.10. *Total cone force (Q_c) and pore pressure mobilization force (u) at the tip of the piezocone*

Q_u is developed in the annular space ($A_c - A_u$) and is equal to:

$$Q_u = (A_c - A_u) \times u \quad [3.34]$$

This leads to the following relation:

$$q_t = \frac{Q_c}{A_c} + \frac{(A_c - A_u) \times u}{A_c} = q_c + u \times (1 - a) \quad [3.35]$$

The piezocone allows us to determine the pore pressure coefficient B_q and is equal to:

$$B_q = \frac{(u - u_0)}{q_t - \sigma_{v0}} \quad [3.36]$$

where σ_{v0} is the total vertical stress in the soil immediately before the test at the depth (z) of the test and is equal to: $\gamma . z$

The mobilization of the pore pressure requires the correction of the two measured parameters: point resistance (q_{cm}) and lateral friction (q_{sm}). Thus, the real corrected values (q_c and q_s) are:

$$q_c = q_{cm} - \Delta q_c \quad [3.37]$$

with:

$$\Delta q_c = \Delta u \times \left(\frac{q_{cm}}{\sigma_{v0}} - 1\right) \quad [3.38]$$

and

$$q_s = q_{sm} - \Delta q_s \quad [3.39a]$$

where:

$$\Delta q_s = \Delta u \times f(\phi_{cu}) \quad [3.39b]$$

ϕ_{cu} is the friction angle under undrained conditions.

3.4.2.4. *Theoretical interpretation of cone penetration tests*

The application of cone penetration tests makes it possible to establish a direct analogy with piles. In Volume 3, Chapter 1, which concerns foundations, we recall and demonstrate Terzaghi's relation, giving the failure stress (q_u) at the base of a foundation.

In a heavy (bulk density γ), frictional (angle of friction φ) and cohesive (cohesion c) soil, the point failure stress (q_u) of a pile driven to a height D in the soil is (general equation [1.28a] in Volume 3, Chapter 1):

$$q_u = \gamma \times D \times N_q + c \times N_c \quad [3.40]$$

where N_q and N_c are the bearing capacity factors, whose numerical values are given by Table 1.8 of Volume 3, Chapter 1.

By analogy, the cone resistance (q_c) can be written as a linear relation as a function of depth (D), that is to say:

$$q_c = a \times D + b \quad [3.41]$$

Relation [3.40] seems to be simple and promising for deducing the two essential soil shear parameters (c and φ) from cone penetration tests, but this analogy unfortunately gets very badly confused with experimental realities, especially in purely frictional soils.

Figure 3.11 shows the analogy between a vertically loaded pile (Figure 3.11(a)) and the cone penetrometer (Figure 3.11(b)).

To better analyze and understand this analogy, let us successively examine the case of two standard materials: sand and clay.

Case of sands

In purely frictional sand ($\varphi \neq 0$), without cohesion ($c = 0$), the relation in [3.40] is simplified and becomes:

$$q_u = \gamma \times N_q \times D \quad [3.42]$$

By analogy, the simplified relation in [3.41] leads to:

$$q_c = a \times D \quad [3.43]$$

with:

$$a = \gamma \times N_q \quad [3.44]$$

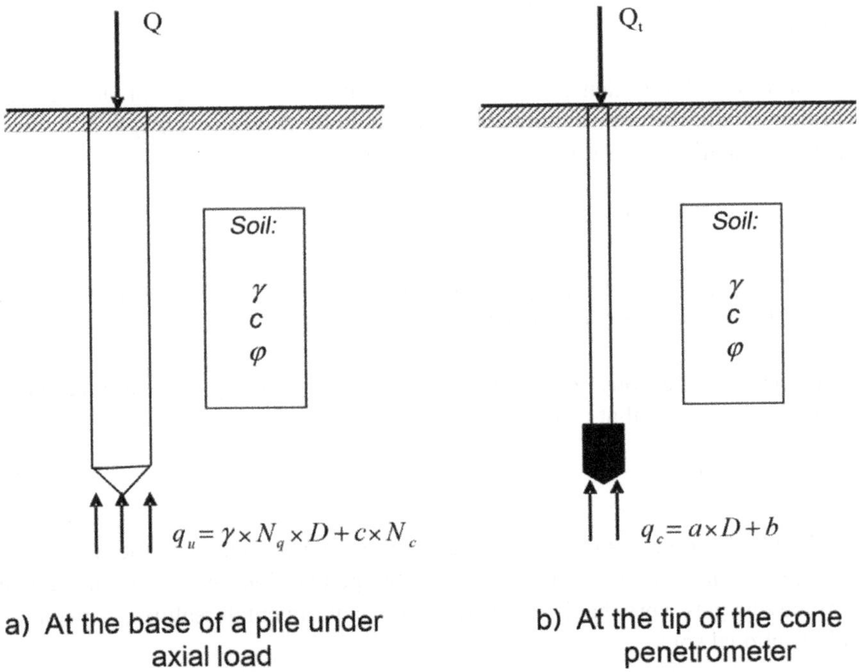

Figure 3.11. *Failure stress: pile-cone penetrometer analogy*

On tests in reduced models, L'Herminier (1967) notes that during penetration, slip lines in the form of logarithmic spirals are mobilized around the penetrometer, and that the internal friction angle (φ) must be corrected in order to avoid high N_q values. In order to bring the calculation data closer to the experimental ones, the author introduces a hypothetical friction angle (φ_s) defined by:

$$\tan \phi_s = \frac{2}{3} \tan \phi \quad [3.45]$$

Other authors (de Beer 1946, 1949; Meyerhof 1956) have tried to give a representation of N_q as a function of φ from the pile-cone penetrometer analogy. However, such results are only slightly significant and can in no way replace more appropriate shear tests in the laboratory (triaxial or direct shear test, laboratory vane test) or on site (vane or phicometer test, presented later in section 3.4.6).

Case of clays

In a purely coherent clay ($c = c_u \neq 0$) and without friction (0), the relation [3.40] is simplified and becomes:

$$q_u = c_u \times N_c + \sigma_{v0} \quad [3.46]$$

where σ_{v0} is the total vertical stress in the soil.

By analogy, the simplified relation [3.41] leads to:

$$q_c = b = c_u \times N_c + \sigma_{v0} \quad [3.47]$$

Relation [3.46] allows us to derive the undrained cohesion c_u of clays by:

$$c_u = \frac{q_c - \sigma_{v0}}{N_c} \quad [3.48]$$

The value of the bearing capacity factor N_c can be deduced from a continuous penetration, with measurement of the tip resistance (q_c), and by applying the corresponding state theorem presented in section 2.3.5 of Volume 2, Chapter 2. In some devices (Dutch cone penetrometers, for example), the tip is provided with a protective skirt which, in addition to the tip, generates a certain lateral friction (q_{sl}) that is mobilized locally on this skirt. This lateral friction can be reasonably considered as a fraction (coefficient β) of the undrained cohesion (c_u) of the soil (Cassan 1987), in other words:

$$q_{sl} = \beta \times c_u \quad [3.49]$$

The total force mobilized at the tip of the straight section of cone A_c is the sum of the force caused by cohesion (equation [3.48]) and the local frictional force mobilized on the lateral surface (A_L) of the skirt (relation [3.49]), that is:

$$A_c \times q_c = (c_u \times N_c + \sigma_{v0}) \times A_c + (\beta \times c_u) \times A_L \quad [3.50]$$

Hence, the undrained cohesion (c_u):

$$c_u = \frac{q_c - \sigma_{v0}}{N_c + \beta \times \frac{A_L}{A_c}} \quad [3.51]$$

With an N_c value of 10 (current value, Table 3.2) and section ratios (A_L/A_c) of 10 (the case of Dutch penetrometers), we obtain the following equality for low cohesions ($\beta = 1$):

$$c_u = \frac{q_c - \sigma_{v0}}{20} \quad [3.52]$$

Based on the Brandt–Caquot formula for the failure stress at the base of a buried circular footing, Begemann (1963) gives a relation that is similar to equality [3.50] with a ratio of 14. Several authors have given relatively different values of the N_c factor. Table 3.2 summarizes the most typical and well-known in the literature.

Type of tip	Author	Value of N_c	Notes
-	L'Herminier	11	-
-	Skempton	10	-
-	Meyerhof	9	-
-	Caquot and Kérisel	7	-
Conical	Bejemann	14	With protective skirt
Conical	Caquot and Kérisel	20	With protective skirt
Conical	-	10–15	-
Dutch	-	15–20	With protective skirt

Table 3.2. *Values of the N_c factor given by several authors*

3.4.2.5. *Concept of critical depth and height*

In a purely frictional soil ($c' = 0$), Biarez (1962) and Grésillon (1970) note that, at medium depth, the slip lines generated around the cylinder of soil in motion go up to the surface. The result is that the point resistance q_c follows a parabolic law in shallow depths (Figure 3.12(a)). The tangent at the origin of the parabola corresponds to a minimum value of the bearing capacity factor (N_{qmin}) for shallow foundations (footings) or semi-submerged foundations (foundation pits). For greater depths, it is generally and hypothetically considered that the point resistance would present a linear part whose slope would correspond, from a "critical height" (h_c) to the maximum value of the bearing capacity factor (N_{qmax}) that is applied for deep foundations (piles or slurry trench piles). Foray (1972) and Puech (1974) confirm these results, but Puech (1974) makes the difference between the critical height h_c and the critical depth, which corresponds to the stabilization level of the point resistance (q_c) as a function of depth (Figure 3.12(a)).

To understand the point resistance (q_c) in sands of varying density, Kérisel (1965) shows that the tip that is mobilized before the saturation plateau decreases as the diameter (B) of the tip increases and that the critical depth (D_c) increases with the diameter of the tip, as shown for dense sands (tight grains) by the evolution of the point resistance as a function of depth for several tip diameters (Figure 3.12(b)).

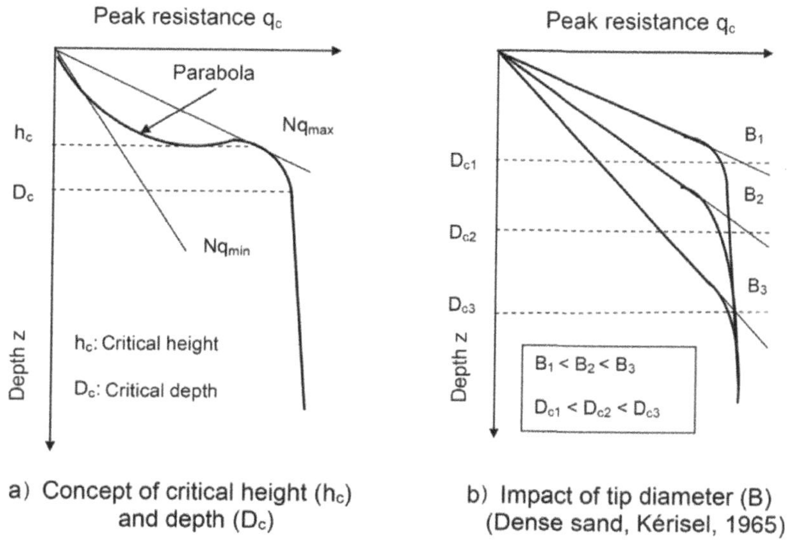

Figure 3.12. *Evolution of the point resistance q_c as a function of depth z*

3.4.2.6. *Soil classification from cone penetration tests*

In cone penetration tests, the soils drilled through by the tip during the drilling operations are not known. In consequent and serious geotechnical investigations and tests, cone penetration tests must be accompanied by appropriate test drillings, preferably core sampling, helical auger drillings, excavations with shovels for short penetrometers (search for solid substratum with real penetration setting) and a calibration on some pressuremeter samplings for a better optimization of the foundations. In France, for about 40 years, the sizing of foundations, has generally been based on the pressuremeter test results obtained by the pressuremeter invented by Louis Ménard (in section 3.4.5, presented below).

Several authors in France and in other countries have tried to classify soils according to the two fundamental parameters measured by the cone penetrometer: q_c and q_s, and the friction ratio F_R (equation [3.27]):

– In France, Sanglerat (1971) links the lateral friction (q_s), measured on the sleeve, to the point resistance (q_c) and gives the empirical ranges of variation between the two parameters, summarized in Table 3.3.

– Bejermann (1965) and Laréal *et al.* (1974) give correlations between the tip (q_c) and lateral friction (q_s) for various soils in a graphical form, as shown in Figure 3.13(a).

– Laréal *et al.* (1974) and Sanglerat *et al.* (1974) establish the ranges of variation of the point resistance (q_c) and the friction ratio (F_R) according to the nature of the soil. Table 3.4 summarizes the variations of the two parameters for several types of soils, measured with the "Andina" cone penetrometer and given by Laréal *et al.* (1974).

– Robertson (2009) establishes a more recent diagram, as illustrated in Figure 3.13(b) in non-standard (q_c, RF%) and standard forms:

[$Q_t = ((q_t - \sigma_{v0})/\sigma'_{v0})$; FR% $= f_s/(q_t - \sigma'_{v0})$ in logarithmic scale]

Soils	Variation (q_s, q_c)	Notes
Peat and soft clays	$\frac{q_c}{30} < q_s < \frac{q_c}{10}$	-
Consistent clays	$\frac{q_c}{25} < q_s < \frac{q_c}{12}$	$q_s = \frac{3q_c}{50}$ (Begemann 1965) $q_s = \frac{q_c}{50}$ (Sanglerat 1971)
Silt and sandy clays	$\frac{q_c}{100} < q_s < \frac{q_c}{25}$	-
Fine and medium sands	$\frac{q_c}{150} < q_s < \frac{q_c}{100}$	-
Coarse sand and gravel	$q_s < \frac{q_c}{150}$	-

Table 3.3. *Values of lateral friction (q_s) and point resistance (q_c) measured with a cone penetrometer (Sanglerat 1971)*

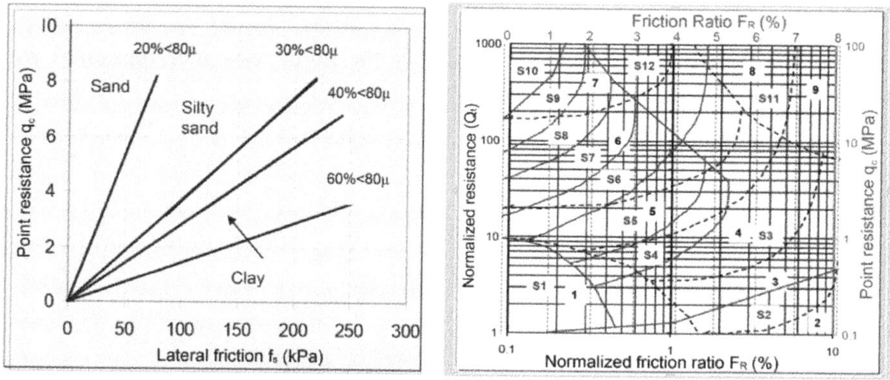

a) Laréal *et al.* (1974) b) Robertson (2009)*

* Standardized Robertson legend (Q_t, F_R):

1: Fine clayey soils; **2**: Organic soils; **3**: Clays and silty clays; **4**: Clays/silts; **5**: Silty sands; **6**: Clean to silty sands; **7**: Sands and gravels; **8**: Dense sands; **9**: Very stiff fine soils.

Figure 3.13. *Summary of correlations between point resistance and lateral friction. For a color version of this figure, see www.iste.co.uk/dhouib/geotechnics1.zip*

Soil nature	Values of q_c (MPa)	Values of F_R (%)
Clays	< 0.8	> 3.0
Sandy silts	$0.4 < q_c < 3.0$	$1.0 < F_R < 7.0$
Silts	$07 < q_c < 2.0$	$1.5 < F_R < 5.0$
Clayey silts	$0.4 < q_c < 5.0$	$1.0 < F_R < 2.5$
Sands	< 1.5	< 2.5
Clayey sands	$0.8 < q_c < 4.0$	$1.0 < F_R < 5.0$
Sandy clays	$0.3 < q_c < 6.0$	$0.5 < F_R < 8.0$
Silty gravels	> 2.0	< 2.0
Clayey gravels	> 1.0	< 5.0

Table 3.4. *Nature of soils and typical values of point resistance (q_c) and friction ratio F_R*

3.4.2.7. *Influential parameters*

Several influencing parameters can modify the point resistance measured with the cone penetrometer. They are generally:

– cone penetration velocity;

– the diameter of the cone tip;

– the shape of the cone tip;

– the continuity of the penetration;

– the discontinuity of the penetration.

Influence of the cone penetration speed

The cone penetration speed is the most significant influential factor with regard to the parameters measured with the cone penetrometer, and this is the case in fine saturated soils as in aquifer sands. Indeed, like in the vane tests (section 3.3.1.6 of this chapter), where the shear strength is influenced by the rotation speed of the reel, the point resistance measured with the cone penetrometer is also sensitive to the cone penetration speed.

Thus, in soft clays, Jézéquel (unknown source, cited by Cassan (1980)), the increase of the penetration speed from 1 to 10 cm/s leads to an increase in the point resistance of 14%–26%, with a normal Gouda tip, and 35% with an electric tip. In stiff clays, it is acknowledged (Cassan 1988) that the influence of the penetration speed is low and is accompanied by a slight increase in the point resistance (q_c).

In saturated silts and aquifer sands with a given grain size, the increase in the drilling speed is accompanied by pore over-pressures and the collapse of the structure of the material, as well as the decrease in point resistance when the density of the soils in situ is lower than the critical density. This is the phenomenon of liquefaction for loose sands and liquefiable silts, or thixotropy, in the case of chalk.

In compact sands, Kérisel (1961, 1965) concludes that the point resistance is not very affected by the drilling speed.

Impact of cone tip diameter

According to the literature, it is generally accepted that in sands, the diameter of the tip does not affect the parameters measured with the cone penetrometer, although, by comparing the point resistances (q_c) measured in tight sands with peak diameters (B) from 45 to 320 mm, Caquot and Kérisel (1967) note that q_c decreases when B increases. However, in fine soils such as clays, the size influence is not negligible. The point resistance decreases when the diameter of the tip increases, due to the elasto-visco-plastic behavior of these materials. Caquot and Kérisel (1966), as well as Kérisel and Adam (1967), thus show that during cone penetration tests and piling of growing sections in Sannoisian green clays, the point resistances decrease.

This observation concerning fine soils such as clays is an element of reflection on the analogy made between the cone penetrometer and the pile.

Influence of the shape of the cone tip

The geometry of the tip is of great significance with regard to the resistance measured with the cone penetrometer. Jézéquel (1969) shows that, for an electric tip from the LRPC of Saint-Brieuc and a Gouda tip equipped with electric gauges of the same diameters, the resistances measured by the latter are higher by 31%–42%. Indeed, contrary to the simple and fixed cone of most cone penetrometers, a cone equipped with a protection skirt like the Gouda cone penetrometer causes parasitic friction on the skirt and overemphasizes the point resistance (q_c).

Impact of the continuity of the penetration

There are two ways to measure the point resistances with the cone penetrometer: continuous measurement with a fixed cone or discontinuous measurement with a mobile cone, where the operator has to stop to measure q_c.

Contrary to the indications of Kerisel (1961) who postulated that, in the case of successive advances of the tip and the lateral envelope, there will be an overemphasis of point resistance, Raedschelders (1961) maintains, in light of tests of continuous cone penetration (fixed cone) in sands, that the values of the point resistance (q_c) are higher than in the case of discontinuous penetration (mobile cone). De Beer (unknown source, quoted by Cassan (1980)) explains that this increase by the mobilization of an artificial load at the tip is due to the lateral friction exerted on the shaft of the penetrometer. According to Sanglerat (unknown source, quoted by Cassan (1980)), this increase is of 8%–15%, and of 10%, according to Jézéquel (unknown source, quoted by Cassan (1980)), for dry sands.

Influence of discontinuity of the penetration

In aquifer clays, silts and sands, Jézéquel (unknown source, cited by Cassan (1980)) recorded point resistances (q_c) that were 28%–36% higher in discontinuous penetration by a normal Gouda tip than in continuous penetration by a Gouda tip of the same geometry equipped with an electric load cell.

Cassan (1988) specifies that, in general, Gouda tips give point resistances of around twice those obtained by electric tips. Jézéquel (unknown source, cited by Cassan 1987) explains these discrepancies by two equal phenomena: friction on the protective skirt and the effect of the discontinuity of the penetration.

3.4.2.8. *Role of the cone penetrometer in geotechnical project studies*

In France, geotechnical study reports (Mission G1 to G5) often include cone penetration surveys, which remain the tool of some specialized soil and foundation agencies.

Cone penetration tests are continuous measurements that make it possible to assess:

– the resistance of the different layers of soil, without detecting its exact nature;

– the possible presence of anomalies: presence of blocks (premature rejection or occasionally high resistance) or of compact substratum (high resistance leading to a real rejection).

However, this reconnaissance tool cannot detect possible anomalies (karst pockets, for example) in a substratum. It can be used to orient the choice of foundations and to deduce the load-bearing capacity of the soils, and must be accompanied by appropriate surveys (core samplings, reconnaissance drilling with auger and/or cutter and pressuremeter tests) to be able to appreciate the geological and mechanical aspect of the soils crossed, whether they are hard or soft. In some cases, geotechnical projects have been resumed and resized with significant added values following anomalies in the “solid” substratum that were not detected by the cone penetrometer, for which all the tests encountered real rejections at the top of substratum (not to be done in Volume 3, Chapter 3).

3.4.3. *Dynamic penetrometer test (PDA and PDB)*

3.4.3.1. *Principle of the test*

The dynamic penetrometer is like a “cone” penetrometer, but the tip is not driven into the soil in a static way, and instead pile-driven by synchronized driving. The “lost” tip is hammered into the soil, from a well-determined drop height, with the help of a driving hammer.

There are two types of dynamic penetrometers:

– the Type-A dynamic penetrometer (French Standard NF P94-114), where the shafts are equipped with mud injection holes at the lower end;

– the Type-B dynamic penetrometer (French Standard NF P94-115), without injection.

The test device is composed of:

– a driving hammer mass m, ranging from 0.052 to 1.5 kN;

– a lost conical tip of diameter B;

– shafts, a driving helmet and various accessories of mass m'.

The drop height (noted as h_d) of the driving hammer varies from 0.2 to 1.0 m. The number of blows (N_d) is noted during the descent of the tip into the soil over a given height Δh_d: generally 0.1 and 0.2 m by European standards. The take up of the tip (noted as e_d and improperly called "elastic" take up) is defined as the ratio of Δh_d to the number of blows N_d, in other words:

$$e_d = \frac{\Delta h_d}{N_d} \quad [3.53]$$

Figure 3.14 shows a schematic cross-section of a Type-A (Figure 3.14(a)) and Type-B (Figure 3.14(b)) dynamic penetrometer test setup.

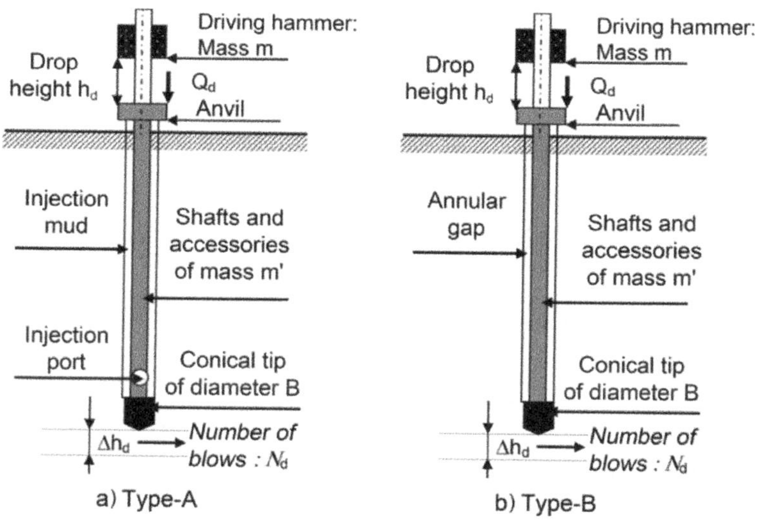

Figure 3.14. *Dynamic penetrometer test device (A and B)*

3.4.3.2. *Theoretical interpretation of the test*

There are generally two approaches to interpreting the dynamic penetrometer test:

– the energy approach;

– the theory of vibrations.

Energy approach

The energy approach consists of analyzing the evolution of the take up of the tip (e_d) at the time of the piling as a function of the dynamic force (Q_d), applied at the head. The total take up (e_d) of the tip is assumed to be the result of two components: the so-called "reversible" take up (noted as e_{dr}) and the so-called "irreversible" take up (e_{di}), that is to say:

$$e_d = e_{dr} + e_{di} \qquad [3.54]$$

As the dynamic force (Q_d) increases, the total take up (e_d) increases. The force-take up curve has a similar shape to the one shown in Figure 3.15. The interior of this curve (OABCO area) represents the energy transmitted during the piling, which allows us to express the dynamic force (Q_d) as a function of the take up, the drop height and the masses by:

$$Q_d \times (e_{di} + \frac{e_{dr}}{2}) = mgh_d \times \frac{m + \varepsilon^2 \times m'}{m + m'} \qquad [3.55]$$

where:

– g is the acceleration of gravity;

– m and m' are, respectively, the masses of the driving hammer, shafts and accessories;

– ε is Newton's coefficient, where: $0 < \varepsilon < 1$.

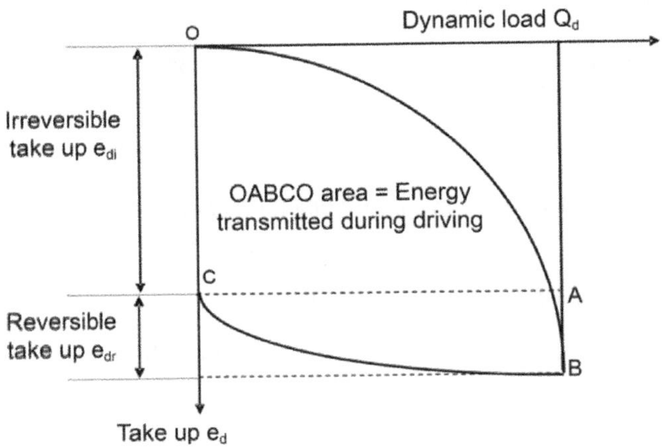

Figure 3.15. *Energy approach: force-take up curve of the dynamic penetrometer test*

Reversible take up (e_{dr}) is low compared to irreversible take up (e_{di}) and can be reasonably ignored. Two assumptions are possible:

1) Bodies of masses m and m' are perfectly elastic: $\varepsilon = 1$

In this case, the dynamic resistance (q_d) at the tip of the straight section A_p is deduced from equation [3.55], which is:

$$q_d = \frac{mgh_d}{A_p \times e_{di}} \qquad [3.56]$$

COMMENT.– This formula is recommended by the European standard.

2) Bodies of masses m and m' are perfectly soft (no rebound): $\varepsilon = 0$

Under this assumption, the dynamic resistance (q_d) is (equation [3.55]):

$$q_d = \frac{mgh_d}{A_p \times e_{di}} \times \frac{m}{m + m'} \qquad [3.57]$$

COMMENT.– This formula, called the "Dutch formula", is the most commonly used.

Vibration theory

Meunier (1974) relates the dynamic point resistance (q_d) to the velocity (wave propagation speed: v_p), to the velocity of the driving hammer at impact (v_m) and to the density of the materials involved. The author establishes the following relation:

$$q_d = \frac{mgh_d}{A_p \times (e_{di} + \frac{\rho'}{\rho_s} + \frac{v_m}{v_p} \times l_m)} \qquad [3.58]$$

where:

– ρ' designates the bulk density of the shafts;

– ρ_s is that of the soil;

– l_m represents the length of the driving hammer.

Considering that the mass of the shafts m' is greater than the mass of the driving hammer, Gonin (1978) expresses the peak dynamic resistance by:

$$q_d = \frac{8}{10} \times \frac{mgh_d}{A_p \times e_{di}} \qquad (m' \geq m) \qquad [3.59]$$

COMMENT.– This formula corresponds to about 20% of the expression of the point resistance obtained in the assumption that bodies of masses m and m' are perfectly elastic ($\varepsilon = 1$, equation [3.56]).

3.4.3.3. *Influential parameters*

Dynamic point resistance can be influenced by several parameters, the most significant of which are:

– the strike rate;

– the drop height;

– the energy loss;

– the shape of the tip;

– the size of the tip.

Strike rate

The recommended strike rates in practice should not exceed 30–60 blows/min. The European standard advises limiting the strike rate to 30 blows/min for all soils. It stipulates that this limit should not be exceeded in the case of clays, but tolerates a value of 60 blows/min in other terrains. The purpose of these recommendations is to avoid excessively fast rates of impact, which are outside the scope of piling and fall into the realm of vibro-impact.

Drop height

The drop height of the driving hammer must be regular during piling operations. This regularity must be guaranteed to within 5%.

Energy loss

Energy losses can totally falsify the results. These losses can result from imperfections in the test equipment, such as:

– the parasitic friction of the driving hammer on the shafts, which tends to reduce the dynamic force at the head;

– the lack of connection between the piling system and the penetrometer itself (shafts and tip).

Size and geometry of the tip

The size and geometry of the tip can have a significant impact on the dynamic resistance at the tip.

3.4.3.4. *Role of the dynamic penetrometer in geotechnical projects*

In France, geotechnical study reports (Missions G1 to G5) often include dynamic penetration samplings for economic considerations which are voluntarily orchestrated by the project owners and passively used by the geotechnicians. The sections of the samplings attached to the report rarely clearly describe the material used (different masses used: anvil and guide of the driving hammer, mass of a shaft, mass of the tip and mass of the tip holder, geometrical characteristics of the tips; area and length, shaft length and drop height).

Like the cone penetration tests, dynamic penetration tests involve continuous measurements, making it possible to assess:

– the resistance of the different layers of the soil, without detecting its exact nature;

– the possible presence of anomalies: presence of blocks (premature rejection or occasionally high resistance) or of compact substratum (high resistance leading to a solid rejection).

The Type-A dynamic penetrometer can be used to help guide the selection and the calculation of foundations when it is accompanied by complementary information resulting from cone penetration tests (Standard NF P94-113), field vane tests (Standard NF P94-112), SPT (NF P94-116) and/or pressuremeter tests (Standard NF P94-110). The Type-B dynamic penetrometer (Standard NF P94-115, AFNOR 1990b) "can be used to help guide the selection of foundations, but does not make it possible to assess the load-bearing capacities", and does not make it possible to calculate the settlements.

Particular attention should be paid to the use of the dynamic penetrometer when the depth of the test exceeds 10 m, especially for the Type-B penetrometer, where parasitic friction between shafts and boring walls accumulates with depth.

The use of the dynamic penetrometer in sands under water calls for the same caution, because there is a liquefaction phenomenon, which makes the dynamic resistance decrease at the tip. In firm clays, there can be a de-structuring of the sheets, as well as a decrease in the dynamic resistance at the tip.

3.4.4. *Standard penetration test (SPT)*

3.4.4.1. *History of the test*

The SPT (French Standard NF P94-116) is a dynamic penetrometer with a tip that is replaced by a "standard" core drill, making it possible to take "undisturbed" samples during the piling, in particular, in sands without cohesion, which are difficult to retrieve.

It was originally invented in the United States by an old chief operator of the "Raymond-Pile" company, who, in 1925, suggested that Terzaghi count the number of blows necessary to drive the core drill he usually used one foot (0.33 m) deeper. Despite several versions around the world, it is the procedure developed in 1925 by Terzaghi that dominates, and on which all practical interpretations of the test are based today.

3.4.4.2. *Principle of the test*

The test consists of:

– cleaning the bottom of the boring;

– driving a "standard" core drill by piling it into the soil to a depth of 0.15 m ("priming phase");

– resuming drilling in two successive steps of 0.15 m each;

– noting the number of blows corresponding to each of the two stages (the real measurement phase).

The piling of the core drill into the soil is done under constant energy, with a free-falling driving hammer with a mass of 0.435 kN and a drop height of 0.76 m. The strike rate set by the European standard and the French Standard NF P94-116 is 30 blows/min.

3.4.4.3. *Test device*

The test device is composed of the following elements:

– boring equipment;

– driving and guiding rig;

– soil surface locating system;

– drill stem and its guidance;

– boring wall holding equipment (recoverable casing or injection mud);

– standard core drill;

– measurement system.

Figure 3.16 describes the SPT setup, which has the following characteristics:

– length: 0.813 m;

– outside diameter: 0.051 m;

– inside diameter: 0.035 m;

– total weight: 0.07 kN.

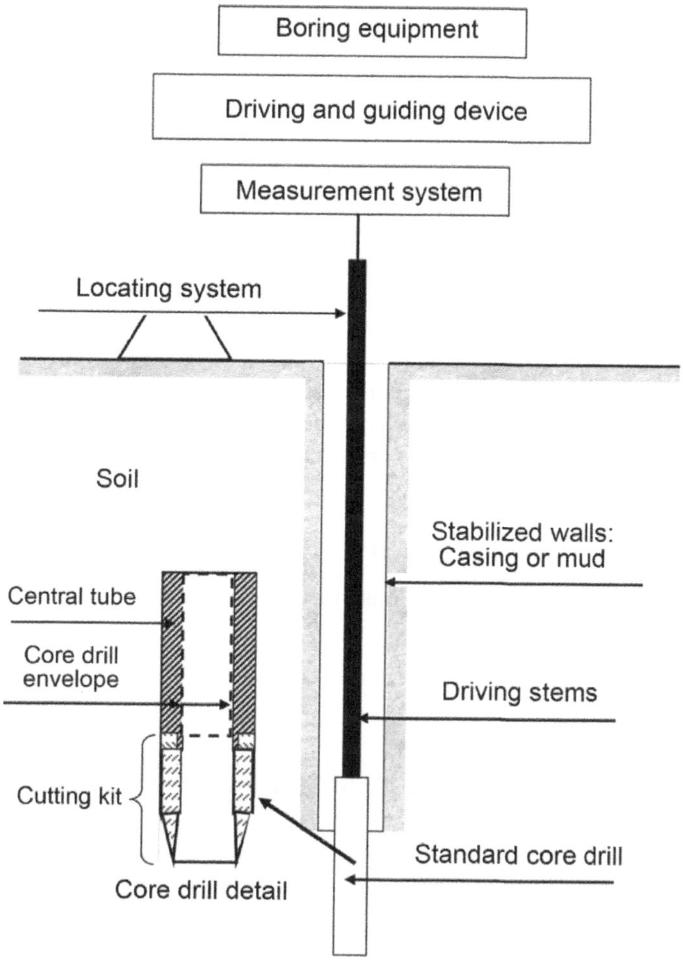

Figure 3.16. *Standard penetration test: SPT*

3.4.4.4. *Measured parameters*

The parameters measured during the piling phases are:

– number of blows N_0 recorded during the 0.15 m piling in order to eliminate the disrupted zone (priming phase);

– number of blows N_1 recorded during a second 0.15 m piling phase (first measurement phase per se);

– number of blows N_2 recorded during a third 0.15 m piling phase (second measurement phase per se).

The number of blows measured (noted N_m) during the core drill test is thus the sum of the two terms, that is to say:

$$N_m = N_1 + N_2 \tag{3.60}$$

COMMENTS.– When the soil is homogeneous, it is clear that $N_1 = N_2$. If not, the values of N_1 and N_2 may be different depending on the degree of heterogeneity of the soil crossed over 0.3 m.

– The British make even more detailed measurements by counting the number of blows for smaller take up heights of 0.075 m.

– In resistant soils, penetration becomes difficult to achieve. For this, the test can be stopped at a conventionally fixed number of blows. Thus, like the European standard, the British standard sets a maximum number of 50 blows/min and the American standard 100 blows/min.

3.4.4.5. *Influential parameters and measurement corrections*

During the core drilling operations, the measured number of blows (N_m) is influenced by two main parameters:

– pore pressure (u);

– vertical stresses (depth).

This requires corrections to be made to N_m in order to take these two parameters into account.

Pore pressure correction

During piling operations, pore pressures are mobilized in the saturated fine sands and underwater silts. This leads to a decrease in the shear strength of the soils.

Terzaghi and Peck (1957) thus recommend correcting the measured number of blows (N_m), when N_m is greater than 15, by the following corrective relation:

$$N_{SPT} = 15 + \frac{N_m - 15}{2} \qquad (N_m > 15) \qquad [3.61]$$

Depth correction (vertical stresses)

Depth impact is related to the increase in vertical stress according to depth (Turnbull and Kaugmann 1956). Teng (1966) considers that the measured number of blows (N_m) is underestimated at shallow depths. The depth correction (z) proposed in this framework by some authors, to obtain the "true" number of blows (N_{SPT}), is as follows:

$$N_{SPT} = N_m \times \frac{350}{70 + \gamma \times z} \qquad (\gamma \times z \leq 280\,\text{kPa}) \qquad [3.62]$$

where γ is the bulk density of the soil (in kN/m^3).

COMMENTS.– This formula is only valid when the vertical stress (γz) is less than or equal to 280 kPa.

– Generally in practice, as a precaution, half the value of N_{SPT} (equation [3.62]) is used when it is greater than $2N_m$.

3.4.4.6. *Practical interpretation of the standard penetration test*

From several tests on all soils, and particularly on sands, for which the use of the standard penetrometer is better adapted than on clayey soils, several correlations have been implemented by the practitioners.

The case of sands

Terzaghi and Peck (1957) give a correlation between the number of blows (N_{SPT}) and the relative density (D_r) of the sands, defined in Table 1.4 of Chapter 1. The correlation between these two parameters is reflected in the graph in Figure 3.17, which, as a function of increasing values of N_{SPT} and D_r, depicts the compactness state of the sands, ranging from loose to very dense.

This graph shows that the response to core drill penetration as a function of relative density (D_r) is greater for very loose to loose sands than for compact to dense sands, and especially for very dense sands. These results, of English-speaking origin, are in agreement with the results of French laboratory tests regarding the definition of the boundaries of various sands as a function of relative density.

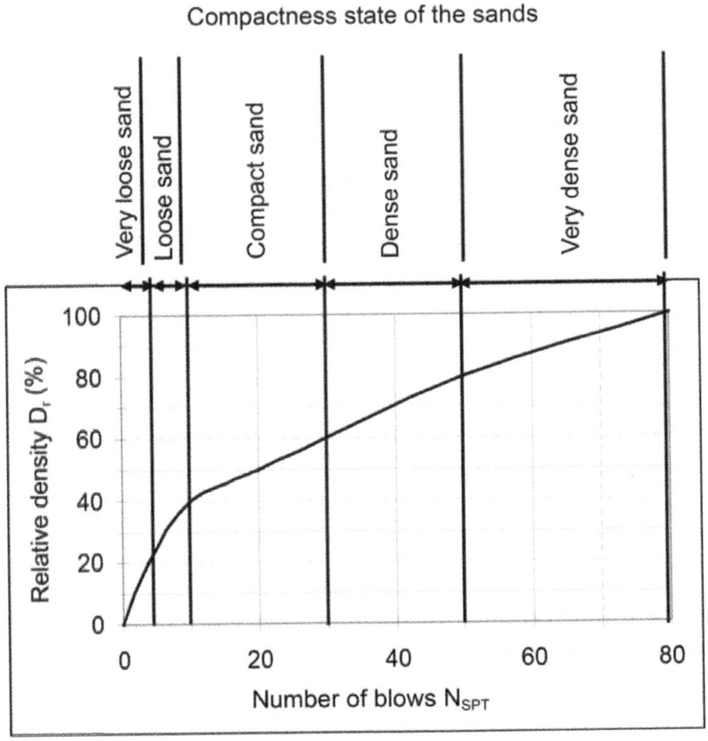

Figure 3.17. *Correlation between the number of blows N_{SPT} and relative density (D_r) of sands (Terzaghi and Peck 1957). For a color version of this figure, see www.iste.co.uk/dhouib/geotechnics1.zip*

Other authors, such as Schultze and Menzenbach (1961), relate the number of blows (N_{SPT}) to the relative density (D_r) and effective stress (σ'_0) of the soil at test depth, and give the following logarithmic relation:

$$Log_e\, D_r = 0 \times 478 \times Log_e(N_{SPT}) - 0.262 Log_e(\sigma'_0) + 2.84 \qquad [3.63]$$

p'_0 in 10^5 Pa

Gibbs and Holtz (1957), in turn, establish a correlation between the number of blows (N_{SPT}) for 0.1 m of core drill take up and the effective earth weight stress (σ'_0) for different values of the relative density (D_r):

– In fine sands and silts with less than 5% fines (passing 80 μm), Meyerhof (1956) establishes an approximate expression for the angle of internal friction as a function of relative density (D_r), which is:

$$\varphi' = 30 + 0.15\,D_r \quad (\varphi' \text{ in degrees and } D_r \text{ in } \%) \qquad [3.64]$$

– In sands and silts with more than 5% fines, the author gives a similar relation:

$$\varphi' = 25 + 0.15\,D_r \quad (\varphi' \text{ in degrees and } D_r \text{ in } \%) \qquad [3.65]$$

These relations are plotted in Figure 3.18 so as to assess the evolution of the angle of internal friction (φ') of the sands as a function of the number of blows (N_{SPT}). However, it should be noted that these relations are of empirical origin. They are to be taken with caution in current practice because they are all based on laboratory tests on sand samples taken with the "SPT" core drill, and have thus inevitably been remolded.

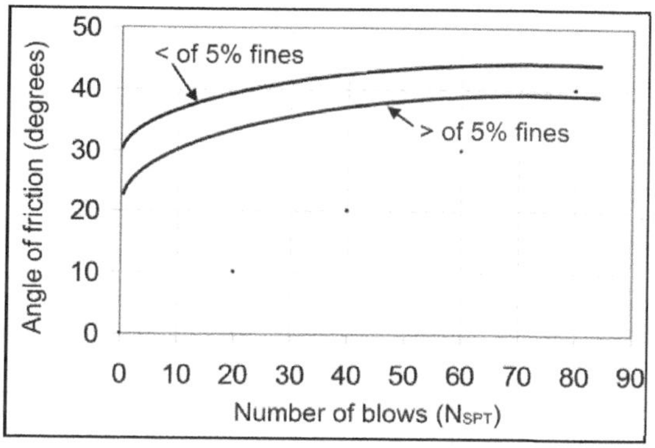

Figure 3.18. *Correlations between the number of blows (N_{SPT}) and the angle of internal friction (φ') of the sands (Meyerhof 1956). For a color version of this figure, see www.iste.co.uk/dhouib/geotechnics1.zip*

Other practical formulas also exist to relate the angle of internal friction to the number of blows for the core drill take up. There is Dunham's formula, which is expressed by:

$$\varphi' = 25 + \sqrt{12 N_r} \quad (\varphi' \text{ in degrees}) \qquad [3.66]$$

and Osaki's:

$$\varphi' = 15 + \sqrt{20 N_r} \quad (\varphi' \text{ in degrees}) \qquad [3.67]$$

The case of clays

In spite of the existence of results and correlations between parameters in the English-speaking literature, the use of the SPT in clays remains very limited and is strongly discouraged. Several reasons play against this test. In clays, pore over-pressures can develop during piling. Parasitic forces due to friction between the sticky clay and the core drill can occur. All these parameters can make the results of the measurements erroneous and unreliable.

In clay soils, Terzaghi and Peck (1951) give correlations between the compressive strengths (R_c) of clay and the number of blows (N_{SPT}). Here are some, quoted by Sanglerat (1965):

– Plastic clays: $$R_c = \frac{N}{40} \ (R_c \text{ in MPa})$$

– Silty clays: $$R_c = \frac{N}{50} \ (R_c \text{ in MPa})$$

– Sandy clays: $$R_c = \frac{N}{75} \ (R_c \text{ in MPa})$$

3.4.4.7. *Role of standard penetration test in geotechnical projects*

Let us recall that, by its very origin, the SPT essentially belongs to English-speaking culture. Its use in France remains relatively limited.

On a practical level in the field of soils and foundations, the test offers some advantages when used in sandy soils. It does indeed make it possible to take samples and identify their nature, to measure the mechanical characteristics in situ, and to carry out identification and shear tests in the laboratory on those same samples. In terms of project sizing, the capabilities developed in France for the calculation of the load-bearing capacity of foundations do not make much use of the SPT.

On the other hand, in seismic zones, the test makes it possible to quantify the risks of liquefaction based on the results of Seed (1976), whose determination capabilities for seismic shear stress are built from this test.

In clay soils, it is strongly discouraged, for obvious reasons. In clays, uncontrolled pore pressures develop during piling, and parasitic frictional forces and sticking phenomena can occur, which can alter the results and make them less reliable.

3.4.5. *Ménard pressuremeter test (MPT)*

3.4.5.1. *History of the pressuremeter and the pressuremeter test*

The MPT (Ménard pressuremeter test) is a French invention made by Louis Ménard (1931–1978) in the 1960s.

Later, the "*Pressiotest*" of the APAGEO company in France was developed, which was another configuration of the Köegler apparatus (1930), with rectangular plates instead of shells that were driven into the soil, like the vane test, and which consisted of a horizontal plate-loading test (section 3.4.7, presented later). Subsequently, the self-drilling pressuremeter of the *Ponts et Chaussées laboratories* was developed in Saint-Brieuc, and was equipped with a special probe called the "self-drilling probe", inside which was placed a drill stem equipped with a helical auger at its end. The latter ensures perforation with an injection mud and a measuring cell without a guard cell, contrary to the Ménard pressuremeter, and is always in contact with the soil. This avoids soft soils from being remolded.

In other countries, the pressuremeter has not had any specific development except the Japanese "Oyo" pressuremeter, which is composed of a single-cell probe.

3.4.5.2. *Analogy of "foundation-pressuremeter probe" behavior*

Louis Ménard's idea is based on the expansion of a cylindrical cavity in a homogeneous and isotropic medium. In terms of a behavioral analogy, consider a foundation anchored in the soil and subjected to a centered vertical load Q. Under this load, a uniformly distributed stress on the soil, noted as p, develops at the base of the foundation. The stability of the foundation is related to the stress that the soil underneath the foundation can provide: this is the intrinsic resistance (noted as p_R) of the soil supporting the foundation. Figure 3.19(a) shows the foundation and the stresses involved.

Let us take the example of a vertical cylindrical boring, in which a radial and uniformly distributed pressure p is applied. The boring remains stable when the pressure p remains lower than the lateral pressure that the soil around the boring can offer; this is the intrinsic resistance of the soil, also noted as p_r. Figure 3.19(b) illustrates a cylindrical boring, subjected to a radial pressure p. It is the principle of the expansion of a cylindrical cavity that inspired Louis Ménard to develop his pressuremeter, which is the most widely used device in the world and whose use in France has increased greatly since the 1960s.

The behavioral analogy between a foundation acting on the soil and the pressure that prevails in a cavity, which can be exerted by a probe, allowed Louis Ménard to develop all the theory and practice that followed for the sizing of foundations. In the following, we give the principle of the Louis Ménard pressuremeter test, the theoretical foundations and the applications of this test.

In current practice, there is the pressuremeter test without cycles (French Standard NF P94-110-1, AFNOR), a test that can be defined as a "monotonic test", and the test with cycles (French Standard NF P94-110-2, AFNOR), which takes the effects of loading–unloading into account, called the "alternating test (or cyclic test)".

Let us look at the two tests.

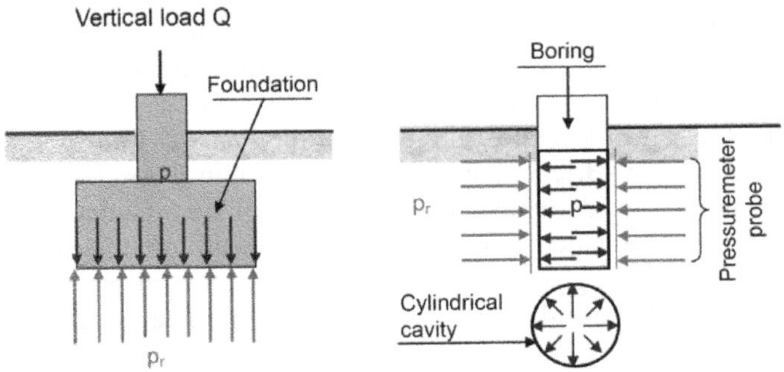

Stability of the foundation: $p < p_r$

a. Foundation under vertical load Q

Stability of the cavity: $p < p_r$

b. Cylindrical cavity under pressure p

Figure 3.19. *Analogy of foundation-pressuremeter probe behavior. For a color version of this figure, see www.iste.co.uk/dhouib/geotechnics1.zip*

3.4.5.3. *Principle of the Ménard pressuremeter test without cycles*

Principle of the test and stages of its implementation

The pressuremeter test without cycles (French Standard NF P94-110-1 of 2000) is carried out according to the following stages:

1) Carry out a boring at variable depths, depending on the strength of the boring walls using a drill stem with a drilling tool (auger, disaggregating tool, core drills, etc.).

2) Bring the drill stem and the drilling tool to the surface.

3) With the help of a drill stem, lower a probe to the depth of the test.

4) Inject pressure into the probe, applied radially in successive increasing steps.

5) Through the decreasing water level in the tubes, record the variations of volume due to the soil deformation around the probe.

6) Conduct the test until the soil around the probe fails.

In terms of behavior, the test goes through the following phases:

– initial phase of boring, where there are zero stresses on the drilling walls;

– (initial) stress recovery phase, also called the "re-compaction phase";

– "pseudo-elastic" deformation phase, where the pressuremeter curve presents a quasi-linear curve;

– creep phase, where the soil starts to undergo a plastic flow;

– failure phase.

In terms of stresses, the different stages result in the pressure state described in Figure 3.20, which is as follows:

– Prior to boring, the soil is in an initial stress state (p_0) that prevails in the soil (σ_{hs}) at the depth of the test, which is (Stage 1 of Figure 3.20(a)):

$$p = p_0 = \sigma_{Hs} \qquad [3.68]$$

Once the boring is drilled, the initial stress (p_0) is nullified on the boring walls (Stage 1 in Figure 3.20(a)) :

$$p = p_0 = 0 \qquad [3.69]$$

– The contact between the probe and the soil passes through a state of stress where the pressure injected ($p = p_1$) into the probe is equivalent to the initial stress (p_0) inside it (equation [3.68]). The progressive increase, in successive stages, of the pressure in the probe leads to deformations in the soil around the pressuremeter probe, where (Stage 2 of Figure 3.20(a)):

$$p \geq p_1 = p_0 \qquad [3.70]$$

In this stage, the pressuremeter curve is linear and the soil is in a pseudo-elastic deformation state (Figure 3.20(b)). Note that the soil is said to be in a state of "pseudo-elastic" deformation because, at this deformation level, the soil cannot be, as we specified in section 2.2 of Chapter 2, linear elastic.

– The quasi-linear part (pseudo-elastic deformation phase) of the pressuremeter curve ends when the pressure p reaches a pressure value (p_2), which verifies the following inequality (Stage 3 in Figure 3.20):

$$p = p_2 \leq p_f \qquad [3.71]$$

p_f is the so-called "creep pressure" of the soil, which corresponds to the bend in the creep curve (dashed curve in Figure 3.20(b)). This curve is constructed from the pair of points (p, ΔV), where ΔV is the change in liquid volume in the central measurement cell between times $t = 30$ s and $t = 60$ s, after the start of a 60 s pressure stage p.

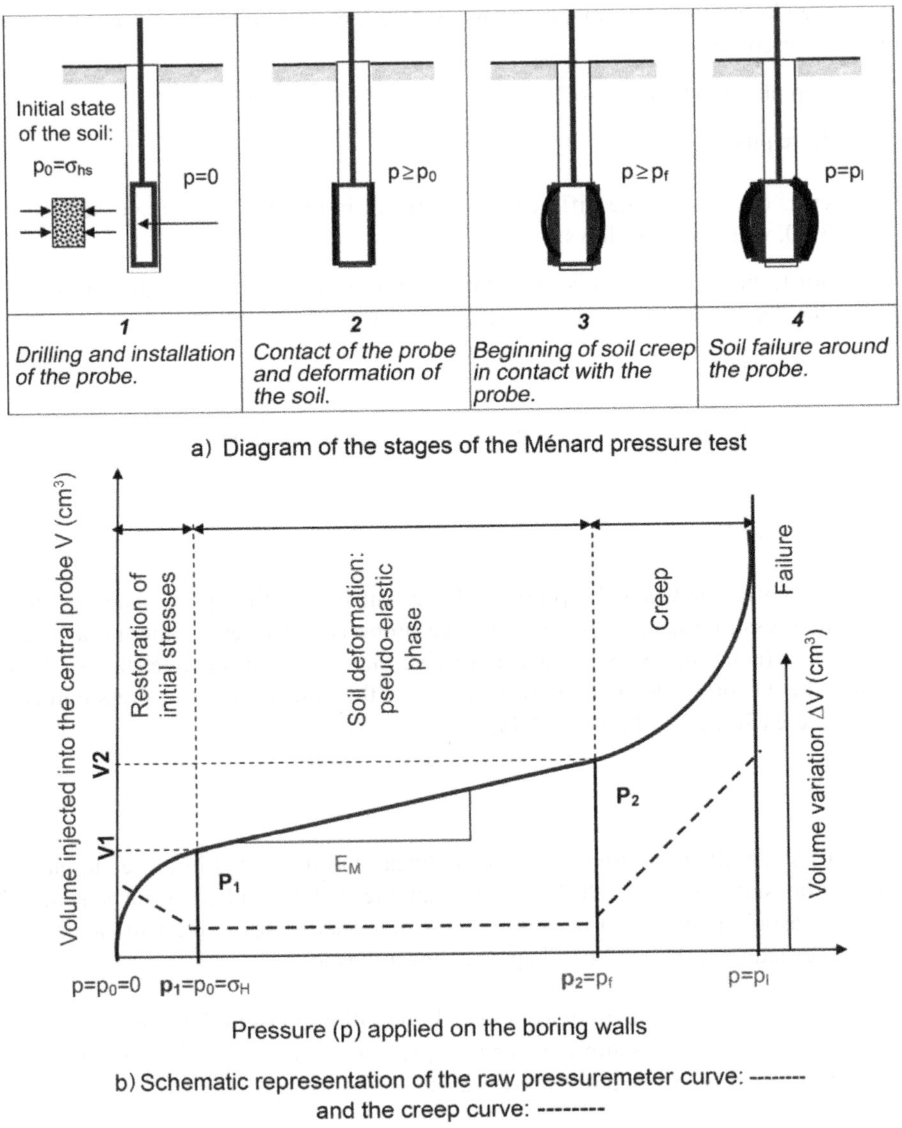

Figure 3.20. *"Monotonic" Ménard pressuremeter test (MPT). For a color version of this figure, see www.iste.co.uk/dhouib/geotechnics1.zip*

– Beyond the creep pressure, the soil undergoes plastic flow around the probe and is eventually broken by a failure pressure called "limit pressure", noted as p_l and where (Stage 4 of Figure 3.20(a)):

$$p = p_l \qquad [3.72]$$

COMMENT.– The limit pressure corresponds to the asymptotic direction of the pressuremeter curve (red solid line) in Figure 3.20(b).

Raw parameters measured

As described in the previous paragraph, the pressuremeter test without cycles measures the following parameters:

– Louis Ménard pressuremeter modulus: E_M

This is the slope of the quasi-linear part of the pressuremeter curve in Figure 3.20(b). It represents the modulus of pressuremeter deformations, written as E_M (Louis Ménard pressuremeter modulus).

– Pressures. These are the two pressures (Figure 3.20(b)):

- creep pressure of the soil around the pressuremeter probe, noted as p_f;

- limit pressure corresponding to the failure of the soil in contact with the probe, noted as p_l.

Measurement time

The pressuremeter test is carried out from several increasing pressure levels that are kept constant. Each pressure level and the volume of liquid injected into the probe are noted at least at the following times (NF P94-110 standard):

Time t = 15 s; t = 30 s and t = 60 s.

The original time of the test is taken at each pressure level, at the moment when the expected pressure is stabilized on the pressure indicator.

Horizontal stress in the soil at the test level

The initial stress (p_0 = σ_{Hs}) in the soil at the test level can be compared to a horizontal stress of soil at rest, and therefore involves the coefficient "K_0" of the soil at rest. It is related to the total vertical stress (σ_{vs}) and the pore pressure (u_e) at the test level (depth z_e), that is to say(Figure 3.21(a)):

$$\sigma_{Hs} = K_{0.}(\sigma_{vs} - u_e) + u_e \qquad [3.73]$$

The term $(\sigma_{vs} - u_e)$ refers to the effective vertical stress in the soil at test depth. For a water table located at a depth z_w in relation to the Natural Terrain (N.T.) and a test projected to a depth z_e, beyond the water table (Figure 3.21(a)), the pore pressure (u_e) at that depth is given by:

$$u_e = \gamma_{w.}(z_w - z_e) \qquad [3.74]$$

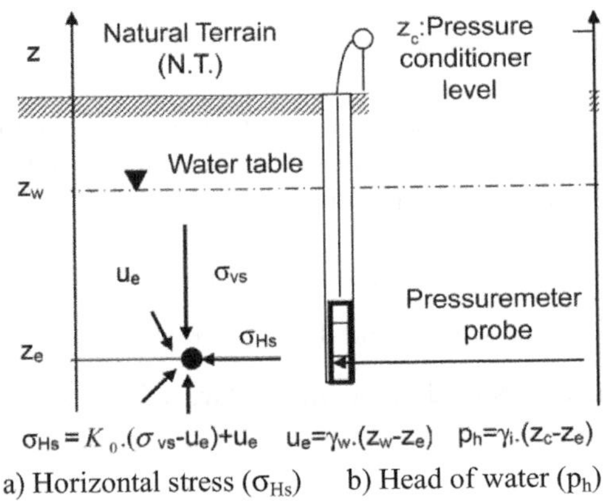

Figure 3.21. *Horizontal stress in soil before test and head of water*

Correction parameters

During the test, the pressure and volume readings used to determine the various parameters are raw. The values of the parameters must therefore be corrected to account for:

– the head of water (noted as p_h) in the soil at the time of the test;

– the resistance of the probe (also called "inertia" of the probe), noted as p_i;

– the expansions of the probe and the tubing (expansion coefficient a, obtained on a metal tube called the "calibration" tube).

Head of water (p_h)

For a monotonic pressuremeter test carried out at elevation z_e, the pressure in terms of the measurement cell (p) is the pressure read on the pressure conditioner, increased by the head of water (p_h), mobilized between the level of the pressure gauge (elevation z_c) and the middle of the probe (test elevation z_e), in other words (Figure 3.21(b)):

$$p_h = \gamma_l \times (z_c - z_e) \quad [3.75]$$

with γ_l being the density of the liquid injected into the central cell of the probe.

Natural resistance of the probe

The test to determine the inherent resistance – also called inertia – of the pressuremeter probe, noted as p_r, is done in the same manner as the real pressuremeter test. The probe is placed in the open air, near and at the same level as the pressure conditioner (or not). The pressure levels (p_r) are worth one-tenth of the maximum applied pressure (p_{rl}) and are maintained for 60 s, until the volume of the injected liquid has reached at least 1.2 V_s (V_s being the conventional initial volume of the central measurement cell).

When the pressure conditioner is placed at the same level as the pressuremeter probe, the head of water in the pipes is equal to zero. If this is not the case, the pressure injected into the probe must be necessarily corrected for this load. Thus, the inertia curve (or "pressure–volume" calibration curve of the central measurement cell) is defined by (Figure 3.22(a)):

$$V = f(p_e) \quad [3.76]$$

where p_e is the natural resistance of the probe and given, after corrections (Figure 3.22(b)), by:

$$p_e = p_r + p_h \quad [3.77]$$

The head of water (p_h) is equal to zero when the probe is placed at the same level as the conditioner ($z_s = z_c$).

The volume (V_r) of the liquid injected into the cell is to be corrected according to the expansion coefficient of the equipment (noted as a). After corrections, the volume (V) of the cell is:

$$V = V_r - a \times p_r \quad [3.78]$$

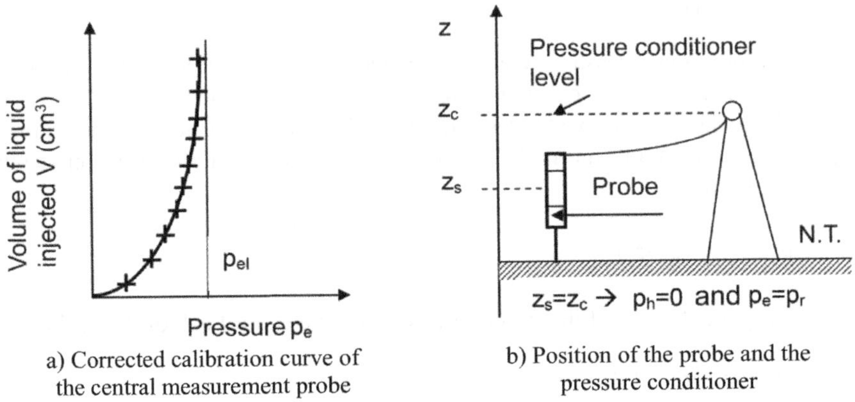

Figure 3.22. *Calibration test of the central probe (probe inertia)*

The expansion coefficient (a) is determined as follows.

Expansion coefficient of the equipment (a)

The expansion coefficient of the probe is determined experimentally, in order to correct the volume in the measurement cell (equation [3.78]) and also to correct the pressuremeter modules (E_M) in very resistant environments (rocks or similar materials).

The probe is introduced into a metal calibration tube and pressurized in stages that are maintained for 60 s. The increments are 0.5 MPa for the probes, reaching 5.0 MPa.

The volume injected at the end of the stage is noted and the pressure–volume curve obtained in the calibration tube is plotted as shown in Figure 3.23. The expansion coefficient (a) is the slope of the experimental line obtained, in other words (Figure 3.23):

$$a = \frac{\Delta V}{\Delta p} \qquad [3.79]$$

This value is expressed in cm^3/MPa. It must not exceed 6 cm^3/MPa for a minimum of 50 m of tubing. The volume at the origin of the pressures (noted as Vc) corresponds to the volume of the liquid injected in the probe, so as to establish its contact with the boring walls.

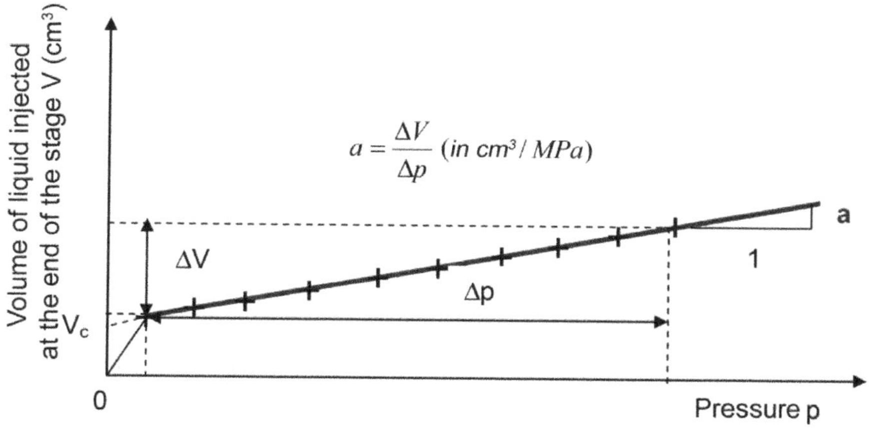

Figure 3.23. *Pressure–volume expansion curve of the pressuremeter probe in a metal calibration tube (expansion coefficient a). For a color version of this figure, see www.iste.co.uk/dhouib/geotechnics1.zip*

Corrections of measured parameters

The raw "pressure–volume" curve is drawn from the measurements before corrections (pressure p_m and volume V_m measured). The corrected pressure–volume curve (corrected pressures p_{cor} and V_{cor}) is obtained by making the following corrections:

– On the pressures:

$$p_{cor} = p_m + p_h - p_e \tag{3.80}$$

where the pressures p_h and p_e are given by equation [3.75] and equation [3.77], respectively.

– On the volumes (equation [3.78]):

$$V_{cor} = V_m - a \times V_r = V_m - \frac{\Delta V}{\Delta p} \times V_r \tag{3.81}$$

Calculation of the main parameters of the pressuremeter test

After the various corrections mentioned above, the main parameters are calculated as presented in the following sections.

Ménard pressuremeter module (E_M)

By definition, the Ménard pressuremeter deformation modulus, noted as E_M, is expressed by:

$$E_M = 2(1+\nu) \times (V_s + \frac{V_1+V_2}{2}) \times \frac{p_2 - p_1}{V_2 - V_1} \quad [3.82]$$

where:

– ν is Poisson's ratio, conventionally equal to 0.33 (long-term value);

– V_s is the volume of the central measurement cell of the pressuremeter probe;

– the pairs (p_1, V_1) and (p_2, V_2) are the pressures and volumes corresponding to the origin and the end of the pseudo-elastic phase (Figure 3.20(b)), respectively.

COMMENT.– Much like European Standard NF EN ISO 22476-2, the French Standard NF P94-110 enables the use of other methods of determining the pressuremeter modulus (E_M), provided that any method used does not give results that differ by more than 5% from the requirements of the standard.

Creep pressure (p_f)

Creep pressure (p_f) is graphically determined from the creep curve. It corresponds to the bend in the creep line (Figure 3.20(b), dotted curve), which separates the "pseudo-elastic" phase from the creep phase. The net creep pressure, p_f*, is the difference between the corrected creep pressure (p_f) and the total horizontal pressure (σ_{Hs}) in the soil at the test level (equation [3.73]):

$$p_f{}^* = p_f - \sigma_{Hs} \quad [3.83]$$

Limit pressure (p_l)

Limit pressure (p_l) is the failure pressure of the soil around the pressuremeter probe. It corresponds to the asymptotic direction of the pressure-volume curve (Figure 3.20(b), continuous curve). In practice, the standard defines it, by convention, as the pressure that causes the doubling of the volume of liquid injected into the central measurement cell. This corresponds to an injected volume (V) of:

$$V = V_s + 2V_1 \quad [3.84]$$

where:

– V_s is the conventional initial volume of the central measurement cell;

– V_1 is the volume corresponding to the origin of the pseudo-elastic phase.

COMMENTS.–

1) Limit pressure (p_l) is defined, by convention, in the old Standard (NF P94-110 of July 1991), as the lower of the two following expressions:

$$p_l = Min\left\{ [-\frac{B}{A} + \frac{1}{A \times (V_s + 2V_1)}], [1.7 p_f - 0.7 \sigma_{Hs}] \right\} \quad [3.85]$$

where A and B are coefficients obtained by the "least squares" method on the experimental values of the curve:

$$\frac{1}{V} = Ap + B \quad [3.86]$$

2) The new Standard NF P94-110, from 2000, has removed this valid requirement. Indeed, according to our experience, it is the second term of identity [3.85] which takes precedence, and the coefficient of 1.7 applied to the creep pressure p_f seems, to us, to constitute a minimal and safe value. Since the publication and application of the new standard, we have seen reports of soils with limit pressures that are strangely identical to the creep pressures. This penalizes projects without reason. For example, for piles, the safety coefficient on the tip goes from 3 to 5.2, not to mention the tests stopped at 2.5 MPa, or even 1.5 MPa in resistant and very resistant environments, contrary to the rules of the standard which clearly requires 5.0 MPa as the final measurement pressure.

3) In a letter dated October 27, 2005, we asked the standardization committee to integrate the rules of Standard NF P94-110 of July 1991 into the European standard being finalized at the time. This request follows a simple logic: to optimize our projects in the face of geotechnical files that are becoming more and more pessimistic, as well as very distant, from projects that demand to be well-known before any geotechnical studies have even been made, and especially to provide the reports of their feasibility.

Like creep pressure, the net limit pressure is defined by the following expression:

$$p^*_l = p_l - \sigma_{Hs} \quad [3.87]$$

3.4.5.4. *Particularities of the test "with cycles"*

The pressuremeter test "with cycles" (also called the "alternating test") is a monotonic test with two unloading stages (Δp). Each stage is conventionally maintained for 15 s. According to Standard NF P94-110-2 (Figure 3.24), its value is:

$$\Delta p = \frac{0.9\,p_f - \sigma_{Hs}}{4} \quad [3.88]$$

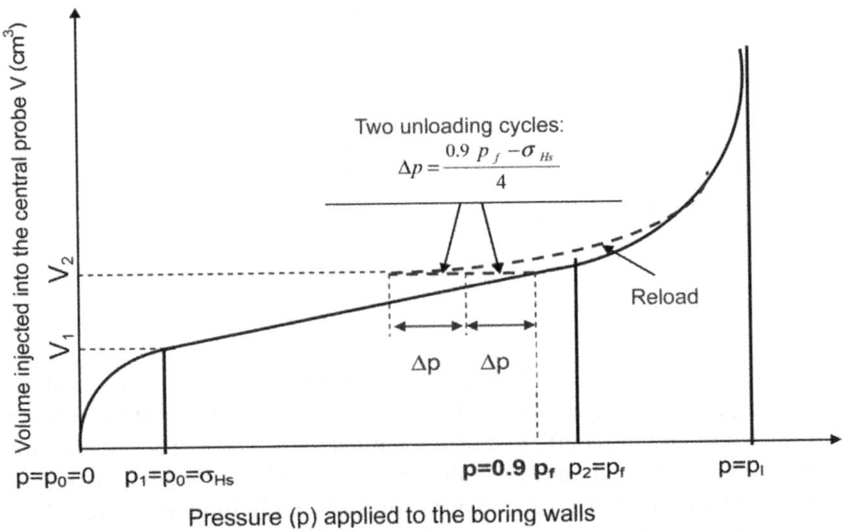

Figure 3.24. *Schematic representation of the two-cycle alternating test. For a color version of this figure, see www.iste.co.uk/dhouib/geotechnics1.zip*

COMMENT ON FIGURE 3.24.– *The selection made by Standard NF P94-110-2 of two unloading cycles and the amplitude retained (equation [3.88]) is not based on any theoretical rule. The test can be useful for repetitive stresses such as vibratory actions, wind and wave forces. These result in small deformations and their duration is long.*

3.4.5.5. *Theoretical interpretation of the test: theory of expansion*

The Ménard pressuremeter test is based on the theory of cylindrical expansion. The pressure (p) is applied uniformly on the boring walls and transmitted to the soil, which is assumed to be a hollow cylinder of radius r_0 (half-diameter of the boring). These are axisymmetric conditions, as presented in section 2.2.1.5 of Chapter 2 (Figure 2.12).

Figure 3.25 shows the boring and the cylindrical expansion mechanism (Figure 3.25(a)), as well as the equilibrium stresses of the cylindrical crown element around the pressuremeter probe (Figure 3.25(b)). In a ($r - \theta$) plane and by projection on the bisector of the angle $d\theta$ of the stresses involved (volume forces neglected), application of the equilibrium equations in generalized cylindrical coordinates [2.47b] leads to (Figure 3.25(b)):

$$\frac{\sigma_\theta - \sigma_r}{r} - \frac{d\sigma_r}{dr} = 0 \qquad [3.89]$$

The boundary conditions of the pressuremeter test are:

– On the boring walls ($r = r_0$), where the radial stress (σ_r) is equal to the applied pressure (p), in other words:

$$\sigma_r = p \qquad [3.90]$$

– At a far distance from the boring walls ($r >> r_0$), where the radial stress (σ_r) is equal to the tangential stress (σ_θ), which, in turn, is equal to the pressure (p_0) prevailing in the soil at test depth, in other words:

$$\sigma_r = \sigma_\theta = p_0 \qquad [3.91]$$

Let us assume that the soil complies with the Tresca criterion (Figure 2.21, equation [2.70] in Chapter 2) in the case where $\sigma_r > \sigma_\theta$, allowing us to write:

$$\sigma_r - \sigma_\theta = 2k \qquad [3.92]$$

The equilibrium equation in [3.89]) then becomes:

$$\frac{d\sigma_r}{2k} + \frac{dr}{r} = 0 \qquad [3.93]$$

whose direct integration, with the boundary conditions defined by equation [3.90] for $r = r_0$, $\sigma = \sigma_0 = p$, leads to:

$$\frac{\sigma_r - p}{2k} = -\log\left(\frac{r}{r_0}\right) \qquad [3.94]$$

Hypothetically, we can consider that the beginning of the plasticization of the cavity around the pressuremeter probe (in other words, $\sigma_r = p_f$) appears for a radius

r_p ($r = r_p$) of the plasticized zone. In this hypothesis, equation [3.94] makes it possible to express the plasticization radius (r_p) by:

$$r_p = r_0 \times e^{\frac{p-p_f}{2k}} \qquad [3.95]$$

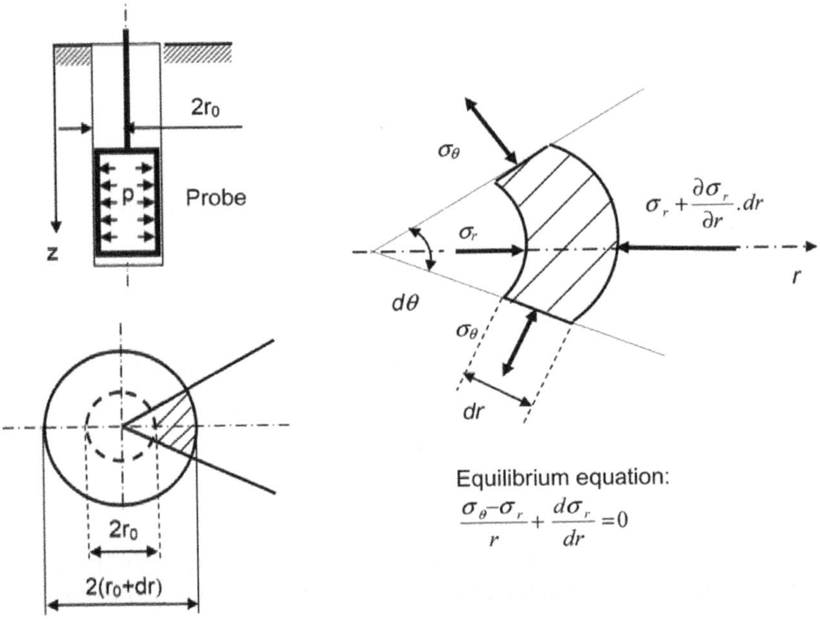

a) Boring and cylindrical expansion in the boring

b) Equilibrium stresses of the cylindrical crown element around the probe

Figure 3.25. *Theory of cylindrical expansion applied to the pressuremeter*

Beyond this, the soil remains elastic around the cylindrical cavity. In this domain, the stress field is defined by the so-called Lamé stress state, that is to say:

$$\begin{cases} \sigma_r = (p - p_0) \times \dfrac{r_0^2}{r^2} + p_0 \\ \sigma_\theta = -(p - p_0) \times \dfrac{r_0^2}{r^2} + p_0 \\ \sigma_z = p_0 \end{cases} \qquad [3.96]$$

The stress field defined by the equations in [3.96] is made up of a spherical part, defined by the mean stress (p_0) given (equation [2.28] of Chapter 2) by:

$$p_0 = \frac{\sigma_r + \sigma_\theta + \sigma_z}{3} \quad [3.97]$$

and of a deviatoric part, which constitutes the other terms of the equalities in [3.96].

Figure 3.26 represents the state of stresses in and around the cylindrical cavity (Figure 3.26(a)), the variations of stresses σ_r and σ_θ as a function of r (Figure 3.26(b)), as well as the boundaries of the elastic and plastic domains. The resulting (plane) deformation field is written as:

$$\begin{cases} \varepsilon_r = \dfrac{du_r}{r} \\ \varepsilon_\theta = \dfrac{u_r}{r} \\ \varepsilon_z = 0 \\ \gamma_{r\theta} = -\dfrac{du_\theta}{r} + \dfrac{u_r}{r} \end{cases} \quad [3.98]$$

Considering a plane deformation state (Figure 2.11(b) and equation [2.35] in Chapter 2), the tangential deformation (ε_θ) can be related to the tangential (σ_θ) and radial (σ_r) stresses by:

$$\varepsilon_\theta = \frac{u_r}{r} = \frac{1-\nu}{E_M}\left[(1-\nu)\times\sigma_\theta - \nu \times \sigma_r\right] \quad [3.99]$$

This leads to the radial displacement u_r:

$$u_r = \frac{1+\nu}{E_M} \times (p - p_0) \times \left(\frac{r_0}{\sqrt{r}}\right)^2 \quad [3.100]$$

The displacement of the boring wall is u_{r0} ($r = r_0$), with:

$$u_{r0} = \frac{1+\nu}{E_M} \times (p - p_0) \times r_0 \quad [3.101]$$

When the stress (p) applied to the boring walls is greater than the total stress (p_0) in the soil in situ before testing, the increase (du_r) in radial displacement is:

$$du_r = \frac{1+\nu}{E_M} \times (r_0 + u_r)\, dp \qquad [3.102]$$

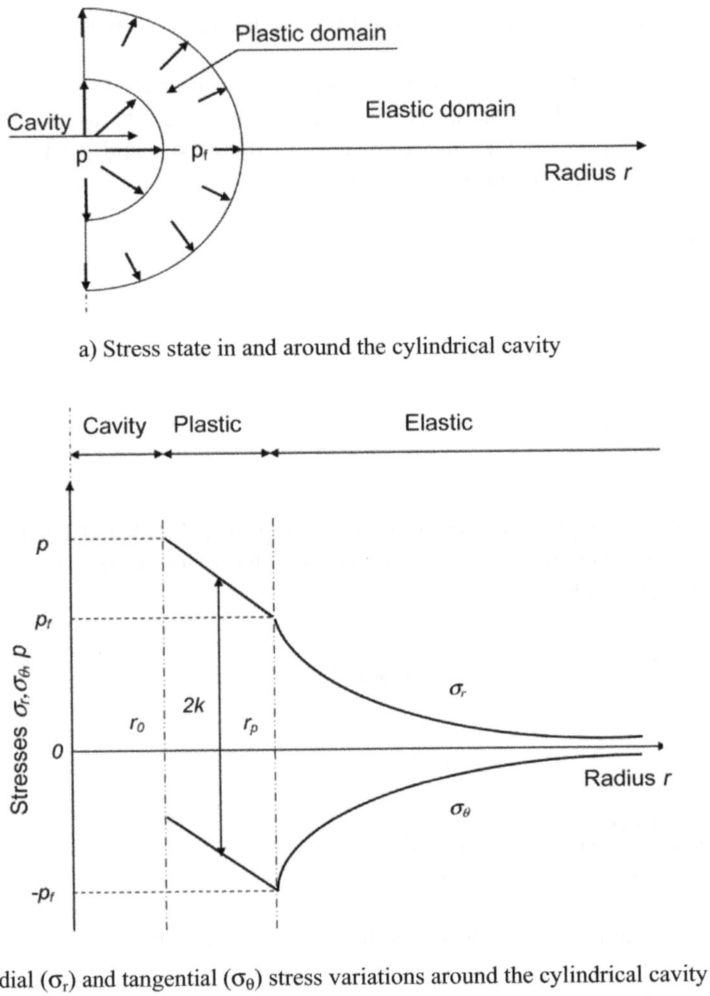

Figure 3.26. *Soil behavior around the pressuremeter probe (Theory of cylindrical expansion in an elasto-plastic medium)*

3.4.5.6. *Relations between pressuremeter parameters*

The main parameters of the pressuremeter test are:

– the Ménard pressuremeter deformation modulus (E_M);

– the creep pressure (p_f);

– the pressuremeter limit pressure (p_l).

In common practice, the following relations are used for the design, sizing and justification of geotechnical projects:

Relation between net limit pressure p_l^* and net creep pressure p_f^*

The p_l to p_f ratio is generally 2. The old Standard NF P94-110 of July 1991 uses a safer ratio of 1.7 (Cassan 1980). According to the literature and our experience, the p_l^* to p_f^* ratio can vary from 1.45 to 2.5 (see Appendix A3.1) for carefully implemented (undisturbed boring walls) and normally monitored tests (choice of extreme points P_1 and P_2 from Figure 3.20, with regard to the creep curve). It is quite acceptable to retain a net pressure ratio:

$$\frac{p_l^*}{p_f^*} = 2 \qquad [3.103]$$

Relation between E_M modulus and net pressuremeter limit pressure p_l^*

The ratio of the Ménard pressuremeter modulus (E_M) to the net pressuremeter limit pressure (p_l^*) depends on the consolidation state of the soil. Ménard (1959a) gives the E_M/p_l^* ratios for different loose soils, summarized in the corresponding Table 3.5, which introduces an additional α coefficient, called a rheological coefficient (or "α Ménard"), according to this ratio.

Indeed, the rheological coefficient characterizes the remolding degree of the boring walls prior to the test, which explains why it is lower for sands and sands and gravels than for clays, the latter being less sensitive to remolding than the former. This means that the E_M/p_l^* ratios are generally lower in sands and sands and gravels than in clays, due to remolding.

For resistant environments (rocks and similar), and depending on the state of the rock, Ménard gives the following α values in Table 3.6.

Soils	Peat		Clay		Silt		Sands		Sands and gravels	
State	$\frac{E_M}{p_l^*}$	α	$\frac{E_M}{p_l^*}$	α	$\frac{E_M}{p_l^*}$	α	$\frac{E_M}{p_l^*}$	α	$\frac{E_M}{p_l^*}$	α
Over-consolidated	-	-	>16	1	>14	2/3	>12	1/2	>10	1/3
Normally consolidated	-	1	9–16	2/3	8–14	1/2	7–12	1/3	6–10	1/4
Under-consolidated	-	-	7–9	1/2	-	1/2	-	1/3		1/4

Table 3.5. *Ratios of Ménard pressuremeter modulus (E_M) to net limit pressure (p_l*) and rheological coefficient ("α Ménard")*

State of the rock	α values
Very fractured rock	1/3
Normal rock	1/2
Very slightly fractured or very weathered rock	2/3

Table 3.6. *α values for rocks*

Relation between the E_M modulus and the conventional soil deformation modulus E (E_s)

The typical relation between the two modules E_M and E (also noted as E_s) is as follows:

$$E = \frac{E_M}{\alpha} \quad [3.104]$$

The selection of α is very delicate because it has an impact on the calculation of the settlements of the structures, as we will demonstrate in the applications of Volume 2, Chapter 2 and Volume 3, Chapter 1 (E/E_M in Standard NF P94-261 in Table 1.7(b)). Attempts to give it a "theoretical" formulation in terms of E_M and p_l* (Baud and Gambin 2014; Baud 2016) do not always yield stable values that conform to the rules of the previous Table 3.5 (α values are sometimes greater than 1.0).

3.4.5.7. *Influential parameters*

The parameters influencing the quality of the pressuremeter test essentially revolve around the quality of the boring, whose impact on the measurements is very significant.

Drilling quality

Depending on the nature of the soil and the strength of the cavity walls, several tools are used (Standard NF P94-110) to carry out the drilling prior to the pressuremeter test. There are semi-destructive tools such as augers, destructive tools such as cutters and "recuperative" tools such as core drills. Table 3.7 summarizes the tools best suited for better drilling quality for several typical soils.

Soils	Tools	Notes
Soft to medium compact clay	TH	Drilling passes limited to 1–2 m (below the water table)
Compact to stiff clay	CD – CAC – TA	-
Loose silts and sands	TH – TA	With mud injection (Bentonite, GSP)
Compact sands	TH – TA	With mud injection (Bentonite, GSP)
Submerged loose sand and gravel	TL	Direct drilling
Compact sand and gravel	TA	With mud injection (Bentonite, GSP)
Rocks or similar	TA	Water or air drilling

TH: Helical auger; CD: Core drill; CAC: Cable core drill; TA: Cutter; TL: Lantern tube (or pile).

Table 3.7. *Typical soils and the most suitable tools for a better drilling quality for the pressuremeter test*

COMMENTS.–

1) The driving of a core drill to make a bore hole has an impact on the quality of the test. On tests carried out in clays, Jézéquel (1968) shows that the E_M pressure modules are underestimated in tests where the boring is made with a driven core drill, compared to those obtained in borings made by rotation (cutting) and the injection of bentonite. According to the author, the limit pressures measured in the tests conducted in core drilling are overestimated, but with a certain diffusion. These results are confirmed by Cassan (1987) in Agen clays. There are no rules in this area. However, it is legitimate to consider that the pressuremeter modulus (E_M), which is very sensitive to the state of the boring walls, can be underestimated if the material structure is destroyed ("piston effect during the withdrawal of the core drill"). We made this observation in the Lyon Stadium (2010), in over-consolidated clays under very compact alluvium.

2) The lantern tube (or pile) is essential for the development of the test cavity in granular soils under water, such as loose sands and gravels. Like Baguelin and

Jézéquel (1972, 1973), Cassan (1987) shows that the limit pressures are little influenced by the use of the lantern tube, whereas the pressuremeter modules are influenced by the effect of soil displacement around the tube. In sands under the water table, the lantern tube destroys the structure of the sands during the driving, by vibration, and reduces the modules through a liquefaction effect. The limit pressures are less disturbed because there is a fast "rebalance" around the lantern tube.

Other parameters, such as depth, anisotropy (orientation of the probe) and the guard cell can have an impact, to a greater or lesser extent, on the pressuremeter measurements. Cassan (1987) uses simplified calculations by Jézéquel, and concludes that the pressuremeter test can be carried out, without risk, at greater or lesser depths. The author believes that the modules are underestimated when the depth increases, but the limit pressures are little affected.

Few pressuremeter tests cover the impact of probe orientation to characterize the influence of soil anisotropy. Cassan (1987) reduces the impact of soil anisotropy on the two main parameters (modulus and limit pressures) of tests carried out in vertical borings, and tests carried out in horizontal borings for tunnels. Jézéquel *et al.* (1968) reach similar conclusions. We believe that this point is not well clarified because the anisotropy of the soil has a direct impact on the modules and limit pressures. There is an issue of scale that the probe, which is not very long in terms of soil mass, cannot integrate.

The influence of the guard cell is, according to Cassan (1987), negligible. It should be remembered that Louis Ménard introduced the guard cell to avoid the measurements only being carried out in the central part of the deformation field, and that this field remains cylindrical. According to Cassan (1987), tests carried out with the Ménard pressuremeter (with guard cell) and the Japanese OYO pressuremeter (without guard cell) lead to similar results.

To better attract the attention of engineers to the importance of pressuremeter testing in the design and justification of geotechnical projects, we present, analyze and comment on pressuremeter profiles, and tests, at the end of this chapter.

3.4.6. *Phicometer shear test*

3.4.6.1. *Principle and test device*

The phicometer shear test is a French invention by Gérard Philipponnat (Philipponnat 1987). It is used to measure both the angle of friction (noted as φ_i) and the cohesion (noted as c_i) of on-site fine or granular soils. In France, the Standard XP P94-120 describes the test and the measured parameters.

Consider a pile of gravel deposited on the soil by a truck. After unloading, the pile of gravel ends up taking the shape of an "Egyptian pyramid", with a horizontal angle (β). This angle is equal to the angle of friction mobilized between grains (φ_i), which is called the "natural angle of repose" (Figure 3.27(a)). When the soil has a certain cohesion, such as clay for example (Figure 3.27(b)), it can be "temporarily" cut vertically. When the clay is saturated, the water prevents any friction between sheets, and only the undrained cohesion (c_u) prevails in the clay and governs its behavior[2].

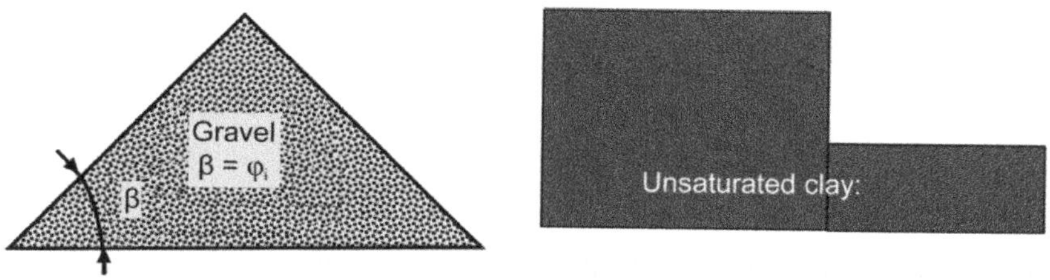

a) Gravel heap in the form of a "pyramid" (equilibrium limit)

b) Vertically cut excavation in unsaturated clay

Figure 3.27. *Role of friction and cohesion in soil. For a color version of this figure, see www.iste.co.uk/dhouib/geotechnics1.zip*

The phicometer shear test setup includes:

In the soil:

– a drill stem for the boring;

– a radially expandable probe, equipped with anchoring rings (teeth) to cling to the ground.

On the soil surface:

– a device for measuring the vertical upward displacement;

– a device for ripping and measuring the ripping force;

– a reaction device and a measurement reference mark;

2 Saturated clay can be cut vertically, like butter out of the fridge, but it will melt very quickly when it is in contact with ambient heat, like the clay of an excavation which, by gradually losing its cohesion when in contact with the air and the sun, collapses. Any excavation must be supported and reinforced in order to guarantee the stability of the excavation itself and the adjoining buildings, and must ensure that the lives of the workers on the site are not put in danger (Volume 4, Chapter 1 deals with the "active-passive pressure earth" theory and retaining walls).

– a pressure system with a pressure–volume controller.

Figure 3.28 describes the test setup.

3.4.6.2. *Implementation of the test*

The test is carried out following the example of the pressuremeter test (see section 3.4.5 of this chapter) according to the following steps (French Standard NF XP P94-120):

1) Carry out drilling passes at varying depths, depending on the strength of the boring walls, using a drill stem with a drilling tool (auger, deagglomeration tool, core drill, etc.).

2) Bring the drill stem and drilling tool up to the surface.

3) Lower the probe to the test depth using the drill stem.

4) In a preliminary phase, expand the probe under a sufficient pressure to make the anchoring rings penetrate the soil to be sheared.

5) During the real test phase (pulling out the probe), inject radial pressures in successive increasing stages into the expandable probe, applying ripping forces at a constant pressure at each stage, so as to shear the soil.

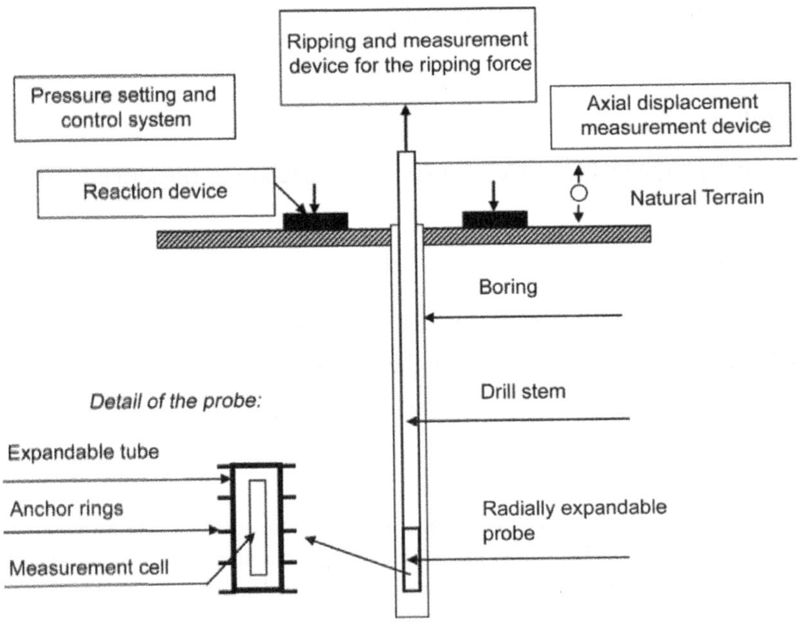

Figure 3.28. *Phicometer: measurement device*

Figure 3.29 shows the stages involved in the implementation of the test (Figure 3.29(a)) and a photo taken on site (Figure 3.29(b)).

Preliminary phase of probe expansion

In the preliminary phase, the measurement cell is expanded to make the anchor rings penetrate the soil to be sheared. The pressure (p_r) measured on the pressure indicator is applied progressively by increments of pressure Δp (Standard NF XP94-120):

– Δp = 50 kPa for soils with a previously estimated pressuremeter limit pressure measured between 0.3 and 1.0 MPa;

– Δp = 100 kPa for more compact soils with a limit pressure exceeding 1.0 MPa.

In this phase, each pressure stage is kept constant for 60 s. At each stage, the pressure p_r and the corresponding injected volume are noted at the following times:

Time t=30 s and t=60 s

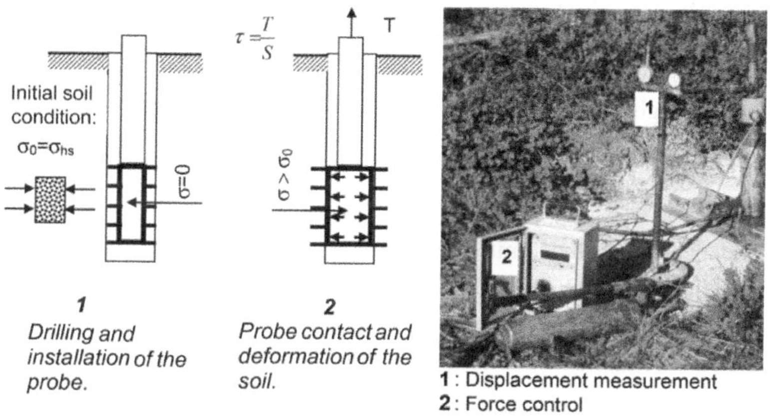

a) Test stages b) Site photo (FONDASOL)

Figure 3.29. *Phicometer shear test. For a color version of this figure, see www.iste.co.uk/dhouib/geotechnics1.zip*

The pressure (p_z) applied to the probe is conventionally determined by:

$$p_z = p_r + p_h \qquad [3.105]$$

where p_h is the hydrostatic pressure given by equation [3.75].

Test phase: expansion and ripping of the probe

The test is performed in increments of pressure that are applied radially in the expandable probe. The whole device (probe and shafts) is subjected to a ripping force at the top in order to shear the soil. Each stage of constant radial pressure and shearing comprises two successive stages:

1) The radial stress applied on the cylindrical surface of the expandable probe is kept constant for 60 s.

2) The application of the ripping force causes the soil to shear around the probe at a constant speed, and the radial stress is kept constant.

In each pressure stage, the probe is pulled out from the soil surface at a constant speed of 2 mm/min. The ripping force is recorded at least every 30 s. The ripping is stopped when:

– either the ripping force has stabilized during three consecutive measurements, spaced out by 30 s;

– or the displacement measured at the head of the drill stem reaches 7 mm.

The phicometer shear test is stopped when:

– either the eight stages provided for by the standard have been completed;

– or the volume injected into the measurement cell has reached 600 cm^3.

The values of the radial pressure increments retained by French Standard XP P94-120 depend on the limit pressure (p_l), estimated beforehand using the pressuremeter. According to the values of p_l, the standard defines eight stages whose values vary:

– between 0.20 and 0.75 MPa for p_l ranging from 0.30 to 1.0 MPa;

– between 0.25 and 1.00 MPa for $p_l \geq 1.0$ MPa.

3.4.6.3. *Measured shear parameters: experimental curves*

The phicometer shear test makes it possible to:

– assess the nature of the soil layers crossed by the drilling through the various drilling passes;

– determine the shear characteristics of the soil in situ from the phicometer tests: cohesion (c_i) and angle of friction (φ_i).

The two shear parameters (c_i and φ_i) are determined on a fitting line on the experimental curve constructed from the pairs (p_c, τ), with p_c being the radial pressure injected into the probe on the x-axis of the curve and τ being the ultimate shear stress corresponding to the radial pressure, plotted on the coordinates of the curve and given by the formula (Figure 3.30):

$$\tau = \frac{T_l}{\pi \times d_s \times l_s} \qquad [3.106]$$

where T_l is the ultimate ripping force, and d_s and l_s are the diameter and length of the probe, respectively.

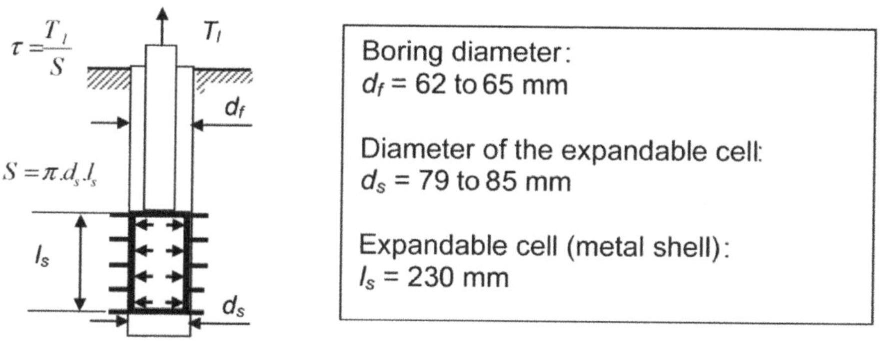

a) Shear surface b) Practical data

Figure 3.30. *Determination of the ultimate conventional shear stress (τ)*

Figure 3.31 shows the characteristic curves of the phicometer test.

Let us note that:

– On the experimental shear characteristic curve: "σ, τ" (Figure 3.31(a)), the cohesion (c_i) of the soil in situ corresponds to the y-axis of the curve and the angle of friction (φ_i) of the soil in situ is the slope of the line.

– The probe expansion curve represents the change, from the beginning to the end of the ripping of volume ΔV (Figure 3.31(b)).

– Figure 3.31(c) shows the loading curve of the volume injected into the measurement cell.

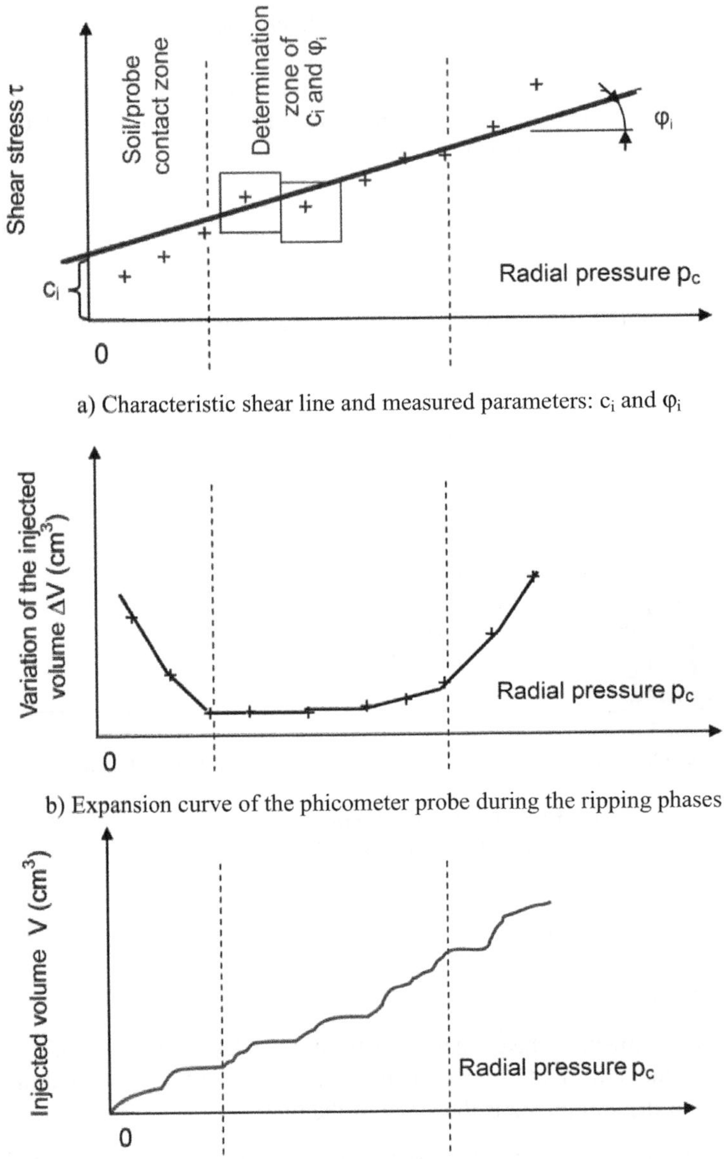

a) Characteristic shear line and measured parameters: c_i and φ_i

b) Expansion curve of the phicometer probe during the ripping phases

c) Evolution of the volume injected into the probe as a function of pressure

Figure 3.31. *Shear curves of the phicometer test. For a color version of this figure, see www.iste.co.uk/dhouib/geotechnics1.zip*

3.4.6.4. *Practical use of measured parameters: empirical correlations*

Phicometer shear parameters were interpreted by Philipponnat and Zerhouni (1993) for use in slope stability studies, through a summary of correlations. In recommending laboratory shear tests for such studies, the authors established precautionary rules for deducing effective shear characteristics, c' and φ', from the phicometer parameters, c_i and φ_i, which were measured on site (Table 3.8).

Phicometer measurements		**Soil type**	**Effective characteristics**	
φ_i (degrees)	**c_i (kPa)**		**φ'_i (degrees)**	**c'_i (kPa)**
≤ 15	<20	Soft clay	17	0
	≥ 20	Slight to stiff clay	17–20	5
15–30	<20	Loose sand	25	0
	≥ 20	Silt	25–30	5–10
>30	<10	Sands and gravels*	30–35	0
	≥ 10	Granular and cohesive soils*	30	5–10
>25	-	Soft rocks**	25–35	10–100

*Difficult test due to blocks; **Inadequate test.

Table 3.8. *Effective shear characteristics correlated from on-site phicometer measurements (Philipponnat and Zerhouni 1993)*

In application 1.7.4 (Volume 4, Chapter 1), the study of a slurry trench wall is presented, whereby the soil in situ are the subject of measurements of the shear characteristics by the phicometer.

3.4.7. *Plate-bearing test*

3.4.7.1. *Behavioral analogy*

The plate-bearing test (or plate-loading test) is used to check the mechanical quality of surface platforms used for road pavements and highways, traffic routes on building sites during construction or in logistics buildings (deliveries and goods transport), slabs in industrial buildings, supermarkets and homes, as well as the rafts of various structures, such as those in sewage treatment plants and various buildings.

Figure 3.32(a) shows the body of a roadway. Figure 3.32(b) shows a human crowd in a party hall on a 0.15 m thick concrete slab, which is itself on a 0.30 m thick compacted sub-base. Figure 3.32(c) illustrates a 0.60 m thick general raft resting on a 0.60 m compacted sub-base, and on which a sludge silo of a wastewater treatment plant is founded. In terms of platform behavior, these three cases constitute a type of load that can be assimilated to a "surface" load, induced on the soil by a plate, whose thickness e is small compared to its diameter d (Figure 3.32(d)).

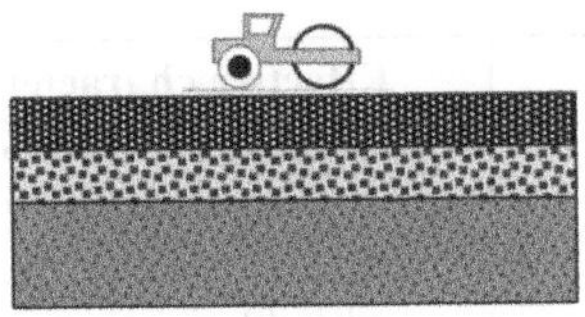

a) Vehicle on the road

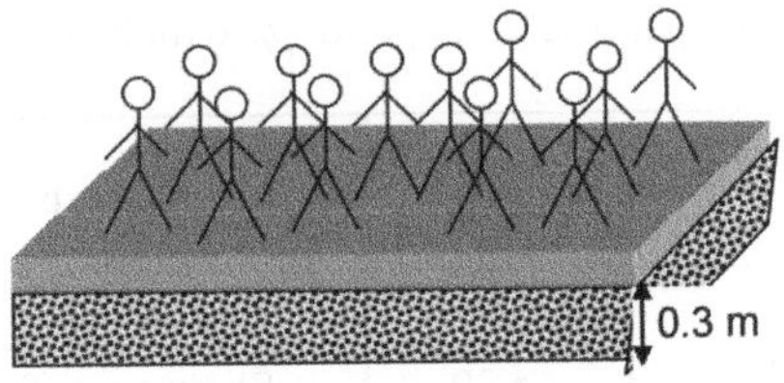

b) Human crowd on a slab
(in a party hall)

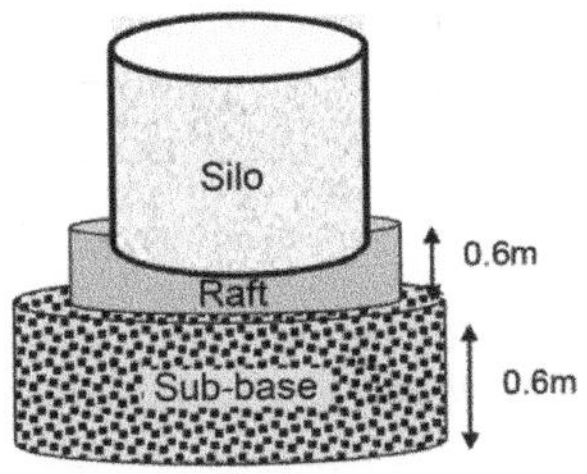

c) Wastewater treatment plant
(STEP) silo on general raft

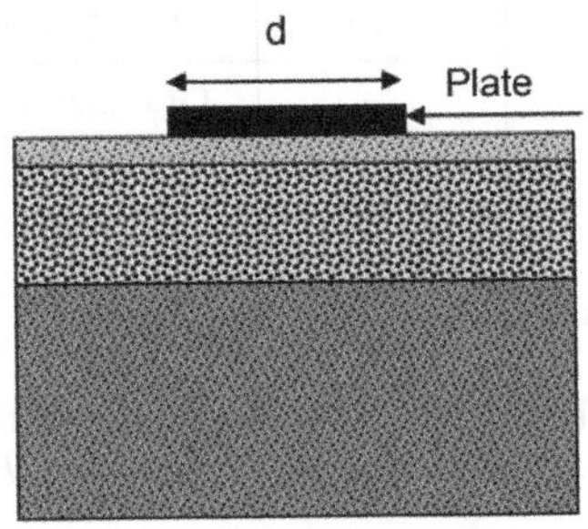

d) Analogy with the plate test

Figure 3.32. *Surface loading at the plate: behavioral analogy. For a color version of this figure, see www.iste.co.uk/dhouib/geotechnics1.zip*

3.4.7.2. *Test device*

The surface plate loading test device includes:

– a plate with a diameter between 0.30 and 0.75 m;

– a girder;

– adjustment chocks;

– a truck for the application of the 180 kN load (or, failing that, a mechanical shovel if the access conditions do not allow it, or even a mini-excavator with a limited load of a few tons, such as 2 tons, allowing K_w to be measured alone, because the stress is limited to 0.070 MPa for a 0.60 m plate).

Figure 3.33 shows a work site photo of a truck, plate, girder, chocks and deflection measuring comparators.

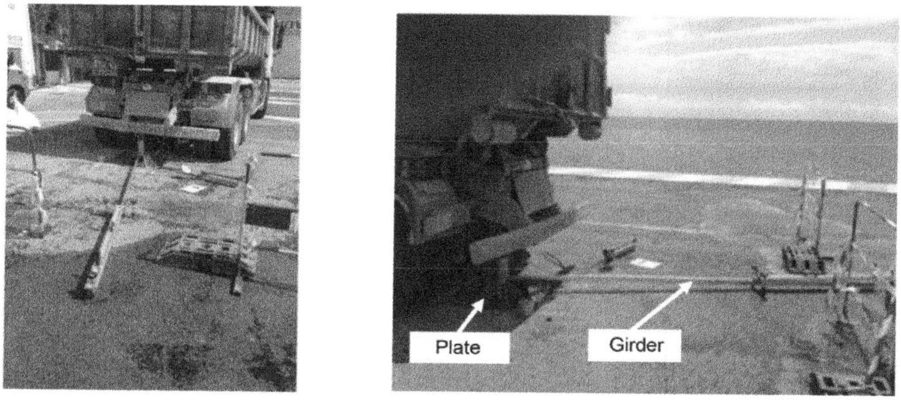

Figure 3.33. *Truck and plate test measuring device (SOL-ESSAIS). For a color version of this figure, see www.iste.co.uk/dhouib/geotechnics1.zip*

3.4.7.3. *Principle and execution of the test*

The plate-loading test consists of:

– placing a sand adjustment layer to flatten the seating surface of the plate;

– installing the test device (plate, girder, chocks and deflection sensors);

– applying a load to remove the settlements of the sand adjustment layer;

– setting all measuring devices to zero (initialization);

– starting the test.

The test is performed through the following stages:

1) Apply a vertical load until there is mean stress (p_K) of 0.07 MPa on the plate and note the corresponding displacement (w_K). This makes it possible to calculate the reaction modulus, called "Westergaard's modulus: K_W".

2) Increase the load to a stress (p_1) of 0.25 MPa and note the corresponding displacement (w_1). This level of stress and displacement makes it possible to determine the deformation modulus of the first load, noted as E_{V1}.

3) Unload the plate and reload to a stress (p_2) of 0.20 MPa and note the corresponding displacement (w_2) to deduce the deformation modulus of the second loading E_{V2}.

Figure 3.34 shows the stages involved in performing the test and the three parameters determined from the experimental curve.

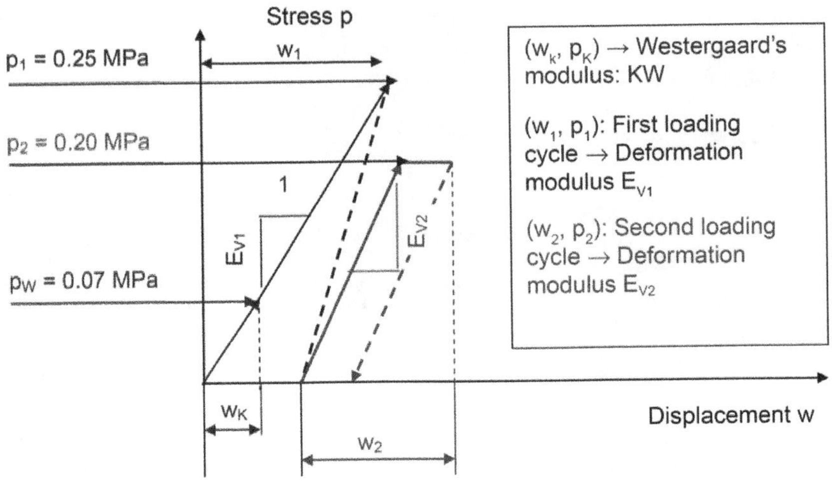

Figure 3.34. *Stages for performing the plate-loading test and basic test parameters. For a color version of this figure, see www.iste.co.uk/dhouib/geotechnics1.zip*

3.4.7.4. *Determination of the test parameters*

The plate-loading test is used to determine the three main parameters shown in Figure 3.34.

Westergaard's modulus: K_W

It is expressed by the pressure p_K, the plate diameter (d) and the displacement w_K as (Figure 3.34):

$$K_W = \frac{p_K}{w_K} \times \frac{d}{d_{75}} \qquad [3.107]$$

where d_{75} is the reference diameter of the 0.75 m diameter reference plate.

Deformation modulus of the first loading cycle: E_{V1}

This modulus is given by (Figure 3.34):

$$E_{V1} = \frac{7d}{10} \times \frac{p_1}{w_1} \qquad [3.108]$$

Deformation modulus of the second loading cycle: E_{V2}

The E_{V2} modulus is expressed as (Figure 3.34):

$$E_{V2} = \frac{7d}{10} \times \frac{p_2}{w_2} \qquad [3.109]$$

Compaction coefficient: k_c

The compaction coefficient (also called the "compression ratio") is the ratio of the two deformation modules:

$$k_c = \frac{E_{V2}}{E_{V1}} \qquad [3.110]$$

This coefficient, which is no longer used today, can conventionally check the quality of soil compaction (subgrade, sub-base, backfill placed in successive layers):

– when $k_c < 2$–2.5, the soil is well compacted (tight soil);

– when $k_c > 2$–2.5, there is a compaction defect (loose soil).

3.4.7.5. *Practical application to roadways and slabs*

When the soil is not very compact, the use of a sub-base is essential. This is the case for a vehicle that cannot move forward on muddy ground, for example. The well-known traditional solution consists of placing planks or spreading gravel on the path of the vehicle in order to allow it to move forward. The thickness of the gravel depends on the quality of the muddy terrain and the weight of the vehicle (light vehicle or truck).

For the installation of roadways, slabs and rafts, the same principle is applied in common practice according to the following steps:

1) stripping of surface soils (topsoil, poor backfill and blocks);

2) compaction of the soil in situ (called "subgrade") and removal of tough spots that appear during compaction;

3) placement of a sub-base on an anti-contaminant geotextile, generally in D2 gravel (with less than 5% fines), carefully compacted over a given height *h*; *h* being a function of the load applied to the structure (overload on slab or raft, weight of traffic on the pavement body) and the quality of the compacted subgrade;

4) installation of the structure.

Considering that the compacted sub-base has a Westergaard modulus (K_s) measured at the plate before the placement of the sub-base, the Westergaard coefficient (K_g) obtained on the sub-base is given by the empirical relation of Claesson *et al.* (1977):

$$K_g = 0.2(\mathrm{h})^{0,45} \times K_s \quad [3.111]$$

where h is the thickness of the sub-base given in mm. The Westergaard coefficient (K_{cf}) of the sub-base is usually set by the project, depending on the nature of the structure, and must be validated by plate tests. Equation [3.111] is used to predict the anticipated thickness (h) of the sub-base to be laid down, in order to obtain the required K_{cf} value.

Let us note, from experience, that on a carefully compacted bottom, a layer of 100 mm of clean gravel (D2 type with less than 5% of fines) generally leads to a K_g increase of 10 MPa.

Figure 3.35 shows a sub-base on a compacted bottom and the relations between plate-loading parameters.

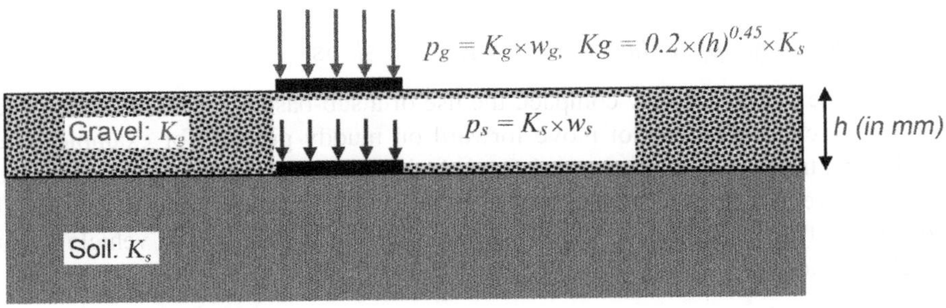

Figure 3.35. *Empirical determination of the height (h) of the sub-base (Claesson* et al. *1977). For a color version of this figure, see www.iste.co.uk/dhouib/geotechnics1.zip*

3.4.7.6. *Example of a field test*

A loading test with a 600 mm diameter plate on the platform of an industrial building site was carried out on a 0.40 m thick D2 gravel layer, resting on untreated but previously compacted silts of the plates before the placement of the sub-base. Figure 3.36 shows the evolution of the deflections as a function of the stresses applied during the two loading-unloading cycles.

The first loading cycle led to the following deflections:

– w_k of 0.48 mm under 0.07 MPa;

– w_1 of 2.08 mm under 0.25 MPa.

Truck load (kN)	Loads							
	14	20	28	42	56	70	Creep	Unload
Stress on plate (MPa)	0.05	0.07	0.10	0.15	0.20	0.25	-	0.00
First cycle (mm)	Deflections							
	0.28	**0.48 (w_k)**	0.72	1.08	1.48	1.88	**2.08 (w_1)**	1.12
Second cycle (mm)	1.56	-	1.80	1.96	2.12	-	2.20	1.24

a) Displacement measurements in both loading cycles

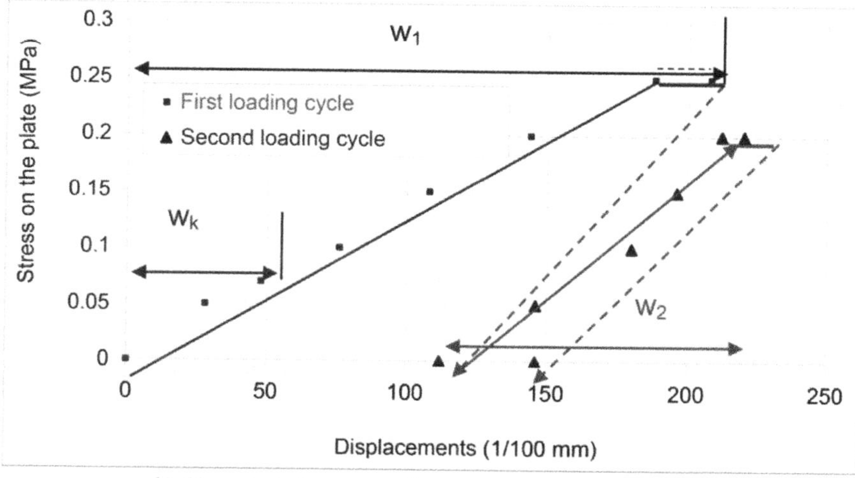

b) Characteristic curve of first and second loading cycles

Figure 3.36. *Example of a plate test on a gravel layer of 0.40 m thickness. For a color version of this figure, see www.iste.co.uk/dhouib/geotechnics1.zip*

The direct application of expression [3.107] and expression [3.108] leads to:

$$K_W = \frac{0.07}{0.48 \times 10^{-3}} \times \frac{0.6}{0.75} = 117\, MPa / m$$

and

$$E_{V1} = \frac{7\times 0.6}{10} \times \frac{0.25}{2.08\times 10^{-3}} = 50\,MPa$$

The second loading cycle results in a modulus E_{V2} (equation [3.109]):

$$E_{V2} = \frac{7\times 0.6}{10} \times \frac{0,20}{(2.20-1.32)\times 10^{-3}} = 94\,\text{MPa}$$

The compaction coefficient k_c (equation [3.110]) is:

$$k_c = \frac{95}{50} \approx 1.9$$

This shows that the gravel was well compacted ($k_c < 2$–2.2).

COMMENT.– There was no plate test on the compacted silts prior to placing and compacting the sub-base. However, the theoretical Westergaard modulus (K_W) is estimated at 35 MPa/m. Indeed, for this value, the direct application of Claesson's practical rule (equation [3.111]) gives a modulus of 104 MPa/m for a gravel thickness (h) of 400 mm.

3.4.7.7. *Order of magnitude of the parameters*

The "common" values of Westergaard's modulus (K_W), measured on various platforms, are summarized in Table 3.9. These modules are indicative and depend on the state of soil compaction and the degree of compaction of the sub-base.

Typical materials	**Westergaard modulus value**
Silt, consistent clays	$K_W \leq$ 20–40 MPa/m
Medium-density sand	K_W from 30–60 MPa/m
Dense sand	K_W from 60–100 MPa/m
Compacted gravel	K_W from 80–150 MPa/m
Treated silts (lime + cement)	K_W from 60–100 MPa/m
Cement gravel	K_W from 100–200 MPa/m

Table 3.9. *"Common" values of the Westergaard K_W modulus for some typical materials*

3.4.7.8. *Set threshold values in geotechnical projects*

In general, Westergaard's modulus (K_W) is used to define the type of platform according to its final destination. It is common practice to use the following values, unless otherwise specified in the contract:

– $K_W \geq 30$ MPa/m for low-load slabs (slab pressure of about 8 kPa), this is the case for houses and buildings used for housing;

– $K_W \geq 50$ MPa/m for common industrial-use slabs (slab pressure of 20 kPa), this is the case for large surfaces (supermarkets, DIY stores);

– $K_W \geq 80$ MPa/m for heavy industrial slabs (slab pressure of about 120 kPa), this is the case for industrial premises and factories with large storage loads.

The DTU 13.3 (Standard NF P11-213 of March 2005) imposes a K_w value of at least 50 MPa. Standards NF P94-117-1 and 117-3 only use modulus E_{V2} and K_W, respectively, without referring to the first loading modulus E_{v1}, nor to the compaction coefficient k_c (E_{v2} to E_{v1} ratio, equation [3.110]); the latter is only of interest to "qualify" the quality of the compaction carried out. Indeed, when it is higher than 2.0, the compaction coefficient k_c can lead to two different situations: either the compaction is poorly done when the two main parameters, E_{V2} and K_W, are low and the sub-base must therefore be remolded, or the intense compaction de-structured the compacted sub-base when these two parameters were high; the subgrade is, in this case, acceptable with a penetrometer test (such as a SPT), if it is necessary to do so. The two measured parameters E_{V2} and K_W can be calibrated on the graph in Figure 3.37.

To define the quality of the technical backfill platforms (all backfills are designated as "technical backfills" when at least: $p_l^* \geq 1.0$ MPa, $E_M \geq 12$ MPa and $q_d \geq 8$ MPa), traffic and roadways, slabs and rafts, it is the deformation modulus E_{V2} that is often used as an acceptance criterion. Depending on the value of E_{V2} and the type of platform (PF), and according to the GTR Guide (LCPC-SETRA 1992a, 1992b), four types of traffic platforms are commonly defined as a function of the values of the modules of the second loading cycle E_{V2}. This data is presented in Table 3.10.

Platform type	**PF1**	**PF2**	**PF3**	**PF4**
E_{v2} value (MPa)	20	50	200	120

Table 3.10. *Platform type according to modulus E_{V2} (according to the GTR Guide)*

We return to the various parameters measured by the plate tests in section 3.6.2.1, relating to the correlations.

3.5. French conventional soil classification

3.5.1. *Classification according to Fascicle 62, Part V*

From the limit pressures (p_l) measured with the Louis Ménard pressuremeter (Standard NF P94-110, see section 3.4.5 of this chapter) and the point resistances (q_c) measured with the cone penetrometer (Standard NF P94-113, see section 3.4.2 of this chapter), Fascicle 62, Part V (MEL 1993) introduces conventional soil categories, which are used in the current semi-empirical methods for designing foundations of geotechnical structures in France.

Table 3.11(a) gives these conventional soil categories retained by Fascicle 62, Part V (MEL 1993).

Soil class		**Soils**	**p_l (MPa)**	**q_c (MPa)**
Clays, silts	A	Soft clays and silts	<0.7	<3.0
	B	Firm clays and silts	1.2–2.0	3.0–6.0
	C	Very firm to hard clays	>2.5	>6.0
Sands, gravels	A	Loose	<0.5	<5.0
	B	Moderately compact	1.0–2.0	8.0–15.0
	C	Compact	>2.5	>20.0
Chalk	A	Soft	<0.7	<5.0
	B	Weathered	1.0–2.5	>5.0
	C	Compact ("sound")	>3.0	(2)
Marls	A	Soft	1.5–4.0	(2)
Marly-limestone	B	Compact	>4.5	(2)
Rocks (1)	A	Weathered	2.5–4.0	(2)
	B	Fragmented	>4.5	(2)

(1): The term "rocks" can include: limestone, shale, granite, etc.
(2): The cone penetrometer is not suitable because it is rejected.

Table 3.11(a). *Conventional soil categories (according to Fascicle 62, Part V)*

3.5.2. *Soil classification according to Eurocode 7*

Without assigning a specific class to soils, contrary to Fascicle 62, Part V (Table 311(a)), the nationally applied Standard NF P94-262 of Eurocode 7 for piles assigns a class to the soil according to its nature, consistency and defined capacity, in addition to p_l and q_c of Table 3.11(a), by the following:

– the consistency index I_c for fine soils defined in section 1.4.4 (Table 1.6);

– the corrected number of blows ($N_{1.60}$), measured with the SPT (see section 3.4.4 of this chapter);

– the undrained cohesion of fine saturated soils, measured with a Vane test, as mentioned in section 3.4.1 of this chapter.

Table 3.11(b) presents the classification of soils according to different criteria, conventionally introduced by the informative Appendix B of the national application standard of Eurocode 7, Standard NF P94-262 (AFNOR 2012).

<table>
<tr><th colspan="2">Soil class</th><th rowspan="2">I_c</th><th rowspan="2">p_l* (MPa)</th><th rowspan="2">q_c (MPa)</th><th rowspan="2">$N_{1.60}$</th><th rowspan="2">c_u (kPa)</th></tr>
<tr><th>Nature</th><th>Compactness</th></tr>
<tr><td rowspan="4">Clays and silts</td><td>Very soft to soft</td><td>0.0–0.50</td><td><0.4</td><td><1.0</td><td>X</td><td><75</td></tr>
<tr><td>Sealed</td><td>0.50–0.75</td><td>0.4–1.2</td><td>1.0–2.5</td><td>X</td><td>75–150</td></tr>
<tr><td>Stiff</td><td>0.75–1.00</td><td>1.2–2.0</td><td>2.5–4.0</td><td>X</td><td>150–300</td></tr>
<tr><td>Very stiff</td><td>>1.00</td><td>≥ 2.0</td><td>≥ 4.0</td><td>X</td><td>≥ 300</td></tr>
<tr><td>Intermediate soils</td><td colspan="6">To be classified according to the precise nature of the intermediate soil (silty sand, clayey sand, sandy clay) and the corresponding classes.</td></tr>
<tr><td rowspan="5">Sands and gravels</td><td>Very loose</td><td>X</td><td><0.2</td><td><1.5</td><td><3</td><td>X</td></tr>
<tr><td>Loose</td><td>X</td><td>0.2–0.5</td><td>1.5–4</td><td>3–8</td><td>X</td></tr>
<tr><td>Moderately dense</td><td>X</td><td>0.5–1.0</td><td>4–10</td><td>8–25</td><td>X</td></tr>
<tr><td>Dense</td><td>X</td><td>1–2</td><td>10–20</td><td>25–42</td><td>X</td></tr>
<tr><td>Very dense</td><td>X</td><td>>2</td><td>>20*</td><td>42–58</td><td>X</td></tr>
<tr><td rowspan="3">Chalk</td><td>Soft</td><td>X</td><td><0.7</td><td><5</td><td>X</td><td>X</td></tr>
<tr><td>Weathered</td><td>X</td><td>0.7–3.0</td><td>5–15</td><td>X</td><td>X</td></tr>
<tr><td>Sound</td><td>X</td><td>>3</td><td>≥ 15*</td><td>X</td><td>X</td></tr>
<tr><td rowspan="3">Marls and marly limestone</td><td>Soft</td><td>X</td><td><1</td><td><5</td><td>X</td><td>X</td></tr>
<tr><td>Stiff</td><td>X</td><td>1–4</td><td>5–15</td><td>X</td><td>X</td></tr>
<tr><td>Very stiff</td><td>X</td><td>>4</td><td>>15*</td><td>X</td><td>X</td></tr>
<tr><td rowspan="2">Rock</td><td>Weathered</td><td>X</td><td>2.5–4</td><td>X</td><td>X</td><td>X</td></tr>
<tr><td>Fragmented</td><td>X</td><td>>4</td><td>X</td><td>X</td><td>X</td></tr>
<tr><td>Legend</td><td colspan="6">X: Unsuitable tool (rejection) or non-measurable parameter in the soil.
*: Value at rejection (author's comment).</td></tr>
</table>

Table 3.11(b). *Conventional classification of soils according to the Eurocodes (Appendix B of Standard NF P94-262)*

COMMENTS.– The classification of soils according to Eurocode 7 in Table 3.11(b) maintains the same classification principle as Fascicle 62, Part V, while retaining variation ranges in limit pressures (p_l) and point resistances (q_c) that are very close, or even identical in certain cases (chalk and rock).

– On the other hand, the classification of the national application standard of Eurocode 7 does not comply with Fascicle 62, Part V, for sound chalk (Class C), marl and marl-limestone, and rocks; the cone penetrometer is unsuitable because it is rejected when encountering these hard materials.

– These inconsistencies are also observed in the calculation rules regarding shallow and deep foundations of the two national application standards of Eurocode 7, respectively Standards NF P94-261 (AFNOR 2013) and P94-262 (AFNOR 2012), as highlighted in the practical applications of Volume 3, Chapter 2 relating to foundations (section 2.1) and piles (application 2.2).

To distinguish between cohesive soils loaded with calcium carbonate $CaCO_3$, Standard NF P94-262 retains (Table B.2.2 of Appendix B) a classification similar to that explained in Table 1.4 of Chapter 1, to which it adds "porous and light chalk", together with limestone.

For intermediate soils derived from the three traditional formations of clay, silt and sand, Demolon (1948) establishes a ternary diagram, based on particle sizes ranging from 2 to 20 μm, to classify silt, sand and clay, and intermediate soils derived from these three formations. This diagram is provided in Figure 1.9 of Chapter 1, and can be used, among other purposes, to complete the classification of intermediate soils, as a function of the percentage of aggregates sized between 2 and 20 μm, in Table 3.11(b), adopted by the national application standard of Eurocode 7, Standard NF P94-262 (Appendix B, AFNOR (2012)).

3.5.3. *Classification according to PS92 rules*

The earthquake-resistant rules that apply to buildings (PS 92, Standard NF P06-013, AFNOR (1995)) adopt a conventional classification, which, in addition to the point resistance (q_c) measured with the cone penetrometer and the limit pressure (p_l) measured with the Louis Ménard pressuremeter, introduces additional data, such as:

– the number of blows (N_{SPT}) measured with an SPT (see section 3.4.4 of this chapter);

– the pressure modulus (E_M) measured with the Louis Ménard pressuremeter;

– the simple compressive strength (R_c) of stiff and cohesive soils;

– the relative density D_r (also called density index I_D), defined in section 1.3.2.4 and Table 1.4 for granular soils;

– the compression index C_c, measured with the oedometer in compressible soils, presented and studied in section 2.2.3 of Volume 2, Chapter 2;

– the velocity of the shear probes (v_T) and longitudinal waves (v_c) under and outside the water table.

Table 3.11(c) presents the classification of soil types according to the PS 92 earthquake-resistant recommendations (PS 92 rules – Standard NF P06-013, AFNOR (1995)).

<table>
<tr><th colspan="2">Soil type</th><th rowspan="2">q_c
(MPa)</th><th rowspan="2">N_{SPT}</th><th colspan="2">Pressuremeter</th><th rowspan="2">R_c
(MPa)</th><th rowspan="2">D_r
(%)</th><th rowspan="2">C_c</th><th rowspan="2">v_T
(m/s)</th><th colspan="2">v_c
(m/s)</th></tr>
<tr><th>Group</th><th>Nature</th><th>E_M
(MPa)</th><th>p_l
(MPa)</th><th>SN</th><th>HN</th></tr>
<tr><td>Rocks</td><td>Healthy rocks and chalk</td><td>X</td><td>X</td><td>>100</td><td>>5</td><td>>10</td><td>X</td><td>X</td><td>>800</td><td>XX</td><td>>2,500</td></tr>
<tr><td rowspan="2">a</td><td>Compact granular soils</td><td>>15</td><td>>30</td><td>>20</td><td>>2</td><td>X</td><td>>60</td><td>X</td><td rowspan="2">>400</td><td>>1,800</td><td>>800</td></tr>
<tr><td>Stiff clay and marl</td><td>>5</td><td>X</td><td>>25</td><td>>2</td><td>>0.4</td><td>X</td><td><0.02</td><td>-</td><td>>1,800</td></tr>
<tr><td rowspan="3">b</td><td>Weathered or fractured rock</td><td>X</td><td>X</td><td>50–100</td><td>2.5–5.0</td><td>1–10</td><td>X</td><td>X</td><td>300–800</td><td>-</td><td>400–2,500</td></tr>
<tr><td>Moderately compact granular soils</td><td>5–15</td><td>10–30</td><td>6–20</td><td>1–2</td><td>X</td><td>40–60</td><td>X</td><td rowspan="2">150–400</td><td>1,500–1,800</td><td>500–800</td></tr>
<tr><td>Cohesive soils moderately consistant and soft chalk</td><td>1.5–5</td><td>X</td><td>5–25</td><td>0.5–2.0</td><td>0.1–0.4</td><td>X</td><td>0.02–0.10</td><td>-</td><td>1,000–1,800</td></tr>
<tr><td rowspan="2">c</td><td>Loose granular soils</td><td><5</td><td><10</td><td><6</td><td><1</td><td>X</td><td><40</td><td>X</td><td rowspan="2"><150</td><td rowspan="2"><150</td><td rowspan="2"><150</td></tr>
<tr><td>Soft cohesive soils and weathered chalk</td><td><1.5</td><td><2</td><td><5</td><td><0.5</td><td><0.1</td><td>X</td><td>>0.10</td></tr>
<tr><td>Table legend</td><td colspan="11">Group a: Soils with good to very good mechanical resistance.
Group b: Soils of moderate mechanical resistance.
Group c: Soils with low (to very low) mechanical resistance.
v_T: shear wave speed, v_c: longitudinal wave speed.
SN: Below the water table.
HN: Beyond the water table.
X: Unsuitable tool (rejection) or non-measurable parameter in the soil.
XX: No such situation in practice.</td></tr>
</table>

Table 3.11(c). *Classification of soil types according to PS 92 (AFNOR 1995)*

3.5.4. *Classification according to Eurocode 8*

To characterize soils with respect to earthquakes, Eurocode 8 uses a restrictive classification of soils as a function of the shear wave velocity, noted as $v_{s,30}$ (m/s), the number of blows N_{SPT} for 30 cm of take up with the standard penetrometer test and the undrained cohesion c_u measured with a field vane test. Table 3.11(d) presents this classification.

<table>
<tr><th>Soil class</th><th>Soil description</th><th>$v_{s,30}$ (m/s)</th><th>N_{SPT} (blows/ 30 cm)</th><th>c_u (kPa)</th></tr>
<tr><td rowspan="2">A</td><td rowspan="2">Rock or other such geological formation with a surface layer of up to 5 m of less resistant material.</td><td>>800</td><td>X</td><td>X</td></tr>
<tr><td colspan="3">$E_M > 100$ and $p_l > 5$*</td></tr>
<tr><td rowspan="2">B</td><td rowspan="2">Stiff deposits of sand, gravel or over-consolidated clay, at least several tens of meters thick, characterized by a progressive increase in mechanical properties with depth.</td><td>360–800</td><td>>50 (sand/gravel)</td><td>>250 (clay)</td></tr>
<tr><td colspan="3">$q_c > 3.5$ (clay) and $q_c > 20$ (sand)*
$E_M = 25–100$*,
$p_l > 1.2$ (clay) and 2–5 (sand)*</td></tr>
<tr><td rowspan="2">C</td><td rowspan="2">Deep deposits of low to medium density sand, moderately stiff gravels or clays of a few dozen to a few hundred meters in thickness.</td><td>180–360</td><td>15–50 (sand/gravel)</td><td>70–250 (clay)</td></tr>
<tr><td colspan="3">$q_c = 1–3.5$ (clay) and 6–20 (sand)*
$E_M =5–25$*,
$p_l = 0.5–1.2$ (clay) and 0.8–2 (sand)*</td></tr>
<tr><td rowspan="2">D</td><td rowspan="2">Cohesive soil deposits of low to medium density (with or without soft coherent layers) or comprising a majority of soft to firm coherent soils.</td><td><180</td><td><15 (sand)</td><td><70 (clay)</td></tr>
<tr><td colspan="3">$q_c < 1$ (clay) and $q_c < 6$ (sand)*
$E_M < 5$*,
$p_l < 0.5$ (clay) and <0.8 (sand)*</td></tr>
<tr><td>E</td><td>Soil profile comprising a shallow layer of alluvium with v_s values of class C or D, and a thickness between about 5 and 20 m, resting on a stiffer material with a $v_s > 800$ m/s.</td><td>-</td><td>-</td><td>-</td></tr>
</table>

S_1	Deposits composed of, or containing, a layer of soft muddy clays of at least 10 m thickness, with a high plasticity index ($I_P > 40$) and significant water content.	<100	-	10–20 (clay)
		$q_c < 0.6$ and $p_l < 0.2$*		
S_2	Liquefiable soil deposits of sensitive clays or any other soil profile not included in classes A to E or S_1.			
Table legend:				
X: Unsuitable tool (rejection). **A**: Rock. **B**: Stiff soils. **C**: Moderately stiff soils. **D**: Not very stiff soils. **E**: Class C or D soils on substratum. **S_1**: Soft clays and muds. **S_2**: Liquefiable sands and sensitive clays.	$v_{s,30} = \dfrac{30}{\sum_1^N \frac{h_i}{v_i}}$ (N soil layers) Thick layer of h_i and speed v_i *Values completed by the AFPS Guide (AmSol 2011) q_c, p_l and E_M in MPa			

Table 3.11(d). *Soil class according to Eurocode 8*

3.6. Correlations between parameters

3.6.1. *Correlations between mechanical parameters measured on site*

Several authors (Costet and Sanglerat 1981; Waschkowski 1983; Cassan 1988; Gonin *et al.* 1992) have attempted to cross-reference mechanical parameters measured by different techniques on the same sites, with a double objective:

1) to verify the correct execution of the tests performed on site;

2) to develop a summary of correlations between these parameters.

The usual correlations applied between the various on-site tests are as follows:

– pressuremeter;

– static and dynamic penetrometers;

– field vane test;

– other means of investigation on site.

These correlations must be taken with much caution and should never be used in the design phase (G2, as defined in Standard NF P94-500). Table 3.11 summarizes the correlations (statistical analysis by Cassan (1988)).

Correlation	Soils					Notes
	Clays	**Silts**	**Sandy clays**	**Sands**	**Sands and gravels**	
$q^*_c/c_u =$	15	-	-	-	-	-
$p^*_l/c_u =$	5.5	-	-	-	-	$p^*_l \leq 0.3$**
$q_c/q_d =$	$\neq 1$	1	0.5–0.9	1	1	No water table
	-	1.2–>3	0.1–0.4	-	1.0–2.6	No water table
$q_d/N_{SPT} =$	0.2	0.2–0.3	0.3	0.4	0.8	-
$q_c/N_{SPT} =$	0.1–0.2	0.2–0.3	0.3–0.4	0.3–0.6	-	-
$N_{SPT}/p^*_l =$	15–20	30	20	20	-	MPa
$N_{SPT}/E_M =$	1.0–1.5	3	-	1.5–2.0	-	MPa
$q^*_c/p^*_l =$	3.0–4.0	6	-	7–10	-	-
$E_M/q^*_c =$	3.0–6.0	3	-	1.0–1.5	-	-
$q^*_d/p^*_l =$	3.0–4.0	-	-	-	-	No water table

$p^*_l = p_l - p_0$, $q^*_c = q_c - p_0$. (p_0 : Horizontal stress in the soil at test depth).

**$c_u = p^*_l/12 + 0.03$ if $p^*_l > 0.3$ MPa, and $c_u = p^*_l/35 + 0.085$ MPa if p^*l varies from 1.0 to 2.5 MPa.

Table 3.12. *Summary of correlations between mechanical parameters measured on site, according to Cassan (1988) (unit in MPa)*

Waschkowski (1983) establishes the following correlative relations between q_d, q_c, p_l and E_M.

Sanglerat (1972) relates the point resistance, measured with a cone penetrometer, (q_c) to the oedometer modulus E_{oed} (see Table 1.6 in Volume 3, Chapter 1).

Soils	Correlative relations		
	q_d/q_c	q_d/p_l	q_d/E_M
Normally consolidated clays, silts and muds, loose or moderately dense sands.	≈ 1	1.4–2.5	0.1–0.3
Over-consolidated clays and silts	1–2	3–5	0.2–0.4
Dense to very dense sands and gravels, silty or clayey sands	0.5–1.0	5–10	0.4–1.5

Table 3.13(a). *Relations between q_d, q_c, p_l and E_M (Waschkowski 1983)*

3.6.2. *Correlations between other parameters*

3.6.2.1. *Plate-bearing test*

Let us recall the basic parameters of the plate test studied in section 3.4.7:

– reaction coefficient: Westergaard modulus: K_W;

– the two deformation modules of the first E_{V1} and second loading E_{V2};

– the compaction coefficient k_c, the ratio of the two modules E_{V1} and E_{V2}.

For the characterization of earthwork platforms, under slab and raft, of technical backfills, the analysis and designing of slab is mainly based on the second loading modulus E_{V2}. This parameter can be associated with the reaction coefficient K_W for the support of earthwork and slab platforms (Table 3.10). The two parameters are not directly related, but a summary of correlations can link them. In common practice, the following connections are practiced for roadways:

Parameters	Variation ranges			
K_W value (MPa/m)	30	50	120	150
E_{V2} value (MPa)	50	55	170	250

Table 3.13(b). *Practical connections between K_w and E_{V2}*

Figure 3.37 shows a large number of on-site measurements for K_W and E_{V2}, a graph that can be used in practice as a measurement “gauge”.

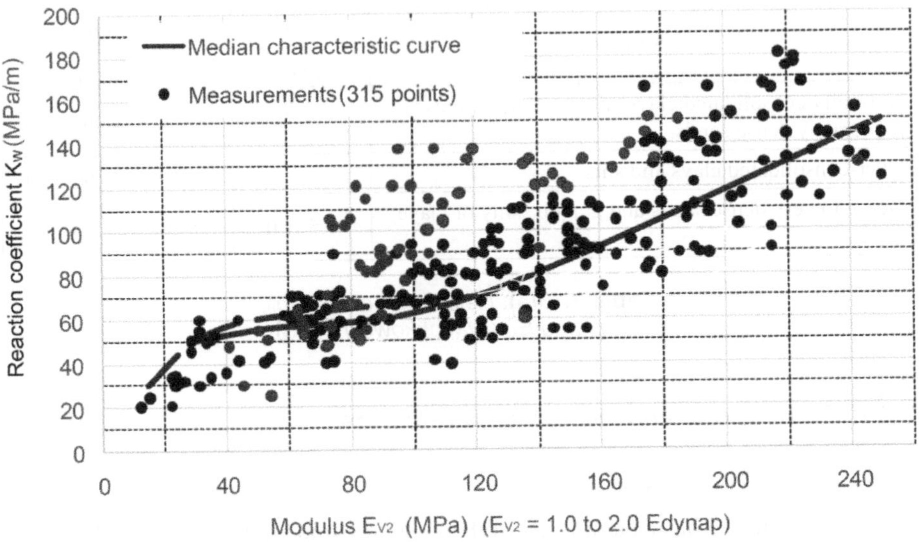

Figure 3.37. *On-site K_W and E_{V2} plate measurements. For a color version of this figure, see www.iste.co.uk/dhouib/geotechnics1.zip*

3.6.2.2. *Correlations with other parameters*

A summary of correlations exists between the mechanical parameters measured during the surface plate-loading test (K_W and E_{V2}) and the dyna-plate (E_{dynap}) measured by other means of investigation (pl^*, E_M and q_c), and the deformation modulus of soil deformation E_s under slab. Table 3.14 summarizes some of these correlations.

Plate test	Pressuremeter	Cone
$K_w = 50$ MPa/m	→ $p_l \approx$ 1.0 MPa, $K_W \approx 3.E_M/.\alpha$ $E_s \approx 0.35.K_W$ (E_s under slab)	-
$E_{V2} =$	3–4 E_M	9 q_c
Dyna-plate test	Coefficient of restitution $R = 0.5$ (50%) → $E_{V2} \approx$ 50 MPa $E_{\text{dynap}} \approx 0.5\text{–}1.0\ E_{V2}$	

Table 3.14. *Other types of summary correlations (in MPa)*

3.6.2.3. *Limit pressure p^*_l – angle of friction of the soil φ*

The correlation between the limit pressure p^*_l in the soil, measured with the Louis Ménard pressuremeter, and the angle of friction of the soil φ is expressed by the following relation (Figure 3.38), which applies for wet or dry soils without intrinsic cohesion:

$$p^*_l = 2.5 \times 2^{\frac{\phi-24}{4}} \qquad [3.112]$$

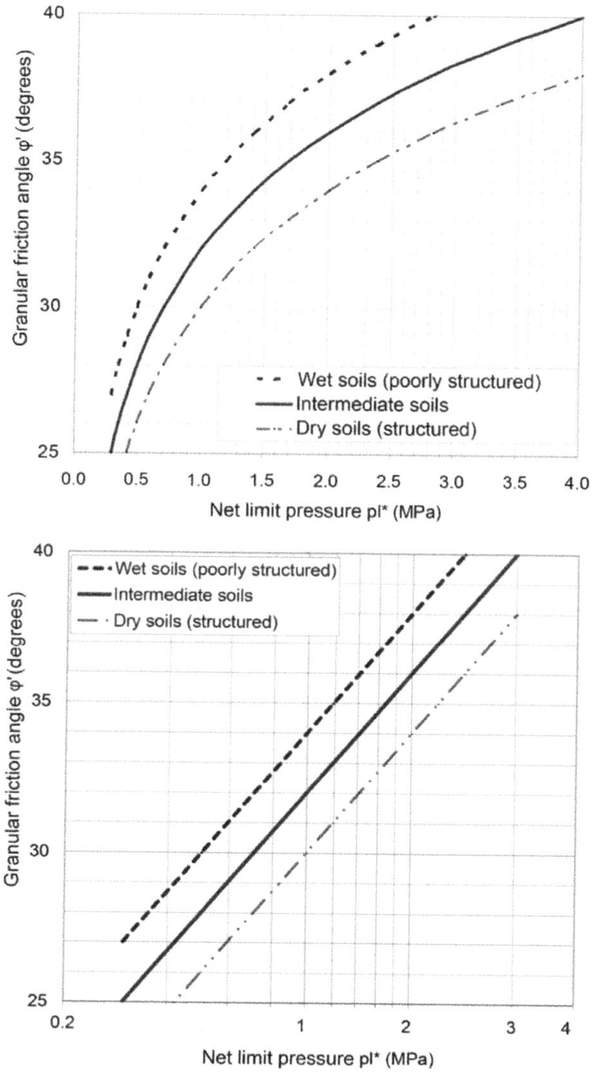

pl*(MPa)	φ (degrees)
0.02-0.05	10
0.05-0.12	15
0.12-0.15	20
0.15-0.18	21
0.18-0.20	22
0.20-0.25	23
0.25-0.30	24
0.30-0.35	25
0.35-0.40	26
0.40-0.50	27
0.50-0.60	28
0.60-0.70	29
0.70-0.85	30
0.85-1.00	31
1.00-1.20	32
1.20-1.40	33
1.40-1.70	34
1.70-2.00	35
2.00-2.40	36
2.40-2.80	37
2.80-3.40	38
3.40-4.00	39

Figure 3.38. *Relation between limit pressure (p_{l^*}) and angle of friction (φ) of the soil. For a color version of this figure, see www.iste.co.uk/dhouib/geotechnics1.zip*

3.7. Practical analysis of typical tests

3.7.1. *Practical example of monotonic pressuremeter tests*

Practical examples of tests, carried out by a specialized geotechnical engineering office in the framework of the project under construction (Mission G12, G2 and G3), are presented and commented on in the following:

– Figure 3.39(a), the pressuremeter curve (called the "pressure–volume curve") in loose soil;

– Figure 3.39(b), the pressure–volume curve of a resistant limestone substratum.

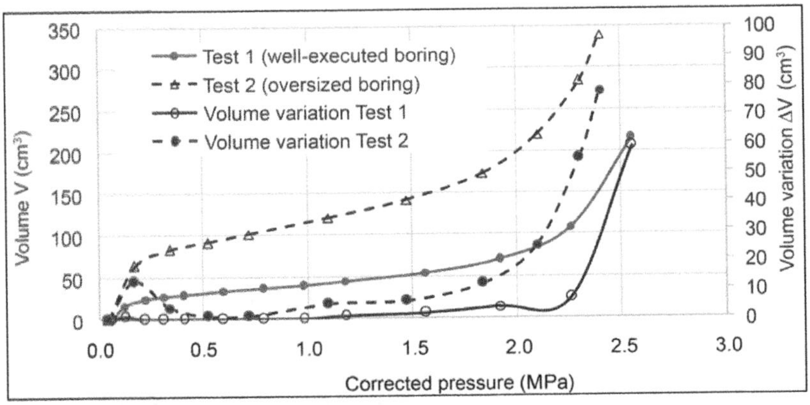

a) Pressure–volume curve in loose soil (FONDASOL)

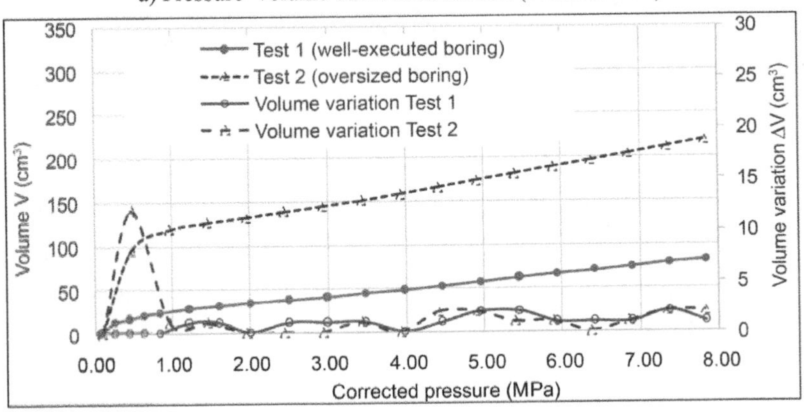

b) Pressure–volume curve in limestone (FONDASOL test for VINCI)

Figure 3.39. *Typical Ménard pressuremeter tests. For a color version of this figure, see www.iste.co.uk/dhouib/geotechnics1.zip*

The pressure–volume curves of the pressuremeter tests in Figure 3.39 lead to the following points:

– Test 1 (Figure 3.39(a)) has 12 measurement stages, ranging from 0.15 (beginning of the test) to 0.3 MPa for the rest of the test; this is a sufficient number (at least eight stages allowed by Standard NF P94-110-1). Having fewer stages can sometimes be explained by the drilling operator's bad assessment of the choice of stages in relation to the compactness of the ground.

– The original volume, about 10/15 cm^3, proves that the boring is well executed, contrary to Test 2 (large boring with a volume of 110 cm^3).

– The pseudo-elastic phase consists of five to eight stages, during which the pressure modulus E_M is calculated (linear part between two points, P1 and P2, before creep starts).

– Both tests show a net creep (around 5 cm^3) from the eighth stage, thus with 3–4 measurement points after the creep bend, with a net creep pressure p_{f*} of 2.4 MPa.

– The net limit pressure p_{l*} of 2.4 MPa corresponds to a creep volume of 60–80 cm^3, and a volume of water injected into the probe of about 300–350 cm^3.

– The p_{l*} to p_{f*} ratio is 1.5 (low).

– Together, these elements and the shape of the pressure–volume curves show that both tests are relatively well executed.

The two tests performed on the resistant limestone substratum of Figure 3.39(b), lead to the following comments:

– The test consists of 16–19 stages of measurement, without achieving creep (remaining zero at 2 cm^3). In other words, the last stage of measured pressure, which is 7.8 MPa, remains below the creep pressure p_f.

– The net limit pressure is, in this case, retained by default, that is to say:

$$\begin{cases} p_{f*} > 7.8MPa, \\ and \\ p_{l*} > 7.8MPa \end{cases}$$

– The 7.8 MPa value, higher than 5 MPa (measurement required by Standard NF P94-110-1), is achieved through tests conducted with a "rock-probe" at the request and on behalf of the company, in order to better optimize the point strength of the foundation piles.

– Finally, the pressure modulus (calculated between P1 and P2, and which is 510 MPa) was corrected by the expansion coefficient, which is 1 cm^3/MPa (<6 cm^3/MPa required by the standard). This correction is necessary for a better assessment of the modules in resistant environments (Dhouib *et al.* 2002) and the calculation of settlements under the works[3] (see Figures 3.46 and 3.57(b)).

Figure 3.40 illustrates a so-called "remolded" test – and therefore uninterpretable and unusable – that should not be removed from the geotechnical analysis and should not be charged.

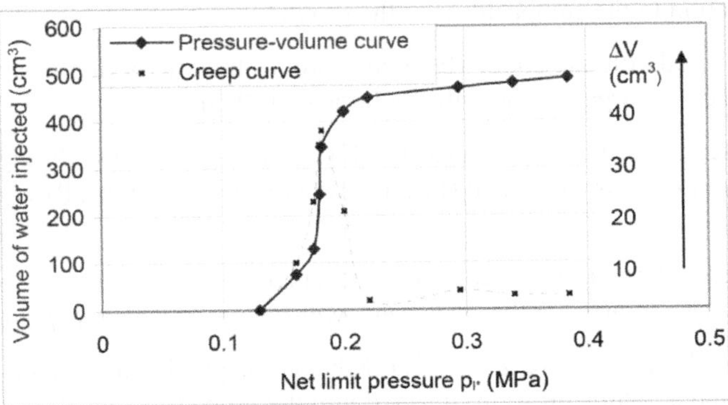

Figure 3.40. *Typical curve of a remolded test (FONDASOL test for VINCI). For a color version of this figure, see www.iste.co.uk/dhouib/geotechnics1.zip*

The remolded test, sometimes named the "non-exploitable test" on pressuremeter profiles provided in the appendices of the geotechnical reports, generally calls for the following comments:

– Sometimes the boring collapses as soon as the drill stem is withdrawn.

– The soil-probe contact can only take place in late stages, that is to say, after several stages of pressure have been injected into the probe, as shown in the graph of Figure 3.40.

– In the example of Figure 3.40, it can be seen that the contact between the probe and the boring walls starts around 400 cm^3 of water, contrary to the well-executed, correct drilling tests of Figure 3.39, where the soil-probe contact is established very quickly (from 50 cm^3 of volume injected into the probe).

3 As an example, in the context of the CMA Tower in Marseille, the average modulus without correction was 180 MPa, and 415 MPa after correction, which allowed the core receiving 110,000 tons to be founded on a general raft with compatible differential settlements between this raft and the neighboring slurry trench piles (CFMS Conference by Jacquard and Dhouib, SMABTP, October 11, 2012).

A landslide generally occurs when water comes into contact with powdery aquifer soil (such as sands and gravels), in very decompressed fine soils with a boring without casing, or when sufficiently thick mud or water with little mud content is used (bentonite, GSP, etc.).

The "remolded" tests can also be the result of poorly executed or poorly calibrated borings, in other words, borings with a "large" diameter or borings drilled in a hurry without any precautions. Good "surveyors" know this well and know when the test is sketchy: "*Boss, I executed a borehole looking at that potato*", in other words, the test was unsuccessful because of the large diameter boring compared to the small 40 mm probe, or because of a thin or insufficient mud (use of casing).

In a remolded test, the pressuremeter modulus (E_M), affected by the first stages, is erroneous because it is underestimated. The limit pressure is to be taken with caution, or even rejected.

In conclusion, the validity of the pressuremeter test depends on the quality of the boring.

Figure 3.41 shows a complete pressuremeter profile, followed by its analysis.

The typical pressuremeter profile in Figure 3.41 requires a thorough reading and makes the following points:

– Semi-destructive drilling with an auger means that the soil is not very resistant, and that the lithological section of the soil is more or less precise, whereas destructive drilling with a cutter, with injection of water and mud (Bentonite, GSP), provides a lithological profile that is very rough and imprecise, and sometimes even erroneous (limestone debris that has been crushed by the cutter can easily be mistaken for sand).

– When the drilling is carried out destructively with a cutter with the injection of water, the continuous recording of the drilling parameters (tool pressure *Po*, drilling speed Via, water injection pressure *Pi*, and rotational torque *Pr* or reflected vibration) is essential to make the pressuremeter profile more "consistent"; in such a case, it is necessary to check the compatibility of the drilling parameters with the limit pressures and the measured pressuremeter modules: if the drilling speed is high, the soil has a low compactness (under-consolidated soil such as backfills and soft soils, natural dissolutions, presence of empty or filled quarries, etc.), which corresponds to low limit pressures and modules. Low speeds correspond to high pressuremeter values.

– The performance of pre-test boring walls can help anticipate foundation work (especially deep foundations); for example, casing the pre-test boring to perform the tests can lead, on a much larger scale, to casing the piles or using thick mud (cased drilled piles or mud drilled piles, see Table 1.11 in Volume 3, Chapter 1).

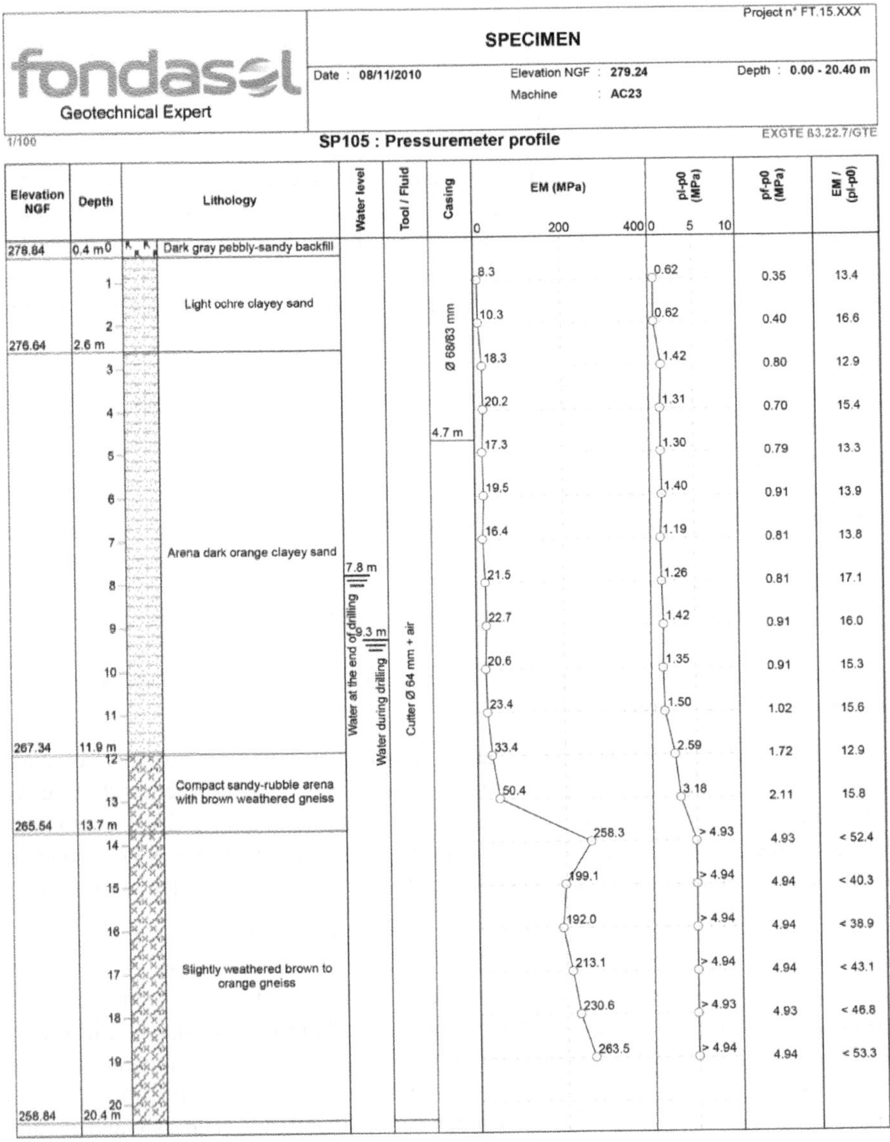

Figure 3.41(a). *Ménard pressuremeter test profile (FONDASOL profile for VINCI). For a color version of this figure, see www.iste.co.uk/dhouib/geotechnics1.zip*

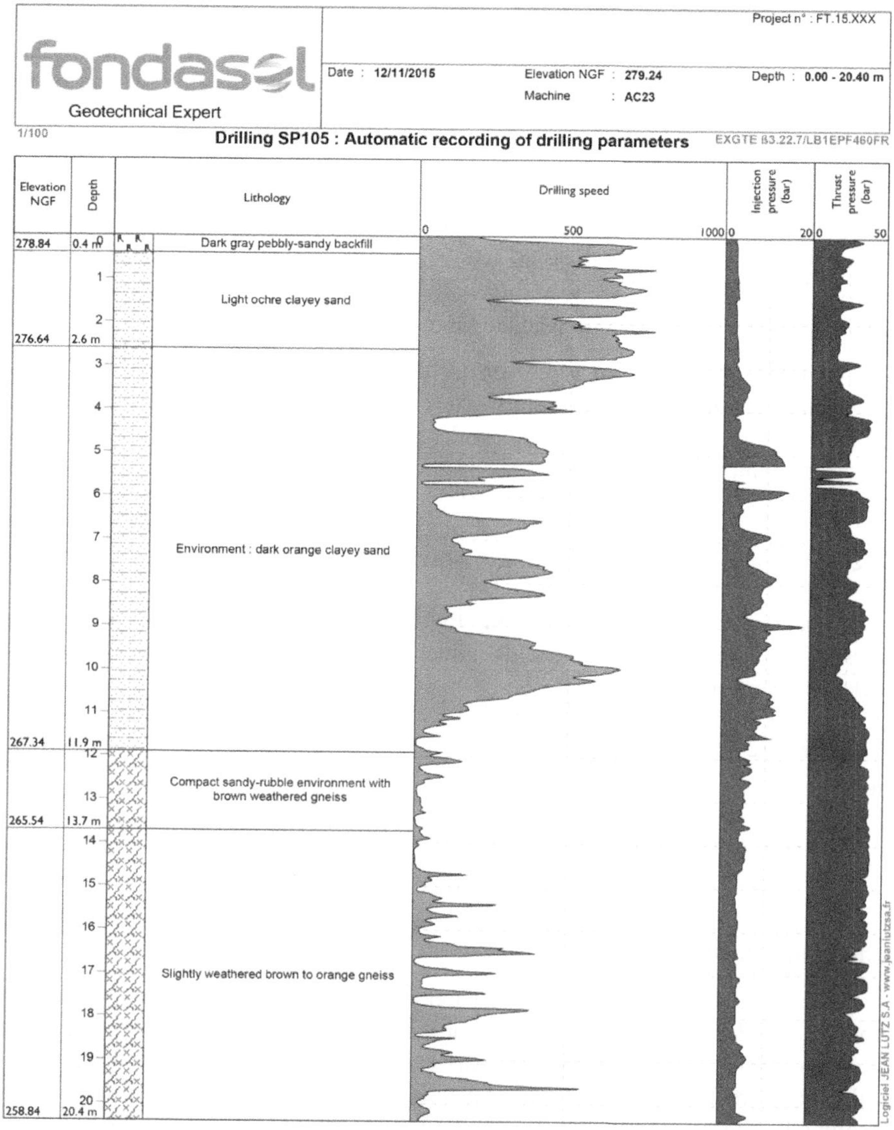

Figure 3.41(b). *Profile for automatic recording of drilling parameters (FONDASOL profile for VINCI). For a color version of this figure, see www.iste.co.uk/dhouib/geotechnics1.zip*

– The verification of the probes used for the tests is necessary for the validity of the tests, in relation to the measured levels of limit pressure: the use of a slotted tube in very soft soils can underestimate the limit pressures, and the pile driving of the slotted tube in clays can normally overestimate the pressuremeter modules.

– When a pressuremeter test is remolded, as shown in Figure 3.40, it is necessary to carefully examine both the test curve (pressure–volume curve) and the corresponding lithological section, while checking the position of the water (non-stabilized level following the drilling operations or the water table level). Indeed, as soon as the soil is under water, the remolding of the boring walls is inevitable, especially in granular soils without cohesion (sands and intermediate sandy soils).

– After the installation of a piezometer tube (Standard NF P94-157) following the presence of water, deduced from the unstable level during drilling operations, a regular monitoring over time before works are carried out is required in order to have a representative water level.

3.7.2. *Practical example of pressuremeter tests with cycles*

The test with cycles (Standard NF P94-110-2) makes it possible to simulate cycles of soil loading and unloading under various repetitive forces (vibrations, shocks and earthquakes).

Figure 3.42 shows two tests with one cycle (Figure 3.42(a)) and with two cycles (Figure 3.42(b)).

The two pressuremeter tests with cycles in Figure 3.42 that were performed on stiff marls (Dhouib *et al.* 2002) call for the following comments:

– The pressure of the first unloading plays an important role in the pressuremeter modulus measured in the test with cycles.

– Indeed, in Figure 3.42(a), the pressuremeter test was conducted in a monotonic manner up to 2.8 MPa, with a pressure around 0.8 p_f (p_f being the creep pressure), but an unloading of up to 0.45 MPa was carried out, and then a reloading with a similar pressure to the unloading. The unloading/reloading loop was closed, the unloading/reloading pressure modulus $E_{D/R}$ was 115 MPa, which was double its E_M counterpart measured in the initial loading, at 61 MPa. This difference is explained by a “stiffening” of the marl soil during the unloading/reloading cycle.

– However, in the test with two cycles in Figure 3.42(b), the first cycle was conducted from a first unloading pressure that was higher than the creep pressure, which makes it possible to observe a clear decrease in the unloading/reloading pressure modules: $E_1 = 60$ MPa in the first cycle and $E_2 = 40$ MPa in the second cycle, modules that are lower than the initial pressure modulus E_M, which is 72 MPa. Beyond the creep pressure, the marl is progressively "destroyed" during the application of the unloading/reloading cycles (see the general graphs in Figure 2.27 of Chapter 2).

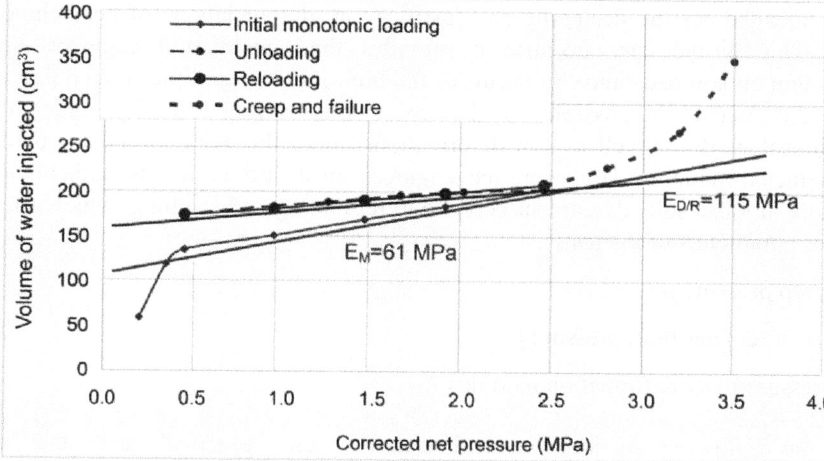

a) Test with one cycle

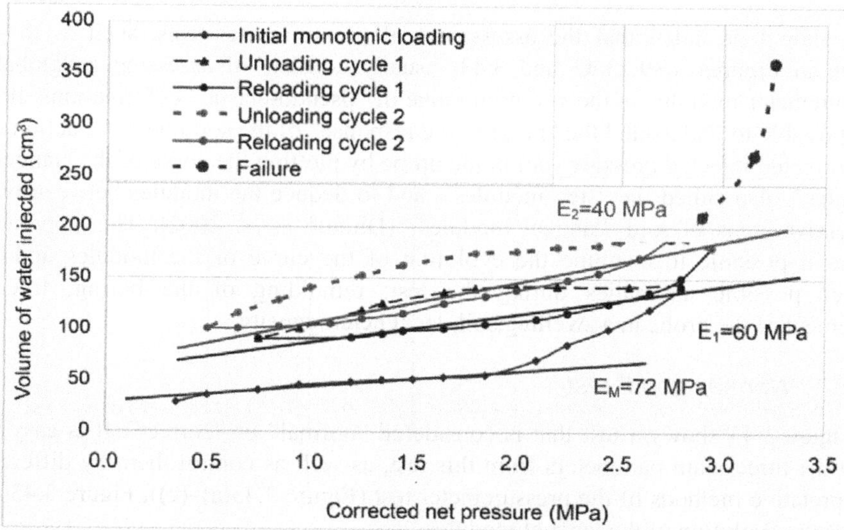

b) Test with two cycles

Figure 3.42. *Pressuremeter tests with cycles (Dhouib* et al. *2002). For a color version of this figure, see www.iste.co.uk/dhouib/geotechnics1.zip*

3.7.3. *Control procedures for pressuremeter tests*

The pressuremeter test is the most substantial test of all of the geotechnical tests in situ intended for the designing and justification of foundations of geotechnical works. This is the case because it provides the construction engineer with information on soil resistance by means of the limit pressure and the creep pressure. Doing this makes it possible to deduce a load-bearing capacity under the geotechnical work, as well as soil deformations under the work, by means of the pressuremeter modulus. However, the engineer must still verify its authenticity, especially in loose soils that are susceptible to remolding and swelling, which affect the three parameters of the test:

– creep pressure p_f;

– pressuremeter limit pressure p_l;

– pressuremeter deformation modulus E_M.

In the following section, we will examine some practical cases that are particularly related to the determination of two main parameters: pressuremeter modulus (E_M) and limit pressure p_l* in a "normal" or "proper" test, a remolded test and a test in soil that swells after drilling.

It should be noted that the assessment of pressuremeter tests, such as those shown in Figures 3.39, 3.42 and 3.43, usually consists of assessing a "global" pressuremeter modulus in the variation range of "pseudo-elastic" deformations. It is also possible to understand the "progressive response" of the soil under the action of the corrected injected pressure (*pc*) in the probe by plotting the curve of the "tangent modules", also called the "step modules", and to deduce the modules between two successive steps of said "tangent modules" (Dhouib *et al.* 2002). This approach makes it possible to examine the evolution of the curve of the modules and to deduce possible anomalies during the test: remolding of the boring, forced placement of the probe in a swelling soil, local heterogeneity.

3.7.3.1. *"Normal: good" test*

Figure 3.43 shows a test that is considered "normal" or "correct". It is easy to draw the three main parameters from this test, as well as control them by different interpretative methods of the pressuremeter test (Figure 3.43(a)–(c)). Figure 3.43(d) shows the evolution of the tangent modules.

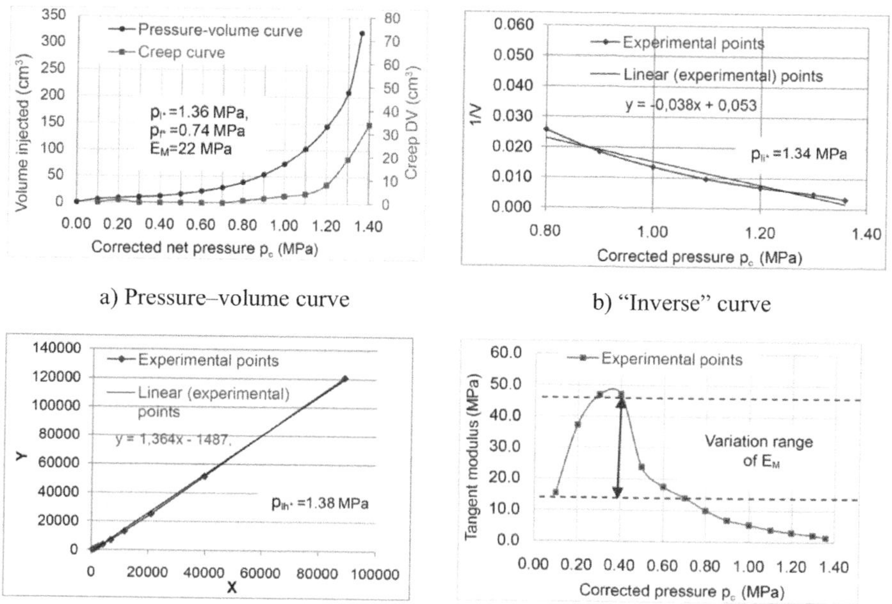

a) Pressure–volume curve

b) "Inverse" curve

c) Hyperbolic extrapolation

d) Curves of tangent modules

Figure 3.43. *Ways to interpret a "good" Ménard pressuremeter test. For a color version of this figure, see www.iste.co.uk/dhouib/geotechnics1.zip*

In Figure 3.43, it is clear that:

– The interpretation of the pressuremeter test, which is considered to be good and brought about by means of the three methods, leads to a perfect consistency of the net limit pressure p_l* with direct determination (Figure 3.43(a)), assessed on the last points of the curve ($p_c > p_2 = p_f$*) by the "inverse curve method": $1/V = Ap + B$ (Figure 3.43(b)) and obtained by the "hyperbolic extrapolation method" by the transformation (X,Y) of Figure 3.43(c). The test is very good, and the traditional ratio $[|p_{li}^* - p_{lh}^*|/p_{lh}^*]$ is 0.03 (<<0.2).

– This test shows that the "overall" pressuremeter modulus E_M, estimated at 22 MPa (Figure 3.43(a)), falls well within the general "bell-shaped" pattern of the tangent modulus curve in Figure 3.43(d), where these modules range from 12 to 48 MPa. Thus, it can be seen that at the beginning of the test, the pressuremeter deformation modulus increases rapidly with the pressure injected into the probe, as the probe compresses the boring walls. This part of the tangent modulus curve can provide important information about soil response at the beginning of the test. During the first stages of the test, the curve stabilizes and passes through a peak before starting to decrease with pressure. In the last stages of the test, the tangent modulus curve decreases rapidly and tends toward an asymptotic direction (with a value that is close to zero), reflecting the total failure of the soil.

3.7.3.2. *"Remolded" test*

A remolded test (Figure 3.40 and Figure 3.44(a)) corresponds to a "remolded" boring, where the boring walls are not vertical and straight, and the tangent modulus curve is very irregular and shows a succession of peaks, as shown in Figure 3.44(b).

A remolded boring may be due to several factors:

– the nature of the soil itself, which is inherently decompressed;

– the presence of water leading to the collapse of the boring walls, especially in the case of granular soils such as underwater sands and gravels;

– a lack of thick mud to keep the boring walls stable after the drilling material has been removed;

– the lack of respect for drilling passes and the associated tool (Table 3.7);

– master drillers use a simple phrase to characterize the remolding of a boring that is not straight and vertical enough: "*My borehole looks like a potato*".

From Figure 3.44(a), showing the pressure–volume curve of the "remolded" test, an "overall" E_M modulus of 4.2 MPa is derived. Analysis of the associated tangent modules in Figure 3.44(b) makes it possible to deduce step modules ranging from 4.5 to 5.5 MPa, close to the overall value of 4.2 MPa. Note that in both assessment procedures of the pressuremeter modulus E_M, it is low and reflects a clear preliminary remolding of the boring walls.

Finally, it is useful to draw the attention of geotechnicians to the quality of the boring, on which the quality of the test naturally depends. This is why it is necessary to have the "pressure–volume" curves attached to the tests, so that the construction engineer can analyze them in detail and choose the "right" parameters before proceeding with the calculation, designing and justification of the foundations of the geotechnical works.

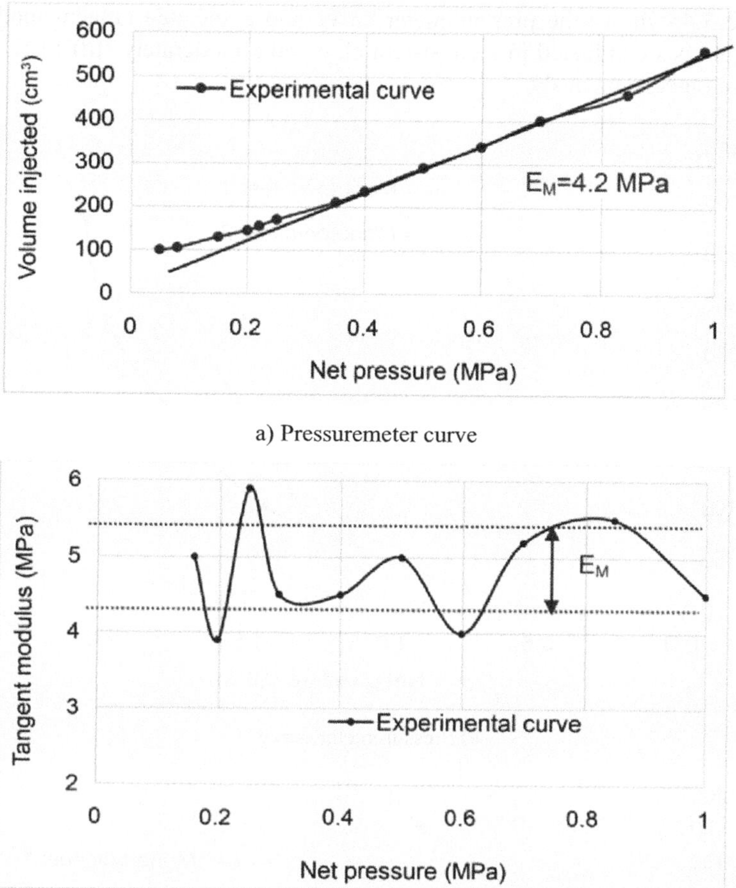

a) Pressuremeter curve

b) Curve of tangent modules

Figure 3.44. *"Remolded" pressuremeter test (Dhouib* et al. *2002). For a color version of this figure, see www.iste.co.uk/dhouib/geotechnics1.zip*

3.7.3.3. *Test in swelling soil*

When the soil swells, the boring walls shrink and the test probe is forced into the soil to perform the test (to learn more about the swelling phenomena, see section 2.4 of Volume 2, Chapter 2).

The phenomena of instantaneous swelling are observed in several soils of swelling clays and marls in France and worldwide.

Figure 3.45 shows the pressuremeter curve and associated tangent modules of two typical tests conducted in a consistent clay and a moderately stiff marl, both of which swell instantaneously.

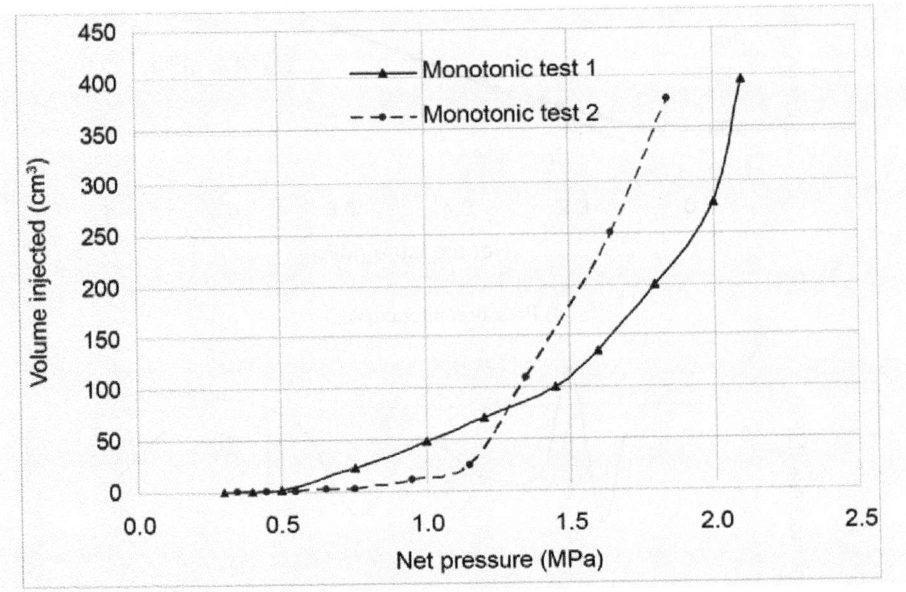

a) Pressuremeter curve

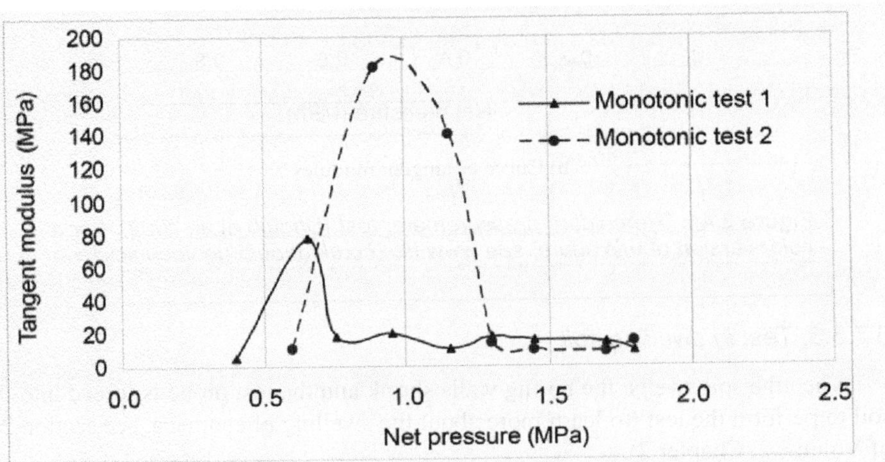

b) Curve of tangent modules

Figure 3.45. *Pressuremeter test in swelling soil. For a color version of this figure, see www.iste.co.uk/dhouib/geotechnics1.zip*

Analysis of Figure 3.45 highlights the following:

– the force of the probe in a swelling soil results in a horizontal pressure–volume curve that only "takes off" from a high pressure level;

– the experimental curve of the tangent modules is characterized by a strong peak, giving extremely high modules with respect to the limit pressures reached;

– in the curve in Figure 3.45(a), the soil failure is sudden and resembles "perfect plastic" behavior (Figure 2.14(d) in Chapter 2).

This example calls for vigilance from geotechnical engineers when interpreting and selecting pressuremeter modules in tests performed in swelling soils. In several projects, we have observed pressuremeter modulus ratios at limit pressures of 20–200, when the limit pressures did not exceed 0.2–0.4 MPa, and the overlying railway embankment suffered significant disruptions (failure before the expected height and very expensive repair with reinforcement of the underlying soil).

3.7.3.4. *Modulus corrections in a resistant environment*

In a resistant environment (limestone, marl and very stiff clay, dense sand and gravel, sound chalk, etc.), it is essential to correct the modules from the expansion coefficient of the equipment, deduced from the expansion curve of the pressuremeter probe in a metal calibration tube of the probe (Figure 3.23).

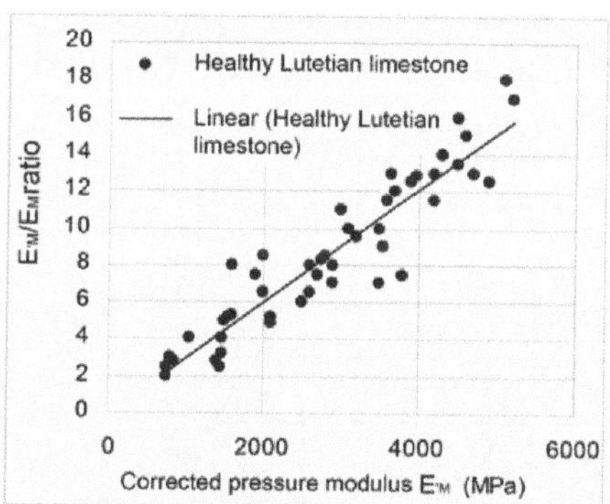

a) Compact and sound limestone

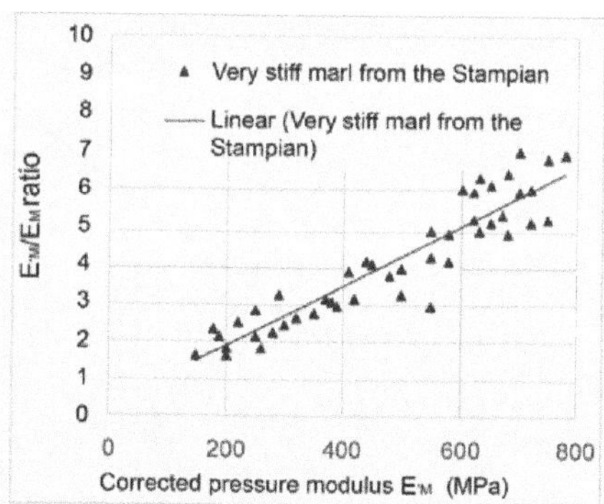

b) Very stiff marl

Figure 3.46. *Correction of pressure modules in resistant soil. For a color version of this figure, see www.iste.co.uk/dhouib/geotechnics1.zip*

Figure 3.46 shows a plot of several dozen measurements of pressuremeter modules in very compact limestone and marl. The evolution of the ratios of the pressuremeter modules E'_M, corrected by considering the expansion coefficient (a)

of the pressure probe (Figure 3.23), to the uncorrected pressuremeter modules E_M, thus shows that the E'_M modules are 2–20 times greater than E_M (Figure 3.46). This correction is necessary in order to be able to justify important geotechnical projects (the example of the tower studied in section 3.1 is a perfect illustration; refer to the data and comments in Figure 3.3 of Volume 3, Chapter 3).

The various examples covered and commented on here show the drilling impact on the quality of the pressuremeter tests, and show that the construction engineer must verify these tests for themselves, instead of using the curves "offered" for free by automatic assessment software that is frequently used.

3.7.3.5. *Soil characterization with the Louis Ménard pressuremeter*

In addition to the various soil and rock classification diagrams, Baud (2005) establishes the diagram in Figure 3.47, which shows how it is possible to classify soils and characterize them by the two main parameters measured with the Louis Ménard pressuremeter: the pressuremeter modulus E_M and limit pressure p_l, with indications on the values of the Ménard rheological coefficient α.

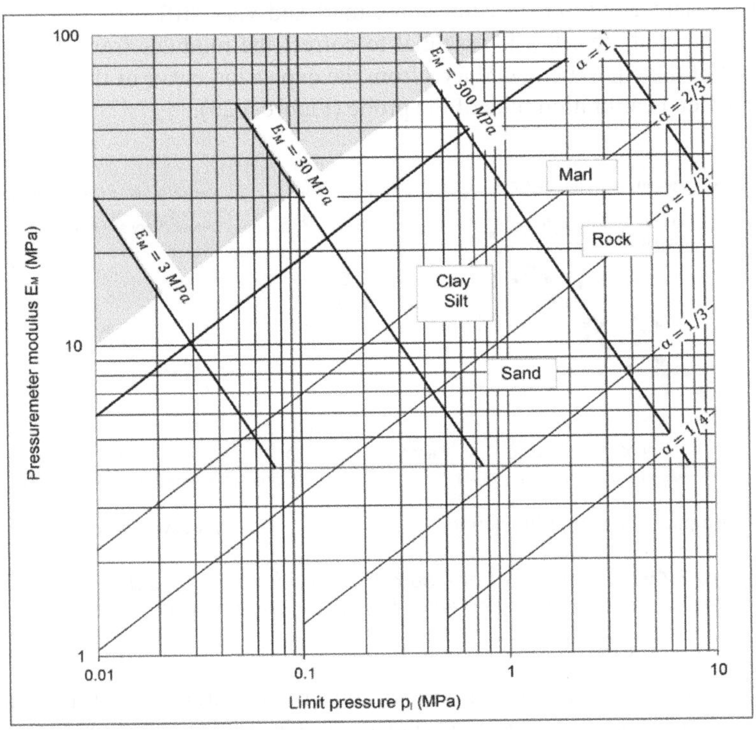

Figure 3.47. *Soil characterization by pressuremeter parameters (Baud diagram)*

The advantage of the Baud diagram (2005) is its simplicity. Its disadvantage lies in its ambiguous distinction of soils in terms of compactness: for sands, one can distinguish loose sands and medium to dense sands; for clays and silts, there are soft silts and clays and stiff clays; with regard to marls, one can find marls that are much less compact than rocks. We will return to the Baud diagram later to integrate experimental points of pressuremeter measurements.

3.7.4. *Practical example of cone penetration tests*

Figure 3.48 shows a cone penetration profile with a normalized electric point (Standard NF P94-113), obtained in the body of an earth dike on sandy-gravelly alluvium, supported by a stiff chalk substratum.

Reading this profile, it should be noted that:

– The point resistances (q_c) are very low in the dike body, ranging from 0 to 3.3 m, around 1.0 MPa, and clearly reflect the very soft character of the sandy silts of which it is composed, with friction ratios (F_R, equation [3.27], noted as R_f on the graph in Figure 3.48) of 1.5%–4%.

– In the sandy-granular alluvium, the point resistances are in a state of flux, varying from 10 to 17.5 MPa, before reaching significant peaks (40 MPa) at the base of the layer, in which the friction ratio F_R remains near-constant, around 1%, up to 8 m in depth.

– The point resistance saturates in the chalk at a height of 50 MPa, corresponding to the rejection on this layer at around 9.6 m in depth.

– The rejections to cone penetration that occur on a stiff substratum do not exclude the possibility that this substratum is the base of strong dissolutions and karstic pockets, which can call into question the sizing of foundations in the substratum. We will give examples at the end of Volume 3, Chapter 3, which are dedicated to foundations.

The presentation of cone penetration results can easily link the profile of point resistances (q_c) with Robertson's diagram (Robertson 2009), which is schematically reproduced in Figure 3.13(b). This very important graph distinguishes the various loose soils as a function of point resistance q_c and friction ratio F_R, which is the ratio of friction f_s to point resistance q_c (equation [3.27]).

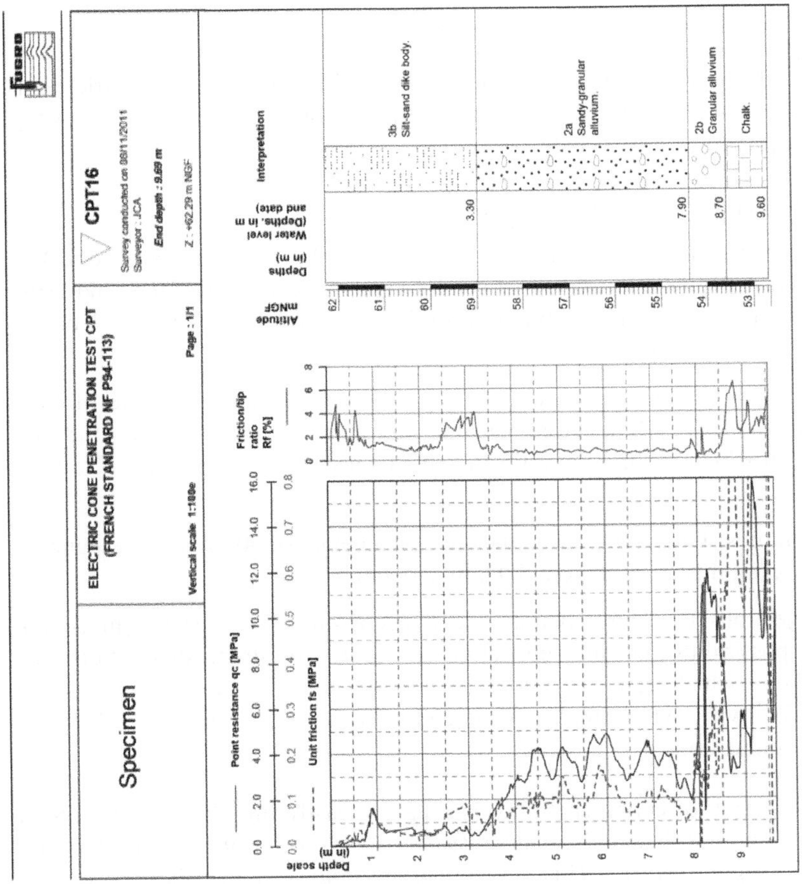

Figure 3.48. *Normalized cone penetration profile (FUGRO document). For a color version of this figure, see www.iste.co.uk/dhouib/geotechnics1.zip*

Figure 3.49 shows another typical profile of point resistance (q_c), measured with a cone penetrometer with a standard electric point (Standard NF P94-113), related to Robertson's diagram in Figure 3.13(b). The analysis of this profile in relation to Robertson's diagram leads to the following:

1) A high concentration of pairs (q_c, F_R) corresponds to clayey silts to silty clays (formation 4 of Robertson's diagram, which designates silts in a general way).

2) A significant fraction is linked to clays, designated on Robertson's diagram as "clay to silty clay" (layer 3).

3) The rest of the formations in Robertson's diagram in Figure 3.49 are dedicated to the very stiff intermediate fine soils, represented by formation 8, and to cemented or dilatant sands (dense sands, Figure 2.17(b)), represented by formation 9.

The two cone penetration profiles in Figure 3.48 and Figure 3.49 illustrate the value of the cone penetrometer, making it possible to:

– carry out continuous measurements without risks of remolding in loose soils of low compactness that are susceptible to remolding;

– establish correlations to distinguish the layers that are "blindly" crossed by penetration operations.

However, like other on-site testing techniques, this method of measuring the resistance of soil in situ suffers from several limitations in its field of application; a field of application that is summarized in the table in Figure 3.58 below.

Another representation of the parameters that is deduced from the cone penetrometer test, equipped with a skirted cone, is the Schmertmann (1978) abacus chart, which is reproduced in the work of Philipponnat and Hubert (2008).

This diagram, represented in Figure 3.50, illustrates the various loose soils, which is the cone penetrometer's purpose, and the corresponding variation ranges of the point resistance (q_c) and the friction ratio (F_R, also noted R_f). It is very simple to use.

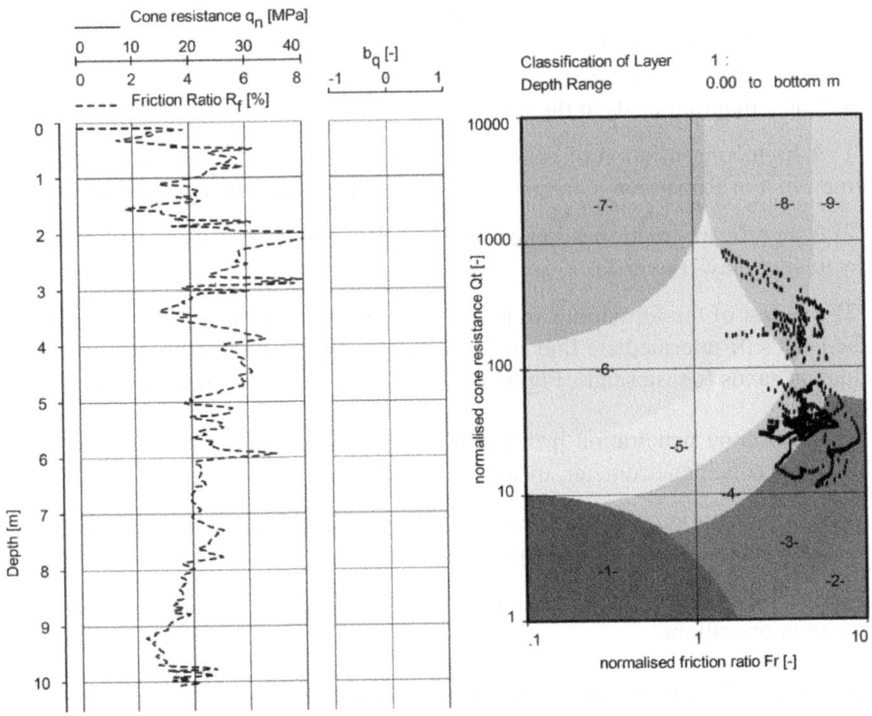

Legend of Robertson's diagram (2009):

1) Fine soils: sensitive clays or silts;

2) Organic soils and peat;

3) Clays: clay to silty clay;

4) Silts: clayey silts to silty clays;

5) Sands: silty sands to sandy silts;

6) Sands: clean sands to silty sands;

7) Sands to granular sands;

8) Very stiff intermediate fine soils;

9) Cemented or dilatant sands (dense).

Figure 3.49. *Point resistance profile q_c associated with Robertson's diagram (FUGRO document). For a color version of this figure, see www.iste.co.uk/dhouib/geotechnics1.zip*

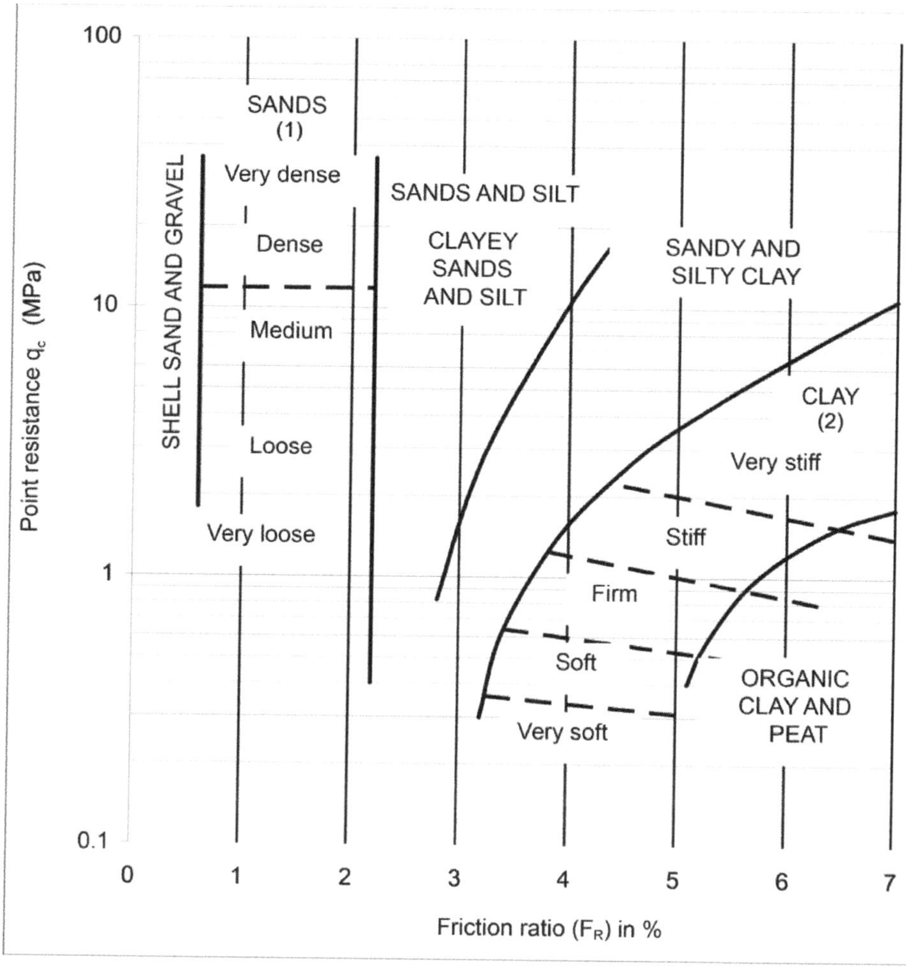

1) For compactness of sands, refer to Table 1.4 in Chapter 1.

2) An uncracked clay, without any organic material.

Figure 3.50. *Schmertmann's abacus chart: skirt cone (1978)*

3.7.5. *Variations and correlations between parameters measured on site*

Figure 3.51 shows the results of five average cone penetration profiles (CPT) and five average SPT profiles in fine clay-silt alluvium.

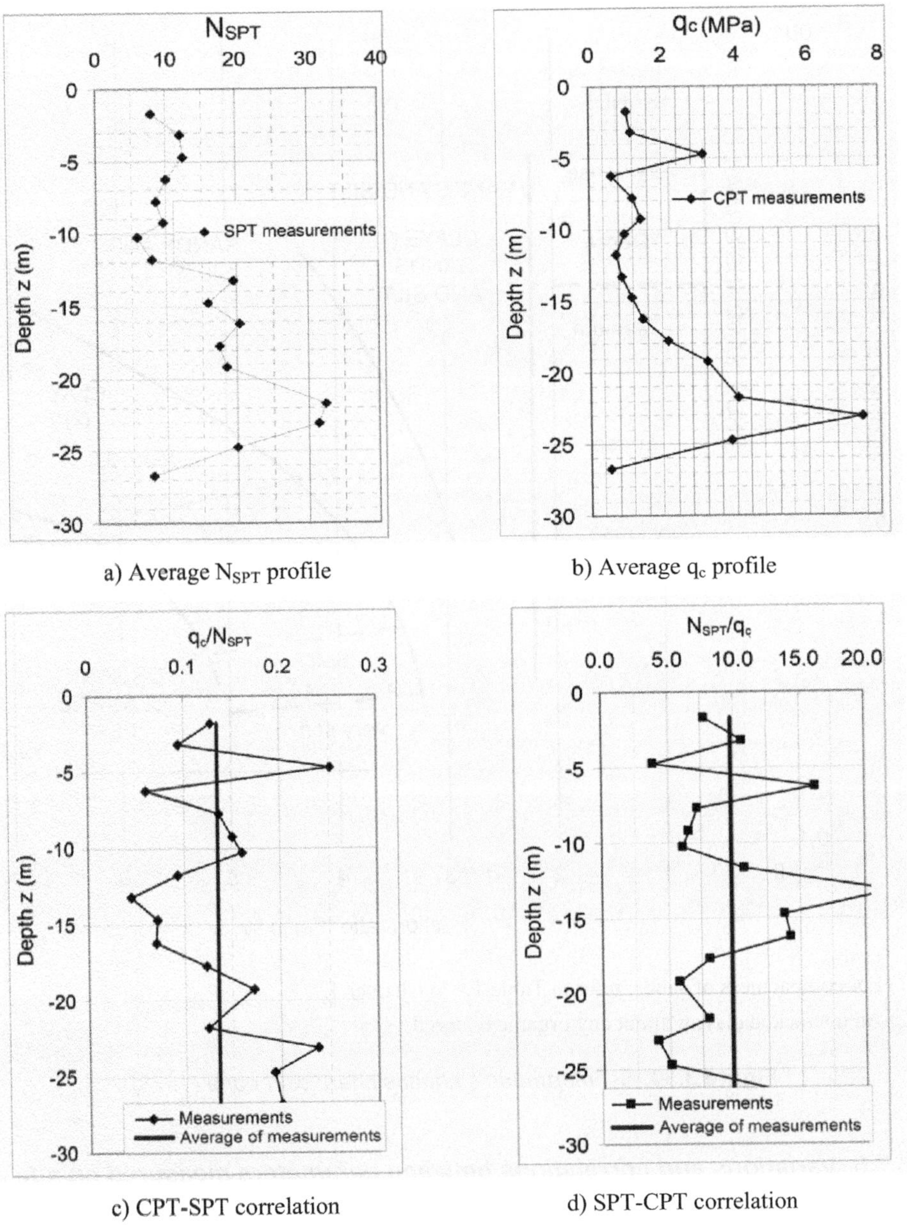

Figure 3.51. *On-site measurements and correlations between SPT and CPT. For a color version of this figure, see www.iste.co.uk/dhouib/geotechnics1.zip*

The graphs in Figure 3.51 call for the following comments:

1) The two average profiles of the number of blows N_{SPT}, measured with the dynamic penetrometer with the "Standard Penetration Test" core drill in Figure 3.51(a), and of the point resistance (q_c), measured with the cone penetrometer (CPT) in Figure 3.51(b), reflect very poor mechanical fine clayey-silt alluvium up to a depth of about 18/20 m, because the measurements show that:

– the number of blows N_{SPT} is around 10 on the first 12 m and 20 up to about 20 m in depth;

– the point resistance q_c varies, on average, between 0.8 and 1.2 MPa, with a "hard point" at 5 m of depth.

2) The two average profiles show a very weak mechanical passage at a depth of 27 m, corresponding to a peaty bed.

3) The correlations between the two parameters q_c (in MPa) and N_{SPT} are more or less in the order of magnitude of the correlations given by Cassan (1988) in Table 3.12, with the following average ratios:

$$\begin{cases} \dfrac{q_c}{N_{SPT}} \approx 0.14 \\ \text{and} \\ \dfrac{N_{SPT}}{q_c} = 9.0 \end{cases}$$

Figure 3.52 shows the results of a compilation of four pressuremeter profiles, compared to an average profile from six cone penetration tests, all carried out in silty-sandy alluvium.

From these pressuremeter and cone penetration profiles, the following can be observed:

– the two profiles of the pressuremeter modulus E_M (Figure 3.52(a)) and cone penetration resistance q_c (Figure 3.52(b)) correspond to decompressed alluvium to a depth of 12 m;

– the correlations between the pressure modules E_M and the point resistances q_c lead to an average $E_M/q_c{}^*$ ratio of 2.8, thus respecting the usual correlations in Table 3.12 for silts and sands ($E_M/q_c{}^*$ ranging from 1.5 to 3);

– the ratios of net point resistances q_c to net limiting pressures $p_l{}^*$ in these tests provide good correlation with a $q_c{}^*/p_l{}^*$ ratio of 2.7, a value compatible with the ratios of E_M modules to point resistances q_c and the usual correlations of Table 3.12.

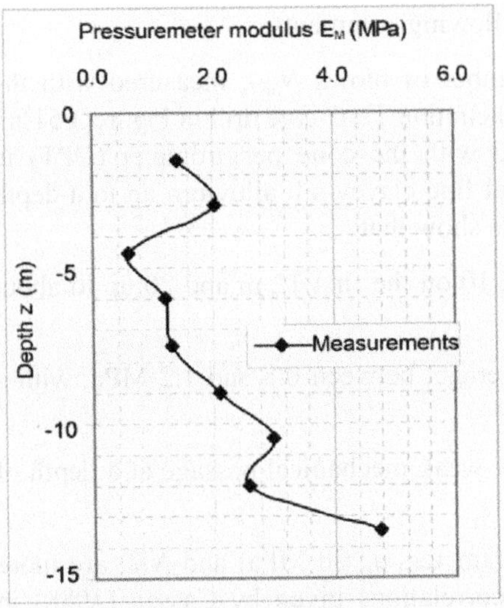

a) Average pressuremeter modulus E_M

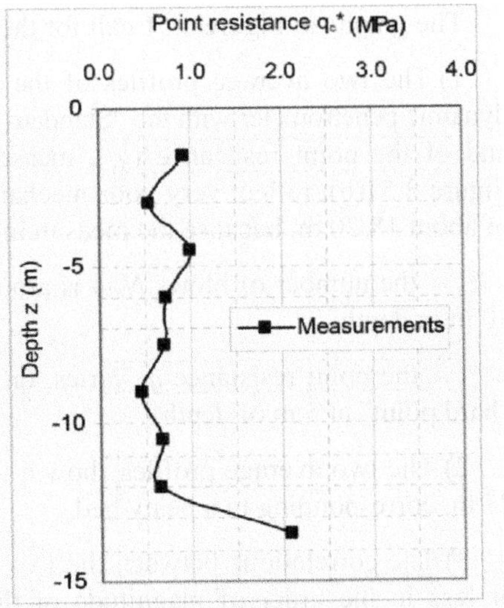

b) Point resistance q_c*

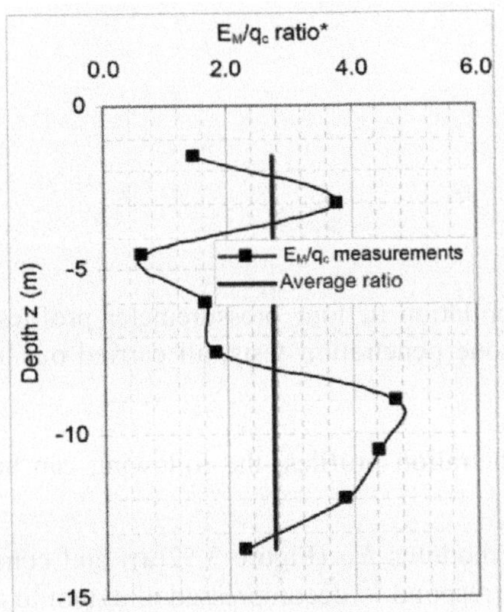

c) E_M-q_c correlation

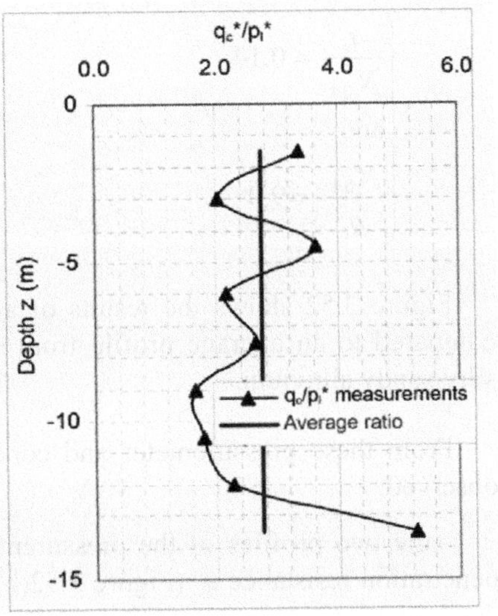

d) q_c*-p_l* correlation

Figure 3.52. *On-site measurements and correlations between pressuremeter (MPT) and cone penetrometer (CPT) tests. For a color version of this figure, see www.iste.co.uk/dhouib/geotechnics1.zip*

The set of tests shown in Figure 3.51 and Figure 3.52 show that:

– when pressuremeter tests are properly performed in a well-calibrated boring, with bare probes in soft soils that are sensitive to remolding, they still provide a good concordance with cone penetration tests;

– the correlations between the parameters presented here corroborate the usual results in Table 3.12.

3.7.6. *Photos of worksites and on-site survey equipment*

3.7.6.1. *Cone penetration material*

Figure 3.53 shows the 180 kN (18 ton) support truck and the electrical cone penetration probes.

a) 180 kN truck

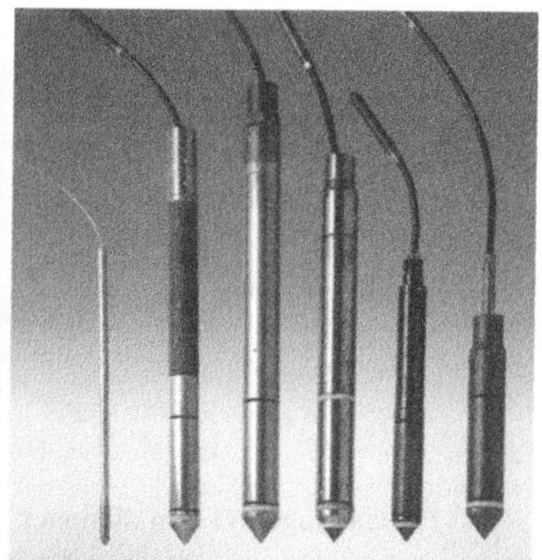

b) Test probes

Figure 3.53. *Truck and cone penetration probes (FUGRO documents). For a color version of this figure, see www.iste.co.uk/dhouib/geotechnics1.zip*

3.7.6.2. *Worksite and drilling and pressuremeter test accessories*

Figure 3.54 shows a crawler drill rig (Figure 3.54(a)) and drilling accessories:

– drill stems (Figure 3.54(b));

– helical auger with a single diameter of 63 mm and a cutter diameter ranging from 51 to 66 mm (Figure 3.54(c));

– typical core drill and a set of core crowns (Figure 3.54(d)).

a) Probe on crawlers

b) Drill stem for pre-drilling

Helical auger 63 mm diameter

Core drill

Cutter 64 mm diameter

c) Helicoidal auger and cutter

Crowns

d) Core drill and crowns

Figure 3.54. *Worksites and equipment for on-site probing and testing (SOL-ESSAIS). For a color version of this figure, see www.iste.co.uk/dhouib/geotechnics1.zip*

COMMENTS.–

1) Semi-destructive drilling with an auger (63 mm diameter), with or without drilling mud, is practiced in soft ground with little to medium resistance. This makes it possible to have a lithological section that is more or less accurate to within 0.5 m.

2) Cutting, either by air drilling or with water injections, with or without drilling mud, is essential in resistant soils. As this is destructive drilling, the lithological section, from drilling debris, is purely summary and may even be erroneous[4].

3) Drilling with a cutter is possible in all soils and rocks, with notable difficulties in flint and millstone blocks, as well as "conglomerates" or "limestone" (blocks cemented in sands and gravels (ancient alluvium)). The cutter can cut a steel shaft, but never a piece of wood (the cutter rebounds without being able to perforate it).

4) A precise lithological section can only be obtained from core drillings, which are sadly lacking in geotechnical investigations nowadays, because the project actors, project owners and company contractors are all looking for "false economies".

5) To illustrate, Figure 3.55(a) shows core samples of drillings that make it possible to accurately identify the nature and state of the soils used for the foundations of the project, and/or crossed by the earthworks (the case of buildings with basements, trench works, etc.).

Figure 3.55(b) illustrates a profile of a 10 m core drilling, which gives important information for the project actors: a precise section to anticipate and adapt the earthworks, the temporary casing height (unstable sandy gravel, as for the drilling of a pile for example), the collection of undisturbed samples (Intact Sample: IS) for laboratory tests, the recovery rate (% core drilling), the stabilized (piezometer installation) or non-stabilized (water in the boring at the end of core drilling) water level, the rock quality (Rock Quality Designation: RQD) in rocky environments, etc.

4 It is possible to confuse sandy limestone that has been crushed by the cutter with sand.

23.0

25.00

a) Core samples in sand and gravel (recovery rate = 50%)

31.50

33.00

b) Stiff clay core (recovery rate = 100%)

c) Limestone cores with varying degrees of fracturing (RQD* = 10%–65%)

(* % of cores longer than 0.1 m in a meter of cores)

Figure 3.55(a). *Cores from core drilling (FONDASOL). For a color version of this figure, see www.iste.co.uk/dhouib/geotechnics1.zip*

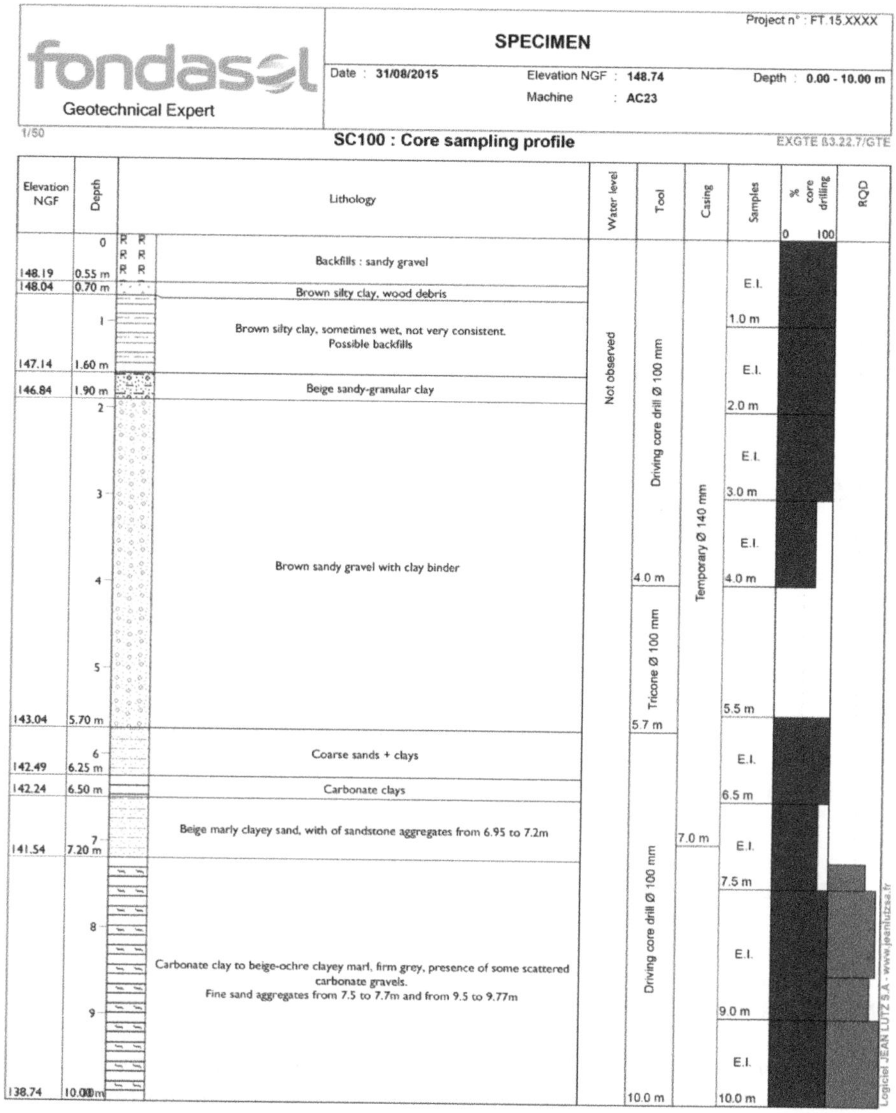

Figure 3.55(b). *Lithological section of a core sample (from FONDASOL). For a color version of this figure, see www.iste.co.uk/dhouib/geotechnics1.zip*

Figure 3.56 shows the accessories for the Louis Ménard pressuremeter test: a bare probe (Figure 3.56(a)), a lantern tube (Figure 3.56(b)), an air cylinder for pressure (Figure 3.56(c)) and the Louis Ménard pressuremeter (Figure 3.56(d)).

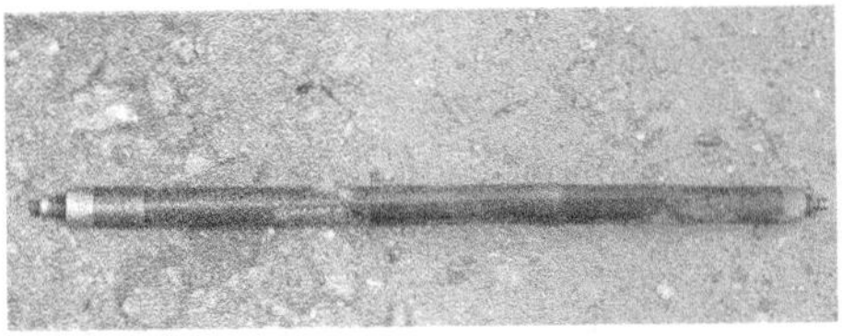

a) Bare probe

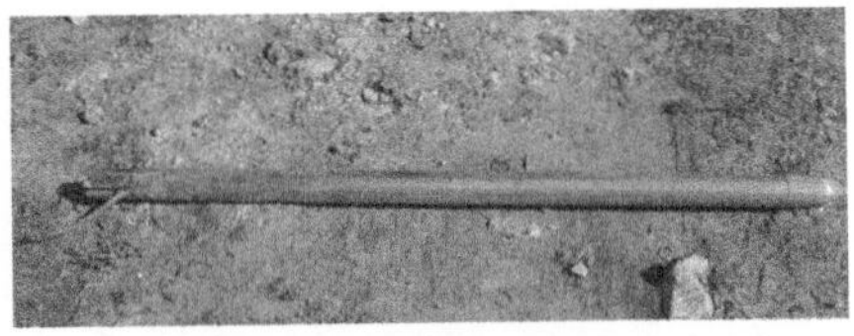

b) Lantern tube (Slotted tube)

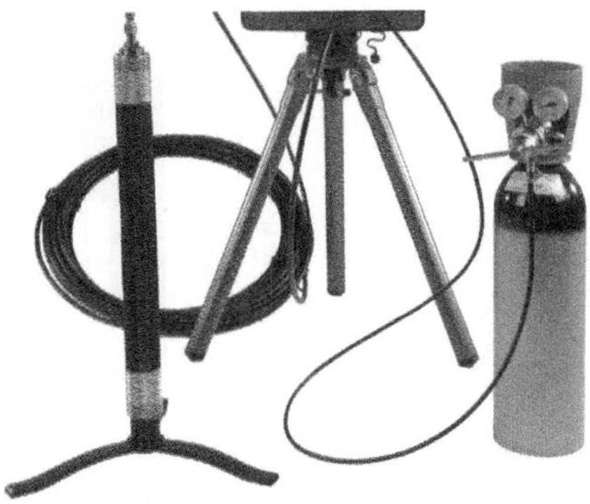

c) Air cylinder (nitrogen) and Geomatech probe

d) Louis Ménard pressuremeter (Geomatech device)

Figure 3.56. *Louis Ménard pressuremeter test accessories. For a color version of this figure, see www.iste.co.uk/dhouib/geotechnics1.zip*

COMMENT ON FIGURE 3.56.–

1) The bare probe, which is 40 mm in diameter and 400 mm long, has a low inertia (0.10–0.15 MPa, Figure 3.57(a), which shows an inertia test) and is well adapted to homogeneous soils without stones, as stones can cause it to shatter.

2) The lantern tube (also called a "slotted tube"), with a higher inertia (0.30–0.45 MPa, Figure 3.57(a)), is used to protect the bare probe against shattering by stones.

3) The probe can be reinforced with metal so that it can be inflated to pressures higher than the regulatory value of 5.0 MPa, imposed by Standard NF P94-110 in resistant environments. In this case, it is a "rock-probe". The use of the rock probe is recommended, sometimes even necessary, in projects with high loads based on piles (see section 1.4 of Volume 3, Chapter 1). As shown in the applications in section 2.2 of Volume 3, Chapter 2 on piles, substantial gains can be made on pile lengths and diameters by optimizing the point force.

4) In resistant soils, probe calibration is mandatory in order to determine the probe expansion coefficient (noted as a, see Figure 3.23). It must be followed by corrections to the modules (Figure 3.46, Dhouib et al. *2002). This procedure, which is simple and does not require significant time or expense, is useful for sizing and justifying geotechnical structures under heavy loads (High-rise buildings (HRB), bridges, dolphins in ports, etc.), as shown in the example presented in the feedback section in section 3.1 of Volume 3, Chapter 3. Figure 3.57(b) shows the calibration test of a bare probe in a metal tube.*

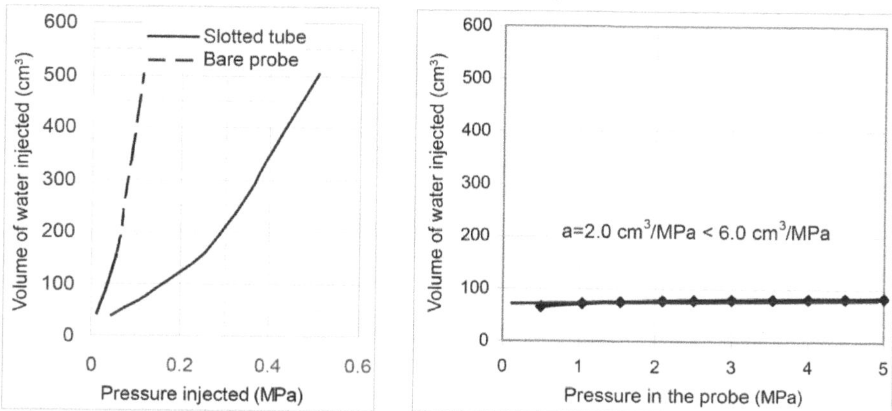

a) Inertia tests b) Probe calibration in a metal tube

Figure 3.57. *Inertia and probe calibration tests. For a color version of this figure, see www.iste.co.uk/dhouib/geotechnics1.zip*

New developments, in terms of the Louis Ménard pressuremeter tests, have recently been implemented in order to be able to perform very high pressure tests, as shown in Figure 3.58. This figure illustrates Geomatech's "HyperPAC" pressuremeter and very high pressure probe (Figure 3.58(a)) and pressure–volume curves, where pressures have been pushed into the probe up to 25 MPa (a test with poorly calibrated boring) and 19 MPa (a test with a well-calibrated borehole) in Figure 3.58(b).

a) "HyperPAC" pressuremeter and very high pressure probe from Geomatech

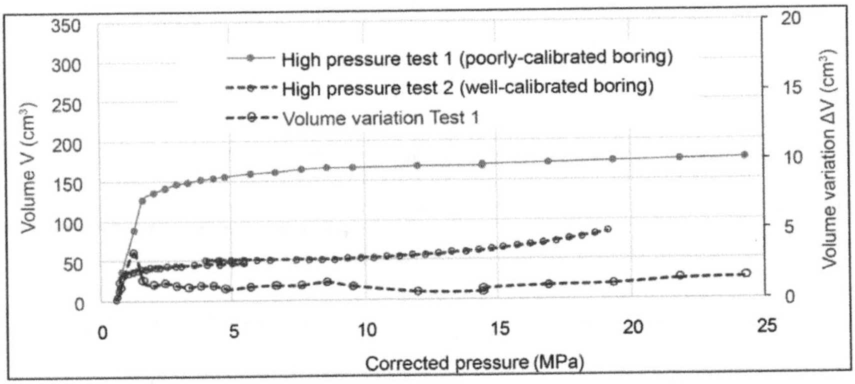

b) Very high pressure tests

Figure 3.58. *Very high pressure tests (Geomatech documents). For a color version of this figure, see www.iste.co.uk/dhouib/geotechnics1.zip*

3.8. Scope of application of on-site testing

In order to simplify this section relating to the scope and field of application of the on-site tests, we present the various investigation tools, the variation ranges of the measured parameters and their field of application to the types of construction works in Figure 3.59 (it is worth also referring to the graph in Figure 3.9 of Volume 3, Chapter 3).

Soil / Tool/Parameter	Soft	Clay and silt			Sands and gravels			Chalk		Marl, marl-limestone and rock
		Soft	Firm to stiff	Stiff clay	Loose	Dense	Very dense	Weathered	Healthy	
Vane test	↔	↔	Pre-drilling	X	X	X	X	X	X	X
c_u (kPa)	10 to 50	20 to 80	80 - 120	X	X	X	X	X	X	X
CPT	↔	↔	↔	↔	↔	↔	X (R)	↔	X (R)	X (R)
q_c (MPa)	<1.0	<1.0	1 to 3	3 to >5	1.5 to 5	10 to 20	>20	2 to 10	X	X
F_R (%)	3 to 8	3 to 6	3 to 6	X	1 to 2	1 to 2	1 to 2	2 to 6	X	X
SPT	↔ (dashed)	↔ (dashed)	↔ (dashed)	X	↔	↔	↔	X	X	X
N_{SPT}	2 to 5	2 to 5	4 to 10	X	1 to 8	20 to 40	>40	X	X	X
MPT	(1) ↔ (dashed)	(1) ↔ (dashed)	↔	↔	↔	↔	↔	↔	(2) ↔	(2) ↔
E_M (MPa)	<5	<5	5 to 15	>20	<5	6 to 15	>15	3 to 10	>30	>50
p_l (MPa)	<0.5	<0.5	0.5 to 1.5	>2.0	<0.5	1 to 2	>2	0.3 to 1.0	>3	>4
Type of work	Earth embankment		Pavilions R+1	R+2 to R+10	(3)	R+2 to R+10, Medium-sized bridges		Small works (4)	HRB, big bridges, dolphins	
Legend	X: Unsuitable tool/non-measurable parameter, (R): Rejection on the hard material, HRB: High-rise building (large towers), (1): Mud drilling and bare probe test, (2): Rock probe and module correction, (3): Improvement (see Volume 4, Chapter 2), (4): Agricultural aggregates, hydraulic nozzle, low embankment, etc.									

Figure 3.59. *Scope of application of on-site testing*

3.9. Test results in typical soils

As a conclusion and in order to enrich this chapter, we will give the results of tests in various soils.

3.9.1. *Pressuremeter tests (Baud diagram)*

Figure 3.60 represents the Baud diagram (2005) from Figure 3.47, filled with test results in clays (Sannoisian green clay), compact sands and gravels, and very compact molassic marls where the limit pressures, stopped at the 5 MPa standard, are given as a default.

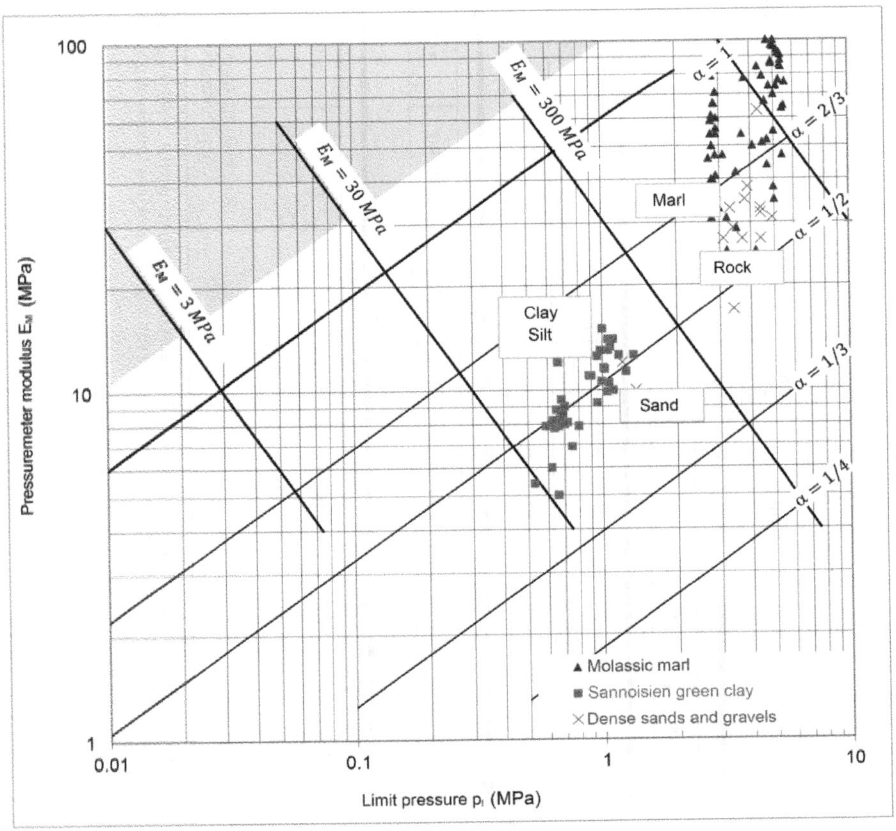

Figure 3.60. *Pressuremeter measurements in various soils linked to the Baud diagram (2005). For a color version of this figure, see www.iste.co.uk/dhouib/geotechnics1.zip*

3.9.2. *Pressuremeter tests in backfills treated with lime*

Figure 3.61 shows plotted values of pressuremeter measurements in sands, classified as B_5 m according to the tables of Appendix 1.1 of Chapter 1, and in crushed marls (classified as R_{22}, according to Appendix 1.1 of Chapter 1), both treated with 2% lime to be used as technical railway embankments.

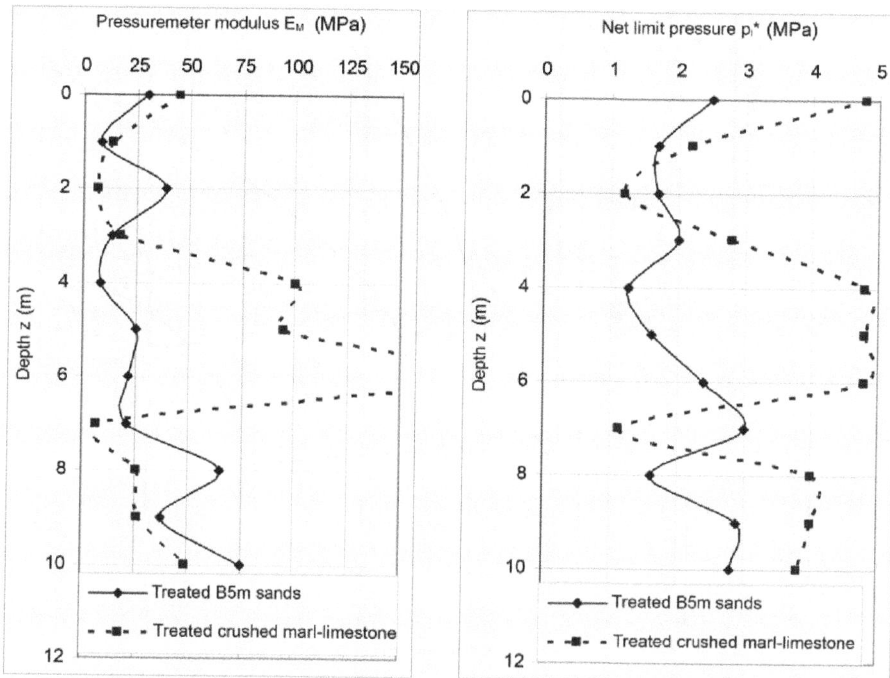

a) Pressuremeter modules E_M b) Net limit pressures p_l*

Figure 3.61. *Pressuremeter measurements in technical backfills in soils treated with 2% lime. For a color version of this figure, see www.iste.co.uk/dhouib/geotechnics1.zip*

3.9.3. *Cone penetration tests*

Figure 3.62 shows the results of cone penetration tests in various soils in the Schmertmann diagram (1978):

- loose sands with mud aggregates;
- compact alluvial sands and gravels;
- lagoon clays with peat aggregates and beds.

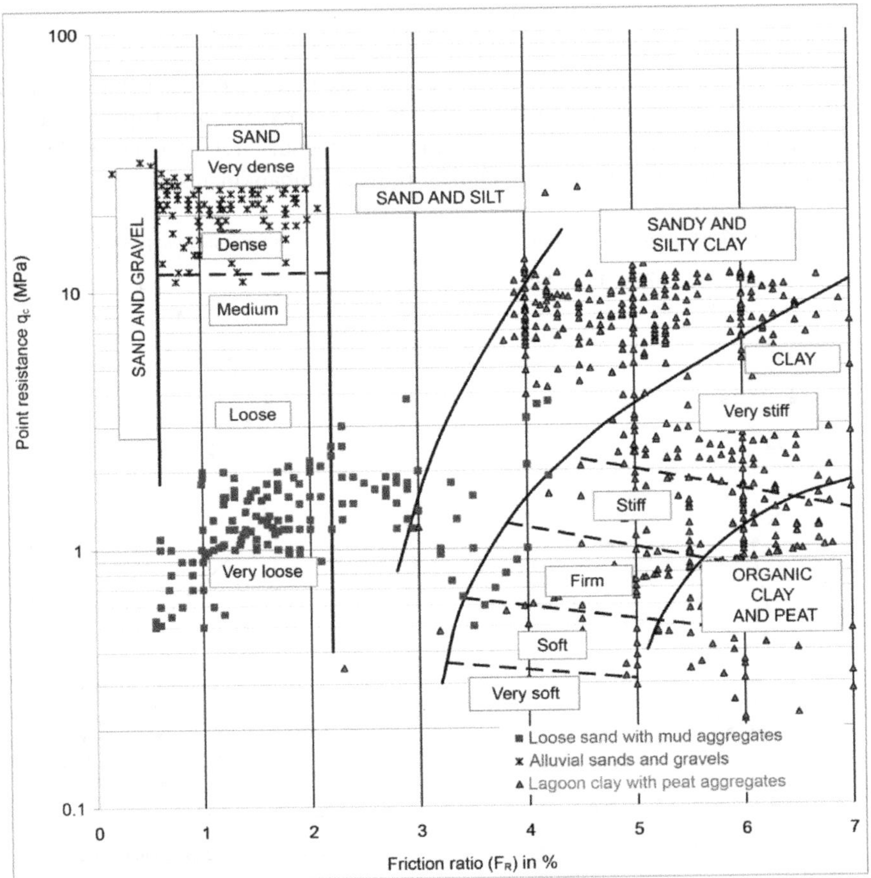

Figure 3.62. *Cone penetration measurements integrated in Schmertmann's diagram (1978) for loose soil. For a color version of this figure, see www.iste.co.uk/dhouib/geotechnics1.zip*

3.9.4. *Common diagram for geotechnical investigations*

Figure 3.63 summarizes the various mechanical data of Figure 3.61 and Figure 3.62, all compiled in a common diagram to characterize the various soils and classify them according to the rules of the table in Figure 3.59, by means of the pressuremeter modules (E_M) and limit pressures (p_l) for very compact molassic marls, on the one hand, and, on the other, point resistances (q_c) and friction ratio (F_R), number of blows (N_{SPT}) and undrained cohesions (c_u) for loose soils.

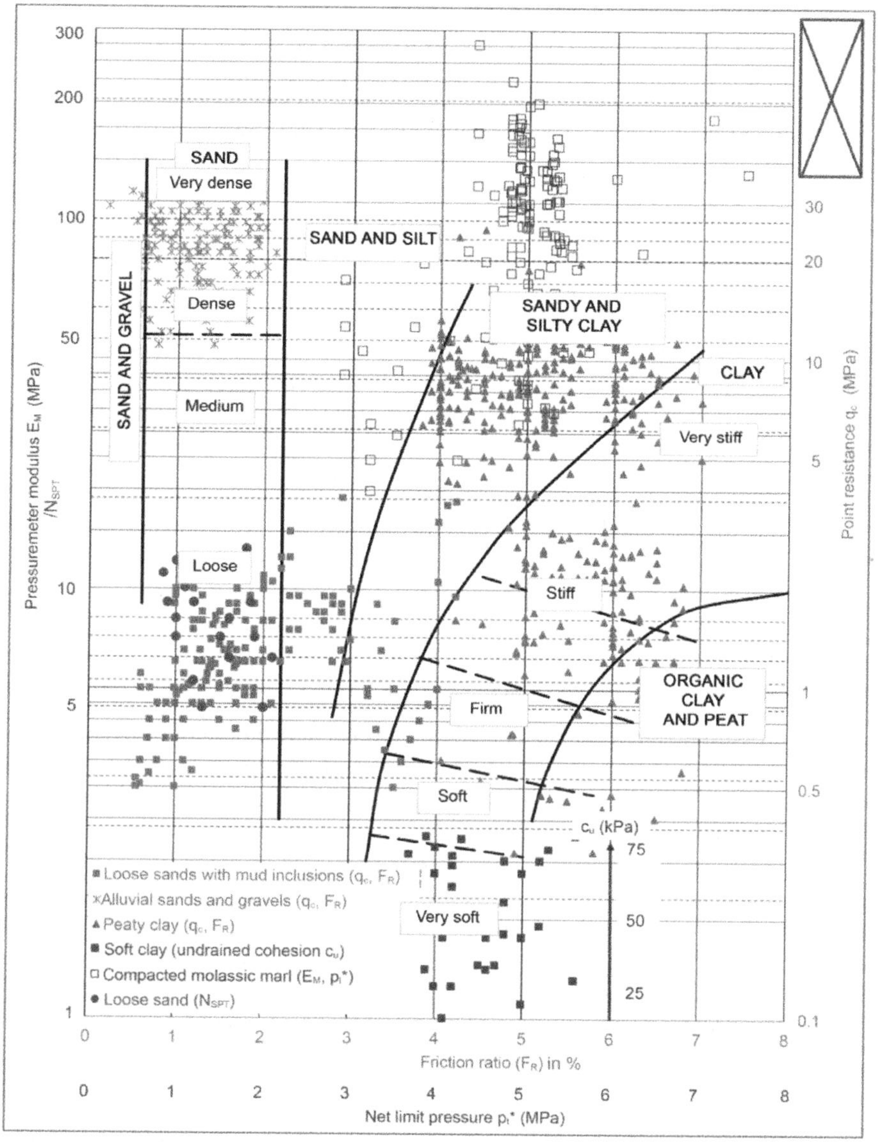

Figure 3.63. *Mechanical characterization of loose soils and rock-like soils by various survey techniques. For a color version of this figure, see www.iste.co.uk/dhouib/geotechnics1.zip*

The diagram in Figure 3.63, linked to the table in Figure 3.59 regarding the scope of the application of on-site testing, dictates the following:

– The field vane test is designed to measure undrained cohesion every 0.5–1.0 m in soft fine soils such as clays, silts and peat (Table 3.1). Consistent with firm clays, pre-drilling with an auger and mud injection is often necessary.

– The cone penetrometer is well adapted to soft soils: soft soil, consistent with firm clay and silt, stiff clay, loose, medium and dense sands, and soft to weathered chalk (Table 3.4). As a continuous measurement, the cone penetrometer offers a high level of accuracy when characterizing soft aggregates in heterogeneous loose soils. Related to core drillings, it gives a reliable image of the geology and the compactness of the soils crossed, and distinguishes the soft and/or organic soil aggregates and beds. The various abacus charts and correlation graphs between parameters (Sanglerat 1971; Larél 1974; Schmertmann 1978; Robertson 2009), derived from this tool (Figure 3.61), support its use in characterizing loose soils. However, this tool is sometimes rejected in compact soils, which limits its use and explains its greater development in Nordic countries (Holland in particular) than in France.

– The use of the standard penetrometer, known as the "Standard Penetration Test" (SPT), is limited to English-speaking countries and used very little in France. Its users want to generalize it to all loose soils, whereas, in reality, it can only be used for granular soils such as sands (Figures 3.17 and 3.63). In addition to the rejection of hard soils, the core drill "sticks" to the clays and influences the measurement (number of blows N_{SPT}).

– The pressuremeter, Louis Ménard's ingenious and productive invention from the 1960s, which was inspired by the expansion of a cylindrical cavity (Figures 3.19, 3.25 and 3.26), is the basic tool for studying, designing and justifying geotechnical projects in France, for obvious reasons: the Louis Ménard pressuremeter makes it possible to measure soil resistance (limit pressure p_l) and its creep (p_f), leading to the determination of the load-bearing capacity of the soil under the work, and also to measure the pressuremeter deformation modulus (E_M), making it possible to check the settlements of the soil under the work. Without coming across any rejection when drilling, it requires well-calibrated and very careful drilling and good understanding, as well as an automatic and continuous recording of the drilling parameters, in order to make the measurements more "consistent".

The diagram in Figure 3.63, associated with the table in Figure 3.59 and combining the Baud (Figure 3.47) and Schmertmann (Figure 3.50) abacus charts, can be easily generalized to loose soils and rock-like soils. Figure 3.64 shows the classification chart of soils according to their nature and the parameters measured by the various tools. The Louis Ménard pressuremeter (MPT): pressuremeter modulus (E_M), limit pressure (p_l) and α Ménard for all soils and rock-like soils. Cone penetration tests (CPT): point resistance (q_c) and friction ratio (F_R or R_f) for loose soils. Standard penetration tests (SPT): number of blows (N_{SPT}) for loose granular soils (sands and gravels) and relative density (D_r) for sands. Field vane test: undrained cohesion (c_u) of soft fine soils.

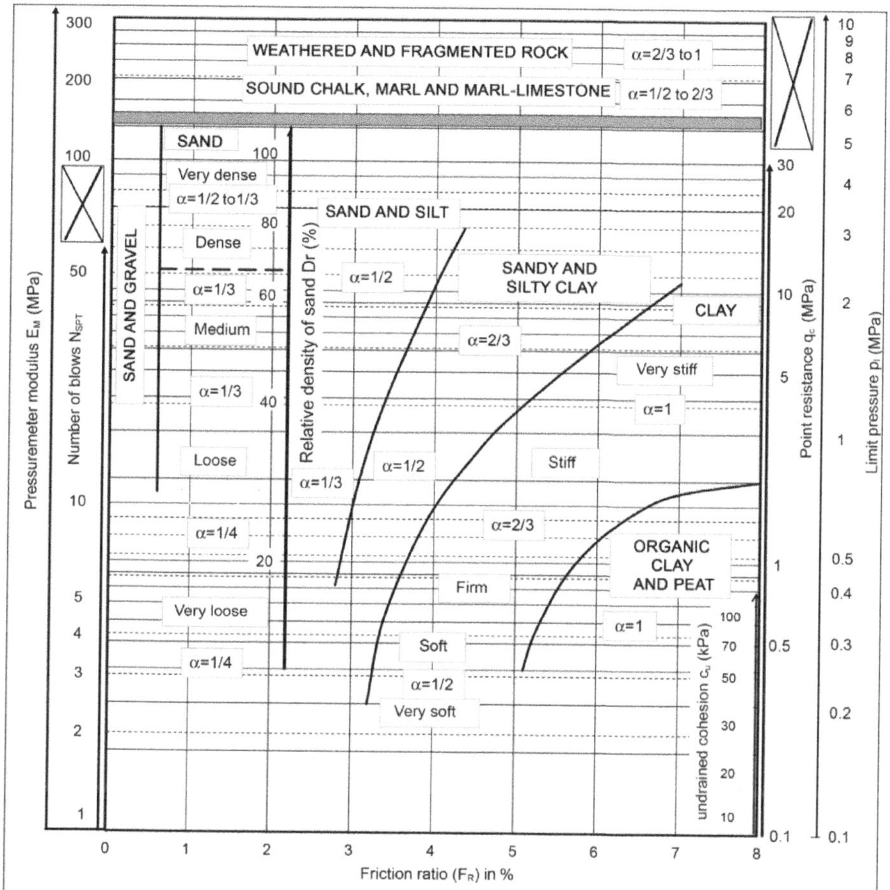

Tool limit (Rejection) Limit of loose and resistant soils and rock

E_M, p_l: Modulus and limit pressure measured with the Louis Ménard pressuremeter (MPT), and α Ménard:
q_c, F_R: Point resistance measured with cone penetrometer (CPT) and friction ratio (F_R)
N_{SPT}: Number of blows with the standard penetration test standard penetration test (SPT)

Figure 3.64. *Mechanical characterization and classification of loose soils and rock-like soils by various survey tools*

3.10. Wise conclusion

We close this chapter, which is of paramount importance for construction projects, ranging from small pavilions to high-rise buildings and giant viaducts, with a few "wise words":

To build projects that leave a mark, you need to:

1) perform good tests in full and unconditional compliance with the rules and standards in effect;

2) analyze and verify the "automated" curves and the measured parameters in detail;

3) select the most representative values possible: minimum and maximum values, averages, standard deviations, invalid (remolded) and singular (hard points) tests, etc.;

4) choose sets of parameters, sweeping all measurements in order to build several typical sections (several "field models").

The rest of this "wise conclusion" is at the end of Volume 3, Chapter 3, which is on geotechnical project foundations.

3.11. Appendix 3.1: Correlations between net creep pressures and net limit pressures

The measurement of the limit pressure in the Louis Ménard pressuremeter test is always accompanied by the choice of the creep pressure, which can give useful information on the correct execution of the test in the field. For this reason, the analysis of the two parameters is of paramount importance in the interpretation and selection of the pressuremeter characteristics used in the sizing, justification and validation of foundations for construction projects.

Figure 3.65 compiles the ratios of net limit pressures (p_l*) to net creep pressures (p_f*) from at least 800 standard pressuremeter tests in sands, sands and gravels, and other relations collected from the literature (Cassan 1980) for sands, clays and silts.

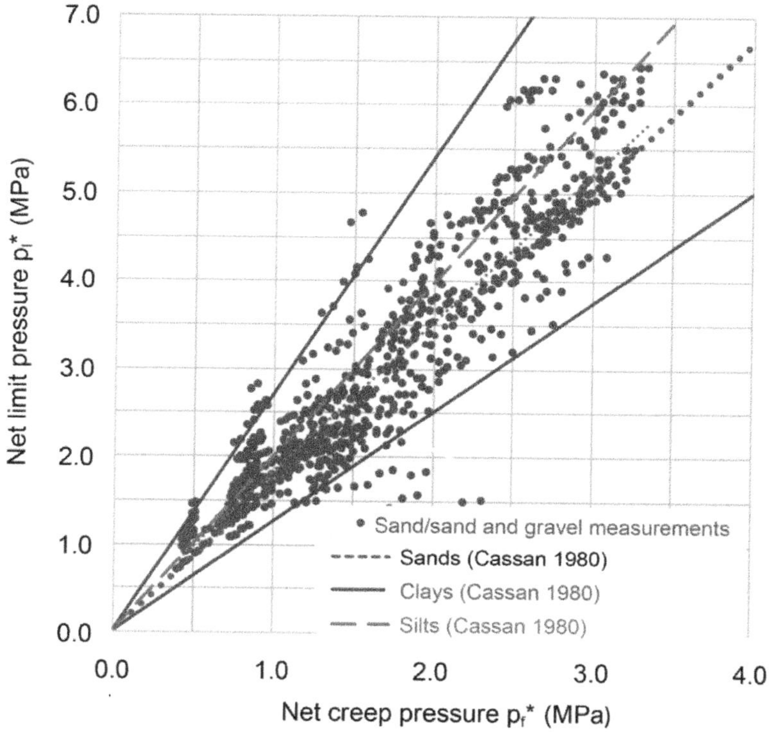

Sand/sand and gravel (Measurements): $\frac{p_{l^*}}{p_{f^*}} \approx 1.70$ (1.45 to 2.50),

Cassan (1980): Sand: $\frac{p_{l^*}}{p_{f^*}} \approx 1.70$; Clays: $\frac{p_{l^*}}{p_{f^*}} \approx 1.69$ (1.25 to 2.70), Silts: $\frac{p_{l^*}}{p_{f^*}} \approx 1.99$

Figure 3.65. *Correlation between limit pressure and net creep (p_l^* and p_f^*). For a color version of this figure, see www.iste.co.uk/dhouib/geotechnics1.zip*

3.12. References

Aas, G. (1965). A study of the effect of vane shape and rate of strain on the measured values of in situ shear strength of clays. *Proceedings of the Sixth International Conference on Soil Mechanics and Foundation Engineering*, Montreal, Canada, 8–15 September.

AFNOR (1990a). NF P94-114 : Essai au pénétromètre dynamique type A. Standard.

AFNOR (1990b). NF P94-115 : Essai au pénétromètre dynamique type B. Standard.

AFNOR (1991a). NF P94-112 : Essai au scissomètre de chantier. Standard.

AFNOR (1991b). NF P94-116 : Essai de pénétration au carottier. Standard.

AFNOR (1991c). NF P94-117 : Portance des plateformes : Partie 1 : Module sous chargement statique à la plaque (EV2). Standard.

AFNOR (1992). NF P94-057 : Analyse granulométrique des sols – Méthode par sédimentation. Standard.

AFNOR (1995a). NF P94-072 : Essai scissométrique en laboratoire. Standard.

AFNOR (1995b). NF P94-091 : Essai de gonflement à l'oedomètre – Détermination des déformations par chargement de plusieurs éprouvettes. Standard.

AFNOR (1995c). Règles PS92 – NF P06-013 (1995) : Règles parasismiques applicables aux bâtiments. Standard.

AFNOR (1996a). NF P94-113 : Essai au pénétromètre statique. Standard.

AFNOR (1996b). NF P94-056 : Analyse granulométrique – Méthode par tamisage à sec après lavage. Standard.

AFNOR (1997a). NF P94-077 : Essai de compression uniaxiale. Standard.

AFNOR (1997b). NF P94-090-1: Essai oedométrique – Partie 1 : Essai de compressibilité sur matériaux fins quasi saturés avec chargement par paliers paliers. Standard.

AFNOR (1997c). XP P94-120 : Essai de cisaillement au phicomètre. Standard.

AFNOR (1999). NF P94-110-2 : Essai au pressiomètre Louis Ménard avec cycles. Standard.

AFNOR (2000). NF P94-110-1 : Essai au pressiomètre Louis Ménard sans cycles. Standard.

AFNOR (2005a). Eurocode 7 : Calcul géotechnique Partie 1 : Calcul géotechnique – NF EN 1997-1. Standard.

AFNOR (2005b). Eurocode 8 : Calcul des structures pour leur résistance aux séismes – NF EN 1998-1 – Partie 1 : Règles générales, actions sismiques et règles pour les bâtiments. September, Standard.

AFNOR (2005c). Eurocode 8 : Calcul des structures pour leur résistance aux séismes – NF EN 1998-1 – Partie 5 : Fondations, ouvrages de soutènement et aspects géotechniques. September, Standard.

AFNOR (2007). Eurocode 7 : Calcul géotechnique Partie 2 : Reconnaissance des terrains et essais – NF EN 1997-2. Standard.

AFNOR (2012). Justification des ouvrages géotechniques. Norme d'application nationale de l'Eurocode 7 – Fondations profondes – NF P94-262. Standard.

AFNOR (2013). Justification des ouvrages géotechniques. Norme d'application nationale de l'Eurocode 7 – Fondations superficielles – NF P94-261. Standard.

AFPS (1993). Guide méthodologique pour la réalisation d'études de microzonage sismique. Methodological guide, AFPS.

AFPS (2004). *Construction parasismique des maisons individuelles aux Antilles. Tome IV : "Recommandations"*. Editions de l'AFPS.

AFPS (2012). *Procédés d'amélioration des sols sous actions sismiques*. Presses de l'Ecole Nationale des Ponts et Chaussées, Paris, France.

Allagnat, D. (2005). *La méthode observationnelle pour le dimensionnement interactif des ouvrages*. Presses de l'Ecole nationale des Ponts et Chaussées, Paris, France.

Amar, S. and Jézéquel, J.-F. (1978). Propriétés mécaniques des sols déterminées en place. *Techniques de l'ingénieur – Traité Construction*, C220, C21.

Arman, A., Poplin, J.K., Ahmad, N. (1975). Study of the vane shear. *Proceedings of the ASCE Conference on in situ Measurement of Soil Properties*, Raleigh, NC, USA.

Baguelin, F. and Jézéquel, J. (1972). Expansion de sondes cylindriques dans les sols cohérents. *Bulletin de liaison des Laboratoires des Ponts et Chaussées*, 61, September–October.

Baguelin, F. and Jézéquel, J. (1973). Le pressiomètre autoforeur. *Annales de l'ITB-TP*, July–August.

Baud, J.-P. (2005). Analyse des résultats pressiométriques Ménard dans un diagramme spectral et utilisation des regroupement statistiques dans la modélisation d'un site. In *ISP5 – Pressio 2005, Volume 1*, Gambin, M., Magnan, J.-P., Mestat, M. (eds). Presses de l'Ecole Nationale des Ponts et Chaussées, Paris.

Baud, J.-P. (2016). Apport de l'essai cyclique à la classification pressiométrique des sols et des roches. *Journées Nationales de Géotechnique et de Géologie de l'ingénieur (JNGGI)*, Nancy, France.

Baud, J.-P. and Gambin, M. (2014). Soil and rock classification from high pressure borehole expansion tests. *Geotechnical and Geological Engineering: An International Journal*, 32, 1397–1403.

de Beer, E.E. (1946). Contributions à l'étude de l'équilibre limite de rupture du sol sous des massifs de fondation transmettant des charges centrées. *Le Génie Civil*, 1 and 15 May and 1 June.

de Beer, E.E. (1949). Quelques exemples d'application des méthodes d'investigation utilisées en Belgique pour la résolution des problèmes de foundation. *Annales de l'ITBTP*.

Begemann, H.K. (1963). The use of the static soil penetrometer in Holland. *New Zealand Engineering*, 18(2).

Begemann, H.K. (1965). The friction jacket cone as an in determining the soil profiles. *Proceedings of the Sixth International Conference on Soil Mechanics and Foundation Engineering*, Montreal, Canada, 8–15 September.

Biarez, J. (1962). Contribution à l'étude des propriétés des matériaux pulvérulents. PhD Thesis, Université de Grenoble, Grenoble, France.

Biarez, J., Cassan, M., Meunier, J., Toutounji, A. (1974). Contribution à l'étude des pénétromètres statiques et dynamiques. *Symposium européen sur les essais de pénétration*. Stockholm, Sweden.

Bjerrum, L. (1972). Stabilité des ouvrages souples. *General Session Report, 5th Soil Mechanics Conference*. Madrid, Spain.

Cadling, L. and Odenstad, S. (1950). The vane borer on an apparatus for determining the shear strength of clay soils directly in the ground. *SGI Proceedings*, 2. Royal Swedish Geotechnical Institute, Stockholm, Sweden.

Caquot, A. and Kérisel, J. (1956). *Traité de Mécanique des sols*, 3rd edition. Gauthier-Villard, Paris, France.

Caquot, A. and Kérisel, J. (1966). *Traité de Mécanique des sols*, 4th edition. Gauthier-Villard, Paris, France.

Cassan, M. (1980). *Les essais d'eau dans la reconnaissance des sols*. Edition Eyrolles, Paris, France.

Cassan, M. (1988). *Les essais en situ en mécanique des sols 1 : Réalisation et interpretation*. Edition Eyrolles, Paris, France.

Claesson, A.I.M., Edwards, J.M., Sommer, P., Uge, P. (1977). Asphalt pavements system: The Schel method. *Proceedings of the 4th International Conference on Structural Design of Asphalt Pavements*. Ann Arbor, MI, USA, 22–26 August.

Costet, J. and Sanglerat, G. (1981). *Cours pratique de mécanique des sols*, 2nd edition. Dunod, Paris, France.

Demolon, A. (1948). Principes d'agronomie. *Tome I : Dynamique du sol*, 4th edition. Dunod, Paris.

Dhouib, A., Caquineau, X., Gambin, M. (2002). Variations des paramètres mesurés par le pressiomètre Ménard. *Symposium International PARAM 2002*, Paris, France, 2–3 September.

Foray, P. (1972). Contribution à l'étude des tassements et de la force portante des pieux. PhD Thesis, Université de Grenoble, Grenoble, France.

Gibbs, H.J. and Hotz, W.G. (1957). Research on determining the density of sand by S.P.T. *Proceedings 4th International Conference on Soils Mechanics and Foundations Engineering*, 1, 35–39.

Gonin, H. (1978). Etude théorique de battage des corps élastiques élancés et applications pratiques. *Annales de l'ITBTP*, 361, May.

Gonin, H., Vandangeon, P., Lafeuillade, M.P. (1992). Etude sur les corrélations entre le Standard Penetration Test et le pressiomètre. *Revue française de géotechnique*, 58.

Grésillon, J.M. (1970). Etude des fondations profondes en milieu pulvérulent. PhD Thesis, Université de Grenoble, Grenoble, France.

Jacquard, C. and Dhouib, A. (2012). Etude de fondations superficielles sur radier général d'une grande tour – modélisation simplifiée et retour d'expérience. *Conférence au Comité Français de Mécanique des Sols et des Fondations*, Paris, France, 11 October.

Jézéquel, J. (1969). Les pénétromètres statiques. Influence du mode d'emploi sur la résistance en pointe. *Bulletin des Liaisons des Ponts et Chaussées*, 36, January–February.

Jézéquel, J., Lemasson, H., Touzé, J. (1968). Le pressiomètre Ménard. *Bulletin des Liaisons des Ponts et Chaussées*, 32, June–July.

Kérisel, J. (1961). Fondations profondes en milieu sableux. *5ème congrès de mécanique des sols*, Paris, France.

Kérisel, J. (1965). Résistance de pointe en milieu pulvérulent de serrage divers. *6ème congrès international de mécaniques des sols*, Montreal, Canada.

Kérisel, J. and Adam, M. (1967). Calcul des forces horizontales appliquées aux fondations profondes dans les argiles et les limons. *Annales de ITBTP*, November.

Laréal, P., Sanglerat, G., Gilley, J. (1974). Comparaison des essais de pénétration effectués avec différents pénétromètres statiques ou dynamique. *Symposium européen sur les essais de pénétration*, Stockholm, Sweden, June.

LCPC Dunod (1973). *Essais en laboratoire et en place de mécanique des sols*. Aide-Mémoire. Dunod, Paris.

LCPC Dunod (1987). *Essai au scissomètre de chantier*. Mode opératoire LCPC Dunod, Paris.

LCPC-SETRA (1992a). Guide technique pour la réalisation des remblais et des couches de forme – Fascicule I : Principes généraux, Technical guide, SETRA, Paris.

LCPC-SETRA (1992b). Guide technique pour la réalisation des remblais et des couches de forme – Fascicule II : Annexes techniques, Technical guide, SETRA, Paris.

L'Herminier, R. (1967). *Cours de mécanique des sols et des chaussées*. Société de diffusion des Techniques du bâtiment et des Travaux Publics, Paris.

Maillard, J-F. (1985). Forages et sondages. *Techniques de l'ingénieur – Traité Construction*, C228, C21.

Mazier, G. (1971). Les essais in situ en rocher et dans les sols meubles. Société Suisse de Mécanique des Sols et des Roches, Fall session.

MEL (1993). Règles de conception et de calcul des fondations des ouvrages de génie civil. 62(V) 93.3 TO, December.

Ménard, L. (1958). La mesure du fluage et l'étude du compactage selon les méthodes pressiométriques. *Congrès National de la Mécanique des Sols*, August, Récife, Brazil.

Ménard, L. (1959a). Notice générale D60 : Règles d'utilisation des techniques pressiométriques et exploitation des résultats obtenus pour le calcul des fondations. *SOLS-SOILS*, 26.

Ménard, L. (1959b). Compressibilité des sols et tassement des fondations. Ministère des armées, *Revue du Génie Militaire*.

Ménard, L. (1960). Application de la précontrainte aux terrains médiocres de foundation. *Bulletin Technique du Génie Militaire*, 1(XCIII).

Ménard, L. (1970). Une méthode économique de fondations sur terrains de remblais gagnés sur la mer. *SOLS-SOILS*, 24.

Ménard, L. and Bourdon, G. (1965). Une méthode économique de fondations sur terrains de remblais gagnés sur la mer. *SOLS-SOILS*, 12.

Ménard, L. and Lambert, P. (1966). Etude expérimentale diun massif de fondation soumis à des vibrations. *SOLS-SOILS*, 17.

Ménard, L., Bourdon, G., Gambin, M. (1969). Méthode générale de calcul d'un rideau ou d'un pieu sollicité horizontalement en fonction des résultats d'essais pressiométriques. *SOLS-SOILS*, 22–23.

Meunier, J. (1974). Contribution à l'étude de l'onde de choc dans les sols. PhD Thesis, Université de Grenoble, Grenoble, France.

Meyerhof, G.G. (1956). Penetration test and bearing capacity of cohesionless soils. *JL Soils Mechanics and Foundations Divisions Proceedings, ASCE*, 82(SM.1, SM.3).

Philipponnat, G. (1987). Le phicomètre : analyse de 200 essais de cisaillement in situ. *Annales ITBTP*, 460, 65–88.

Philipponnat, G. and Hubert, B. (2008). *Fondations et ouvrages en terre*. Edition Eyrolles, Paris, France.

Philipponnat, G. and Zerhouni, M.I. (1993). Interprétation de l'essai au phicomètre. *Revue française de géotechnique*, 65, 3–28.

Pilot, G. (1972). Rupture d'un remblai sur sols compressibles. *Commentaire Bulletin de Liaison des Laboratoires des Ponts et Chaussées*, 61, September–October.

Puech, A. (1974). Calcul de la force portante d'un pieu isolé. PhD Thesis, Université de Grenoble, Grenoble, France.

Rat, M. (1974). Reconnaissance des sols. *Techniques de l'ingénieur – Traité Construction*, C224, C21.

Robertson, P.K. (2009). Guide to cone penetration testing for geotechnical engineering. *Canadian Geotechnical Journal*, 46(11), 1337–1355.

Roy, M. and Leblanc, A. (1986). Facteurs influençant la mesure au scissomètre. *Bulletin de Liaison des Laboratoires des Ponts et Chaussées*, 144, June–July.

Sanglerat, G. (1965). *Le pénétromètre et la reconnaissance des sols*. Dunod, Paris, France.

Sanglerat, G. (1971). Pénétration statique et dynamique. Le pénétromètre statique et les prévisions des tassements. *Recueil des conférences du Séminaire du C.A.S.T. de l'INSA de Lyon sur les essais in situ*.

Sanglerat, G., Nhiem, T.V., Séjourne, M., Andian, R. (1974). Direction soil classification by static penetromters with special friction sleeve. *European Symposium on Penetration Testing*, June. Stockholm, Sweden.

Schlosser, F. (2006). *Eléments de Mécanique des Sols*. Edition des Presses de l'Ecole Nationale des Ponts et Chaussées, Paris, France.

Schmertmann, J.H. (1970). Static cone to compute static settlement over sand. *Journal of the Soil Mechanics & Foundations Engineering Division*, 96, 1011–1043.

Schmertmann, J.H. (1978). *Guidelines for Cone Penetration Test Performance and Design – US Department of Transportation*. Federal Highway Administration (FHA), Washington, DC, France.

Schmertmann, J.H., Hartman, J.P., Brown, P.R. (1978). Improved strain factor influence diagrams. *Journal of the Soil Mechanics & Foundations Engineering Division*, 104, 1031–1035.

Schultze, E. and Menzenbach, E. (1961). L'essai de pénétration standard et la compressibilité des sols. *Congrès international de mécanique des sols*, Paris, France.

Seed, H.B. (1976). Evaluation of soil liquefaction effects on level ground during earthquakes. *State of the Art. Liquefaction Problems in Geotechnical Engineering, ASCE Annual Convention*. Philadelphia, PA, USA.

Skempton, A.W. (1948). Vane test in the alluvial plains of River Firth near Grangemouth. *Géotechnique*, 1(2), 111–125.

Tancelin, M.E. (1952). Mesure des pressions de terrain dans les mines de fer de l'Est. *Annales de l'ITBTP*, 58, October.

Teng, W.C. (1966). *Calcul des fondations et mur de soutènement*. Edition Eyrolles, Paris, France.

Terzaghi, K. (1951). *Mécanique des sols appliquée*. Dunod, Paris, France.

Terzaghi, K. and Peck, R. (1957). *Mécanique des sols appliquée*. Dunod, Paris, France.

Tincelin, M.E. (1952). Mesure des pressions de terrains dans les mines de fer de l'Est. *Annales de l'ITBTP*, 58, October.

Waschkowski, E. (1983). Le pénétromètre dynamique. *Bulletin de Liaison des Laboratoires des Ponts et Chaussées*, 125.

French, European and ISO Standards in the Field of Geotechnics

S.1. Introduction

French standards are noted as NF in order to distinguish them from European (written as NF EN) or international (written as NF EN ISO) standards.

The XP reference indicates the experimental character of the standard on which the NF standards are imposed.

The EN reference indicates that the standard comes from the European Committee for Standardization (CEN).

The standards preceded by the ISO reference are the work of the International Organization for Standardization (ISO).

The applications of the Eurocodes are translated into transcripts by the national standards bodies of each country: for example, Eurocode 7 – Part 1 is transcribed as DIN EN 1997-1 in Germany, NBN EN 1997-1 in Belgium, NF EN 1997-1 in France and BS EN 1997-1 in the United Kingdom.

National application standards of (Eurocode 7) are updated in each country.

Under the decree of January 26, 1984, the standard is considered as "*a reference*" in France. It is therefore neither a law nor a regulation. *Its application becomes compulsory when one party, among other signatories of the same contract, has referred to it.*

The application of a standard may also be mandatory when it is related to the prevention of risks and damage to people and property.

The choice of an NF, EN or ISO standard in the contract is the responsibility of the project owner. However, it is normally up to the project manager, who is considered the "expert" and "conductor" of the project, to advise and guide the project owner toward the choice of one standard over another.

The standards indicated in the "*project*" framework may be referenced in the contracts, subject to precautions in their use.

Finally, the same standards and rules in effect must be applied to the various components of the construction project (soil and foundations, structure).

S.2. General standards: expertise and qualification of assessment and test companies

Official references	*Classification index*	*English title*	*Date of publication*
NF X50-110		Quality of expertise – General requirements of competence for an expertise.	05-2003
NF ISO 31000		Risk management – Principles and guidelines.	01-2010
FD X50-045		Quality of expertise – Possible use of Standard NF X50-110 (General requirements of competence for an expertise).	01-2010
FD X50-046		Quality of expertise – Recommendations for the application of Standard NF X50-110:2003 (General requirements of competence for an expertise).	02-2011
NF S70-003-1 *Mandatory*		Works in the vicinity of networks – Part 1: Prediction of damage and its consequences.	07-2012
NF EN ISO 22475-1	P94-510-1	Geotechnical assessment and tests – Methods of sampling and piezometer measurements – Part 1: Technical principles of the works.	
ISO TS 22475-2		Geotechnical assessment and testing – Boring sampling methods and piezometer measurements – Part 2: Qualification criteria for companies and personnel.	09-2006

ISO TS 22475-3		Geotechnical assessment and testing – Boring or excavation sampling methods and piezometer measurements – Part 3: Assessment of the compliance of companies and personnel.	09-2007
NF S70-003-2		Works in the vicinity of networks – Part 2: Techniques for detecting without excavation.	12-2012
PR XP S70-003-4		Works in the vicinity of networks – Part 4: Special clauses in works contracts.	02-2014
NF P94-500	P94-500	Geotechnical engineering missions – Classifications and specifications.	11-2013

S.3. Soils – Assessments and tests: laboratory tests

S.3.1. *French standards (NF)*

Official references	*Classification index*	*English title*	*Date of publication*
NF ISO 11074	X31-002	Soil quality. Vocabulary.	03-2006
XP P94-010	P94-010	Geotechnical glossary – Definitions – Notations – Symbols.	01-12-1996
XP P94-011	P94-011	Description. Identification. Soil naming – Terminology. Elements of classification.	01-08-1999
NF P94-040	P94-040	Simplified method for the identification of the 0/50 mm fraction of a granular material – Determination of the grading and of the methylene blue value.	01-10-1993
XP P94-041	P94-041	Grading identification – Wet sieving method.	01-12-1995
XP P94-047	P94-047	Determination of the weight content of organic matter in a material – Calcination method.	01-12-1998
NF P94-048	P94-048	Determination of carbonate content – Calcimeter method.	01-10-1996
NF P94-049-1	P94-049-1	Determination of water content by weight of materials – Part 1: Microwave desiccation method.	01-02-1996
NF P94-049-2	P94-049-2	Determination of water content by weight of materials – Part 2: Hot plate or heating panels method.	01-02-1996

NF P94-050	P94-050	Determination of the water content by weight of materials – Steaming method.	01-09-1995
NF P94-051	P94-051	Determination of the Atterberg limits – Limit of liquidity in the bowl – Limit of plasticity at the reel.	01-03-1993
NF P94-052-1	P94-052-1	Determination of Atterberg limits – Part 1: Liquid limit – Cone penetration method.	01-11-1995
NF P94-053	P94-053	Determination of the density of fine soils in a laboratory – Cutting kit, mold and water immersion methods.	01-10-1991
NF P94-054	P94-054	Determination of the density of solid particles in soils – Water pycnometer method.	01-10-1991
NF P94-055	P94-055	Determination of the weight content of organic matter in soil – Chemical method.	01-12-1993
NF P94-056	P94-056	Sieve analysis – Dry sieving method after washing.	01-03-1996
NF P94-057	P94-057	Sieve analysis – Sedimentation method.	01-05-1992
XP P94-058	P94-058	Determination of the state of decomposition (humidification) of organic soils – Von Post test.	01-10-1993
NF P94-059	P94-059	Determination of the minimum and maximum densities of non-cohesive soils.	01-11-2000
XP P94-060-1	P94-060-1	Desiccation test – Part 1: Conventional determination of the shrinkage limit on the 400-micrometer aggregate of a material.	01-12-1997
XP P94-060-2	P94-060-2	Desiccation test – Part 2: Actual determination of shrinkage limit on an undisturbed sample.	01-12-1997
NF P94-061-1	P94-061-1	Determination of the density of a material in situ – Part 1: Point gamma densimeter method (direct transmission).	01-10-1996
NF P94-061-2	P94-061-2	Determination of the density of a material in situ – Part 2: Membrane densimeter method.	01-04-1996
NF P94-061-3	P94-061-3	Determination of the density of a material in situ – Part 3: Sand method.	01-04-1996
NF P94-061-4	P94-061-4	Determination of the density of a material in situ – Part 4: Method for coarse materials (D_{max} « 50 mm).	01-12-1996
NF P94-062	P94-062	Density measurement in situ – Double gamma probe boring.	01-08-1997

NF P94-068	P94-068	Measurement of the methylene blue adsorption capacity of a soil or rock material – Determination of the methylene blue value of a soil or rock material by the spot test.	01-10-1998
NF P94-070	P94-070	Tests with the triaxial revolution device – Generalities – Definitions.	01-10-1994
NF P94-071-1	P94-071-1	Soils: Assessment and testing – Straight shear box test – Part 1: Direct shear.	01-08-1994
NF P94-071-2	P94-071-2	Straight shear box test – Part 2: Alternating shear.	01-08-1994
NF P94-072	P94-072	Vane test in the laboratory	01-09-1995
NF P94-074	P94-074	Tests with the triaxial revolution device – Apparatus – Preparation of the specimens – Unconsolidated undrained test (UU) – Consolidated undrained test (C_u + U) with pore pressure measurement – Consolidated drained test (CD).	01-10-1994
NF P94-077	P94-077	Uniaxial compression test.	01-12-1997
XP P94-090-1	P94-090-1	Oedometer test – Part 1: Compressibility test on near-saturated fine materials with step loading.	01-12-1997
XP P94-091	P94-091	Oedometer swelling test – Determination of deformations by loading several specimens.	01-12-1995

S.3.2. *European standards (NF EN)*

Official references	***Classification index***	***English title***	***Date of publication***
NF EN ISO 14688-1	P94-400-1	Geotechnical assessment and testing – Naming, description and classification of soils – Part 1: Naming and description.	01-05-2003
NF EN ISO 14688-1/A1	P94-400-1/ A1	Geotechnical assessment and testing – Naming, description and classification of soils – Part 1: Naming and description – Amendment 1.	20-12-2013
NF EN ISO 14688-2	P94-400-2	Geotechnical assessment and testing – Naming, description and classification of soils – Part 2: Principles for classification.	01-04-2005
NF EN ISO 14688-2/A1	P94-400-2/ A1	Geotechnical assessment and testing – Naming, description and classification of soils – Part 2: Principles for classification – Amendment 1.	20-12-2013

XP CEN ISO/TS 17892-1	P94-512-1	Geotechnical assessment and testing – Laboratory testing of soils – Part 1: Determination of water content.	01-08-2005
XP CEN ISO/TS 17892-2	P94-512-2	Geotechnical assessment and testing – Laboratory tests on soils – Part 2: Determination of the density of fine soil.	01-08-2005
XP CEN ISO/TS 17892-3	P94-512-3	Geotechnical assessment and testing – Laboratory tests on soils – Part 3: Determination of solid particle density – Pycnometer method.	01-08-2005
XP CEN ISO/TS 17892-4	P94-512-4	Geotechnical assessment and testing – Laboratory tests on soils – Part 4: Determination of grading distribution.	01-08-2005
XP CEN ISO/TS 17892-5	P94-512-5	Geotechnical assessment and testing – Laboratory tests on soils – Part 5: Oedometer step loading test.	01-08-2005
XP CEN ISO/TS 17892-6	P94-512-6	Geotechnical assessment and testing – Laboratory tests on soils – Part 6: Cone penetration test.	01-08-2005
XP CEN ISO/TS 17892-7	P94-512-7	Geotechnical assessment and testing – Laboratory tests on soils – Part 7: Uniaxial compression test on fine soils.	01-08-2005
XP CEN ISO/TS 17892-8	P94-512-8	Geotechnical assessment and testing – Laboratory tests on soils – Part 8: Unconsolidated triaxial test.	01-08-2005
XP CEN ISO/TS 17892-9	P94-512-9	Geotechnical assessment and testing – Laboratory tests on soils – Part 9: Triaxial compression test on consolidated saturated soils.	01-08-2005
XP CEN ISO/TS 17892-10	P94-512-10	Geotechnical assessment and testing – Laboratory tests on soils – Part 10: Direct shear test.	01-08-2005
XP CEN ISO/TS 17892-11	P94-512-11	Geotechnical assessment and testing – Laboratory tests on soils – Part 11: Determination of permeability at constant and decreasing variable load.	01-08-2005
XP CEN ISO/TS 17892-12	P94-512-12	Geotechnical assessment and testing – Laboratory tests on soils – Part 12: Determination of Atterberg limits.	01-08-2005
NF EN 933-8	P18-622-8	Tests to determine the geometric characteristics of aggregates. Part 8: Assessment of fines. Sand equivalent.	01-12-2004
NF EN 1367-1	P18-653-1	Tests for determining the thermal properties and weathering of aggregates. Part 1: Determination of freeze-thaw resistance.	01-08-2007

S.4. Soils – Assessments and tests: on-site surveys and tests

S.4.1. *French standards (NF)*

Official references	*Classification index*	*English title*	*Date of publication*
NF P94-063	P94-063	Compaction quality control – Constant energy dynamic penetrometer method – Principle and calibration method of penetrodensitographs – Exploitation of results – Interpretation.	01-06-2011
NF P94-105	P94-105	Compaction quality control – Dynamic variable energy penetrometer method – Principle and calibration method of the penetrometer – Exploitation of the results – Interpretation.	01-04-2012
NF P94-110-1	P94-110-1	Ménard pressuremeter test – Part 1: Test without cycle.	01-01-2001
XP P94-110-2	P94-110-2	Soils: Assessment and test – Ménard pressuremeter test – Part 2: Test with cycle.	01-12-1999
NF P94-112	P94-112	In-situ vane test.	01-11-1991
NF P94-113	P94-113	Cone penetration test.	01-10-1996
NF P94-114	P94-114	Dynamic penetration test, Type A.	01-12-1990
NF P94-115	P94-115	Dynamic penetrometer survey, Type B.	01-12-1990
NF P94-116	P94-116	Penetration test with a core drill (SPT).	01-10-1991
NF P94-117-1	P94-117-1	Platform load-bearing capacity – Part 1: Modulus under static plate loading (EV2).	01-04-2000
NF P94-117-2	P94-117-2	Platform load-bearing capacity – Part 2: Modulus under dynamic loading.	01-10-2004
NF P94-117-3	P94-117-3	Platform load-bearing capacity – Part 3: WESTERGAARD reaction coefficient under static plate loading.	01-12-2008
NF P94-119	P94-119	Piezocone test.	01-12-1995
XP P94-120	P94-120	Phicometer shear test.	01-12-1997
XP P94-123	P94-123	Boring logging – Neutron probe method.	01-09-1999

NF P94-130	P94-130	Pumping test.	04-2000
NF P94-131	P94-131	Lugeon water test.	09-1994
NF P94-132	P94-132	Lefranc water test.	10-2000
NF P94-157-1	P94-157-1	Piezometeter measurements – Part 1: Open tube.	03-1996
NF P94-157-2	P94-157-2	Piezometeter measurements – Part 2: Pore pressure measurement probe.	03-1996

S.4.2. *European standards (NF EN)*

Official references	*Classification index*	*English title*	*Date of publication*
NF EN ISO 22475-1	P94-510-1	Geotechnical assessment and testing – Methods of sampling and piezometer measurements – Part 1: Technical principles of the works.	01-03-2007
NF EN ISO 22476-1	P94-521-1	In-situ testing – Part 1: Electrical cone and piezocone penetration tests.	*Project*
NF EN ISO 22476-2	P94-521-2	In-situ testing – Part 2: Dynamic penetration test.	01-07-2005
NF EN ISO 22476-2/A1	P94-521-2/ A1	In-situ testing – Part 2: Dynamic penetration test – Amendment 1.	01-04-2012
NF EN ISO 22476-3	P94-521-3	In-situ testing – Part 3: Core penetration test.	01-07-2005
NF EN ISO 22476-3/A1	P94-521-3/ A1	In-situ testing – Part 3: Core penetration test – Amendment 1.	01-04-2012
NF EN ISO 22476-4	P94-521-4	In-situ testing – Part 4: Louis Ménard pressuremeter test.	*Project*
NF EN ISO 22476-5	P94-521-5	In-situ testing – Part 5: Flexible dilatometer test.	*Project*
NF EN ISO 22476-6	P94-521-6	In-situ testing – Part 6: Self-drilling pressuremeter test.	*Project*
NF EN ISO 22476-7	P94-521-7	In-situ testing – Part 7: Stiff dilatometer diameter test.	*Project*
NF EN ISO 22476-8	P94-521-8	In-situ testing – Part 8: Rejected pressuremeter test.	*Project*
NF EN ISO 22476-9	P94-521-9	In-situ testing – Part 9: Field vane test.	*Project*

NF EN ISO 22476-10	P94-521-10	In-situ testing – Part 10: Weight survey test.	05-2005
NF EN ISO 22476-11	P94-521-11	In-situ testing – Part 11: Flat dilatometer test.	05-2005
NF EN ISO 22476-12	P94-521-12	In-situ testing – Part 12: Cone penetration test with mechanical tip cone.	01-10-2010
NF EN ISO 22282-1	P94-523-1	Geohydraulic testing – Part 1: General rules.	18-01-2014
NF EN ISO 22282-2	P94-523-2	Geohydraulic testing – Part 2: Water permeability test in an open tube boring.	18-01-2014
NF EN ISO 22282-4	P94-523-4	Geohydraulic testing – Part 4: Pumping tests.	04-01-2014
NF EN ISO 22282-5	P94-523-5	Geohydraulic testing – Part 5: Infiltration test.	05-01-2014
NF EN ISO 22282-6	P94-523-6	Geohydraulic testing – Part 6: Water permeability test in a closed tube boring.	06-01-2014

S.5. Soils – Standards and benchmarks for geotechnical design and sizing

S.5.1. *French standards and benchmarks*

Official references	*Classification index*	*English title*	*Date of publication*
DTU 13.11	P11-211	DTU 13.11 – Shallow foundations.	01-03-1988
DTU 13.11	P11-211/A1	DTU 13.11. Amendment 1 to the technical specifications (DTU P11-211).	01-06-1997
P11-211 COMPIL		DTU 13.11 – Shallow foundations – Compiled text of Standard DTU 13.11 of March 1988 and its amendment 1 of June 1997.	06-1997
P11-212	P11-212	DTU 13.2 – Deep foundations for buildings.	01-09-1992
NF P11-212-2	P11-212-2	DTU 13.2 – Building works – Deep foundation works for buildings – Part 2: Special clauses.	01-11-1994
NF P11-213-1	P11-213-1	DTU 13.3 – Slabs. Design, calculation and execution. Part 1: Technical specifications for slabs for industrial or similar use.	03-2005
NF P11-213-1/A1	P11-213-1	DTU 13-3 – Slabs. Design, calculation and execution. Part 1: Technical specifications for slabs for industrial or similar use. Amendment 1.	05-2007

NF P11-213-2	P11-213-2	DTU 13.3 – Slabs. Design, calculation and execution. Part 2: Technical specifications for slabs for non-industrial use or similar.	05-2007
NF P11-213-2/A1	P11-213-2	DTU 13-3 – Slabs. Design, calculation and execution. Part 2: Technical specifications for slabs for non-industrial use or similar. Amendment 1.	05-2007
NF P11-213-3	P11-213-3	DTU 13.3 – Slabs. Design, calculation and execution. Part 3: Technical specifications for slabs of individual houses.	05-2007
NF P11-213-3/A1	P11-213-3	DTU 13-3 – Slabs. Design, calculation and execution. Part 3: Technical specifications for slabs of individual houses. Amendment 1.	06-2007
NF P11-213-4	P11-213-4	DTU 13.3 – Slabs. Design, calculation and execution. Part 4: Special clauses.	04-02-2004
NF P11-221-1	P11-221-1	DTU 14.1 – Building works – Casing works – Part 1: Technical specifications.	01-05-2000
NF P11-221-2	P11-221-2	DTU 14.1 – Building works – Private contracts – Casing works – Part 2: Special clauses.	01-05-2001
XP P36-201/A1		DTU 40.5 – Rainwater drainage works. Technical specifications. Experimental Standard XP P36-201 of November 1993. Amendment XP P36-201/A1 of December 1997.	12-1997
DTU 13.12	P11-711	DTU 13.12 – Rules for the calculation of shallow foundations (completed by erratum of November 1988).	01-03-1988
NF EN 12794+A1	P19-802	Precast concrete products – Foundation piles.	01-07-2007
NF EN 14991	P19-817	Precast concrete products – Foundation elements.	01-07-2007
NF EN ISO 13793	P50-745	Thermal performance of buildings – Thermal design of foundations to avoid frost heaves.	01-05-2001
NF P06-001		Bases of calculation of constructions – Operating loads of buildings. Approved in June 1986.	06-1986
NF P06-002		DTU – NV 65 Rules. Rules defining the effects of snow and wind on constructions and annexes – January 2001. Approved in April 2000 – January 2001 edition.	01-2001

NF P06-006		N84 Rules, modified in 1995. Action of snow on constructions. Erratum n°2, July 1997. Amendment P06-006/A1, April 2000.	04-2000
NF P06-013	P06-013	DTU – Seismic construction rules – PS Rules applicable to PS92 buildings. Approved in December 1995. Ministerial order of May 29, 1997 (JORF of June 3, 1997): "Classification and seismic construction rules applicable to buildings in the so-called normal risk category".	03-06-1997
NF P06-013/A1	P06-013	DTU – PS92 Rules. Approved in February 2001. Modifications to NF P06-013. NF P06-013 revision; NF P06-013/A1: Approved in February 2001.	02-2001
NF P06-013/A2	P06-013	Seismic construction rules – PS Rules applicable to buildings. Approved in November 2004.	11-2004
NF P06-014	P06-014	Seismic construction rules – Seismic construction of individual houses and similar buildings – Revised PS-MI89 Rules 92 – Field of application – Design-Execution. Approved in March 1995.	03-1995
NF P06-014/A1	P06-014	Seismic construction of individual houses and similar buildings – Revised PS-MI89 Rules 92 – Field of application – Design-Execution. Approved in February 2001.	02-2001
XP P16-603		DTU 64-1 Implementation of autonomous sanitation systems. Technical specifications. Experimental standard, August 1998.	08-1998
Règles BAEL 91		Technical rules for the design and calculation of reinforced concrete structures and constructions, according to the limit states method. March 1992 – Amendment in February 2000.	02-2000
Fascicule 62, Titre V		Technical rules for the design and calculation of foundations for civil engineering structures (will be issued as soon as the French standards complementary to Eurocode 7 are published).	1993

S.5.2. *Eurocodes*

Official references	*Classification index*	*English title*	*Date of publication*
NF EN 1990/1991	P06-100-1	Eurocodes 0 and 1 – General rules – Basis of calculation of structures.	03-2003
NF EN 1992	P18-716	Eurocode 2 – General rules and calculation of concrete structures.	10-2005
NF EN 1993	P22-311	Eurocode 3 – General rules and calculation of steel structures.	10-2005
NF EN 1994	P22-391	Eurocode 4 – General rules and calculation of steel-concrete structures.	12-1997
NF EN 1995	P21-711	Eurocode 5 – Generalities and design, and calculation of timber structures.	11-2005
NF EN 1996	P10-611	Eurocode 6 – General calculation of masonry structures.	02-1998
NF EN 1997-1	P94-251-1	Eurocode 7 – Geotechnical calculation – Part 1: General rules.	06-2005
NF EN 1997-1/A1	P94-251-1/ A1	Eurocode 7 – Geotechnical calculation – Part 1: General rules.	04-2014
NF EN 1997-1/NA	P94-251-1/ NA	Eurocode 7 – Geotechnical calculation – Part 1: General rules – National Annex to NF EN 1997-1; 2005.	09-2006
NF EN 1997-2	P94-252	Eurocode 7 – Geotechnical calculation – Part 2: Ground assessment and testing.	01-09-2007
NF EN 1998-1	P06-030-1	Eurocode 8 – Calculation of structures for earthquake resistance – Part 1: General rules, seismic actions for buildings.	12-2000
NF EN 1998-1/NA	P06-030-1/ NA	Eurocode 8 – Calculation of structures for earthquake resistance – Part 1: General rules, seismic actions for buildings – National Annex to NF EN 1998-1; 2005 – General rules, seismic actions and rules for buildings.	12-2000
NF EN 1998-2	P06-032	Eurocode 8 – Calculation of structures for earthquake resistance – Part 2: Bridges.	12-2000
NF EN 1998-2/NA	P06-032/NA	Eurocode 8 – Calculation of structures for earthquake resistance – Part 2: Bridges – National Annex to NF EN 1998-2 – Bridges.	12-2000

NF EN 1998-5	P06-035-1	Eurocode 8 – Calculation of structures for earthquake resistance – Part 5: Foundations, retaining structures and geotechnical aspects.	12-2005
NF EN 1998-5/NA	P06-035-1/ NA	Eurocode 8: Calculation of structures for earthquake resistance – Part 5: Foundations, retaining structures and geotechnical aspects – Annex to NF EN 1998-5: 2005 – Foundations, retaining structures and geotechnical aspects.	10-2007
NF EN 1999-1	P22-151	Eurocode 9 – General rules and calculation of aluminum structures.	01-2001

S.5.3. *National application standards of Eurocode 7*

Official references	***Classification index***	***English title***	***Date of publication***
NF P94-261	P94-261	Justification of geotechnical works – National application standards of Eurocode 7 – Shallow foundations.	15-06-2013
NF P94-262	P94-262	Justification of geotechnical works – National application standards of Eurocode 7 – Deep foundations.	01-07-2012
NF P94-270	P94-270	Geotechnical calculation – Retaining structures – Reinforced embankments and in-site nailed walls.	01-07-2009
NF P94-281	P94-281	Justification of geotechnical works – National application standards of Eurocode 7 – Retaining structures – Walls.	*Project*
NF P94-282	P94-282	Geotechnical calculation – Retaining structures – Curtain walls.	03-2009
NF P94-290	P94-290	Geotechnical calculation – Earthworks.	*Project*

S.6. Soils – Tests on foundation elements of the works

S.6.1. *French standards (NF)*

Official references	*Classification index*	*English title*	*Date of publication*
NF P94-150-1	P94-150-1	Soils: Assessment and testing – Static test of isolated pile under axial load – Part 1: In compression.	01-12-1999
NF P94-150-2	P94-150-2	Soils: Assessment and testing – Static test of isolated pile under axial load – Part 2: In tension.	01-12-1999
NF P94-151	P94-151	Soils: Assessment and tests – Static test of isolated pile under transverse forces.	01-10-1993
XP P94-152	P94-152	Soils: Assessment and testing – Dynamic axial loading test of a deep foundation element.	01-12-1997
NF P94-153	P94-153	Soils: Assessment and testing – Static tie-rod test.	01-12-1993
NF P94-156	P94-156	Soils: Assessment and testing – Inclinometer measurements.	01-101995
NF P94-160-1	P94-160-1	Soils: Assessment and testing – Auscultation of a foundation element – Part 1: Transparency method.	01-10-2000
NF P94-160-2	P94-160-2	Soils: Assessment and testing – Auscultation of a foundation element – Part 2: Reflection method.	01-11-1993
NF P94-160-3	P94-160-3	Soils: Assessment and testing – Auscultation of a foundation element – Part 3: Parallel seismic method (MSP).	01-05-1993
NF P94-160-4	P94-160-4	Soils: Assessment and testing – Auscultation of a foundation element – Part 4: Impedance method.	01-03-1994
XP P94-160-5	P94-160-5	Soils: Assessment and testing – Auscultation of a foundation element – Part 5: Gamma ray nuclear scattering methods.	01-12-1999
XP P94-202	P94-202	Soils: Assessment and testing – Soil and rock sampling – Methodology and procedures.	01-12-1995
NF P94-222	P94-222	Soil reinforcement – Reinforced soil structures with extensible and flexible strips – Static extraction test in place of inclusion.	01-08-1995

NF P94-232-1	P94-232-1	Soil reinforcement – Reinforced soil structures with extensible and flexible reinforcement – Part 1: Extraction test in place of a reinforcement strip, by steps.	01-02-2001
NF P94-242-1	P94-242-1	Soil reinforcement – Static nail pull-out test, subjected to an axial tensile force – Constant displacement speed test.	01-03-1993
A05-251	Removed in 2007	Soil corrosion – Corrosiveness assessment – Buried steel structures (sheet piles and piles)	03-1990
A05-252	A05-251	Soil corrosion – Galvanized or non-galvanized steels in contact with natural backfill materials (soil: Reinforced Earth type of embankments).	07-1990

S.6.2. *European and international standards (NF EN ISO)*

Official references	***Classification index***	***English title***	***Date of publication***
NF EN 125051-1	A05-412-1	Protection of metallic materials against corrosion. Risks of soil corrosion – Part 1: Generalities.	09-2003
NF EN 12051-2	A05-412-2	Protection of metallic materials against corrosion. Risks of soil corrosion – Part 2: Ferrous materials with little or no alloy.	09-2003
EN ISO 22477-1	P94-150-1	Static loading test on pile in compression.	*Project*
EN ISO 22477-2	P94-150-2	Static loading test on pile in tension.	*Project*
EN ISO 22477-3	P94-151-3	Lateral static loading test on pile.	*Project*
EN ISO 22477-4	P94-152	Dynamic loading test on pile in compression.	*Project*
EN ISO 22477-5	P94-153	Test on tie-rods.	*Project*
EN ISO 22477-6	P94-222	Test on nails.	*Project*
EN ISO 22477-7	P94-232	Test on improved soil.	*Project*

S.7. Soils – Execution of works and special geotechnical works

S.7.1. *French standards (NF)*

Official references	*Classification index*	*English title*	*Date of publication*
NF P94-325-1	P94-325-1	Execution of special geotechnical works – Gabion structures made of double-twisted hexagonal wire mesh – Part 1: Off-water structures.	01-03-2004
NF P94-325-2	P94-325-2	Execution of special geotechnical works – Gabion structures in double-twisted hexagonal mesh – Part 2: In-water structures.	01-11-2006
NF P95-106	P95-106	Works of art – Repair and reinforcement of concrete and masonry structures – Specifications for the foundations of the works.	01-08-1993
NF P95-301	P95-301	Avalanche protection equipment – Passive point anchors in soft soil – Pull-out test method.	01-08-1994
NF P95-303	P95-303	Avalanche protection equipment – Screen, rack – Design specifications.	01-12-1992
NF P95-304	P95-304	Avalanche control equipment – Avalanche nets – Design specifications.	01-12-1992
NF P95-305	P95-305	Avalanche control equipment – Snow barrier – Design specifications.	01-12-1992
XP P95-306	P95-306	Certification of rockfall protection screens – Testing procedures.	*Project*
NF P95-307	P95-307	Rockfall protection equipment – Terminology.	01-12-1992
NF P95-308	P95-308	Rockfall protection equipment – Net screens.	01-12-1992
NF P98-050-1	P98-050-1	Underground telecommunications structures for pedestrian and vehicular traffic areas – Part 1: Telecommunications chambers – Specifications, testing, marking.	01-04-2006
NF P98-050-2	P98-050-2	Underground telecommunications structures for pedestrian and vehicular traffic areas – Part 2: Closure devices (frames and plugs) – Specifications, testing, marking.	01-05-2006
NF P98-331	P98-331	Roads and outbuildings – Trenches: Opening, backfilling, repairing.	01-02-2005
NF P98-332	P98-332	Roads and outbuildings – Rules of distance between buried networks and rules of proximity between networks and vegetation.	01-02-2005

XP P98-333	P98-333	Roads and outbuildings – Small trenches.	01-06-2009
NF S70-003-1	S70-003-1	Works near networks – Part 1: Prevention of damage and its consequences.	01-07-2012
NF S70-003-2	S70-003-2	Works near networks – Part 2: Techniques for detection without excavation.	01-12-2012
NF S70-003-3	S70-003-3	Works near networks – Part 3: Georeferencing of works.	23-05-2014

S.7.2. *European standards (NF EN)*

Official references	***Classification index***	***English title***	***Date of publication***
NF EN 1536	P94-310	Execution of special geotechnical works – Bored piles.	01-10-2010
NF EN 12699	P94-311	Execution of special geotechnical works – Piles with soil discharge.	01-03-2001
NF EN 14199	P94-313	Execution of special geotechnical works – Micropiles.	01-09-2005
NF EN 1538	P94-320	Execution of special geotechnical works – Slurry trench walls.	01-10-2010
NF EN 1537	P94-321	Execution of special geotechnical works – Tie-rods.	07-09-2013
NF EN 12063	P94-322	Execution of special geotechnical works – Sheet piling.	01-08-1999
NF EN 14475	P94-326	Execution of special geotechnical works – Reinforced embankments.	01-01-2007
NF EN 14490	P94-328	Execution of special geotechnical works – Nailing.	01-09-2010
NF EN 12715	P94-330	Execution of special geotechnical works – Injection.	05-10-2000
NF EN 12716	P94-331	Execution of special geotechnical works – Columns, panels and soil-cement structures created by jet.	01-10-2001
NF EN 14679	P94-334	Execution of special geotechnical works – Treated soil columns.	01-09-2005
NF EN 15237	P94-337	Execution of special geotechnical works – Vertical drains.	01-05-2007
NF EN 14653-2	P93-503-2	Trench shoring components – Part 2: Testing and assessment.	01-08-2005
NF EN 14731	P94-340	Execution of special geotechnical works – Improvement of soil masses by vibration.	01-12-2005

S.8. Rocks – Tests

S.8.1. *French standards (NF)*

Official references	*Classification index*	*English title*	*Date of publication*
NF P94-064	P94-064	Dry density of a rock element – Hydrostatic weighing method.	01-11-1993
NF P94-066	P94-066	Fragmentation rate of rocky materials.	01-12-1992
NF P94-067	P94-067	Degradation rate of rocky materials.	01-12-1992
XP P94-202	P94-202	Rocks – Soil and rock sampling – Methodologies and procedures.	01-12-1995
XP P94-402	P94-402	Rocks – Glossary – Definitions – Notations – Symbols.	01-04-2002
NF P94-410-1	P94-410-1	Rocks – Tests for determining the physical properties of rocks – Part 1: Determination of water content by weight – Steaming method.	01-05-2001
NF P94-410-2	P94-410-2	Rocks – Tests for determining the physical properties of rocks – Part 2: Determination of density – Geometric and water immersion methods.	01-05-2001
NF P94-410-3	P94-410-3	Rocks – Tests for determining the physical properties of rocks – Part 3: Determination of porosity.	01-05-2001
NF P94-411	P94-411	Rocks – Determination of the velocity of ultrasonic wave propagation in a laboratory – Transparency method.	01-04-2002
XP P94-412	P94-412	Rocks – Determination of the resistance index to penetration by a drill.	01-04-2001
NF P94-420	P94-420	Rocks – Determination of uniaxial compressive resistance.	01-12-2000
NF P94-422	P94-422	Rocks – Determination of tensile resistance – Indirect method – Brazilian test.	01-01-2001
NF P94-423	P94-423	Rocks – Determination of triaxial compressive resistance.	01-03-2002
XP P94-424	P94-424	Rocks – Direct shearing along a rock discontinuity – Test under a constant force, perpendicular to the discontinuity surface.	01-03-2003
NF P94-425	P94-425	Rocks – Determination of Young's modulus and Poisson's ratio.	01-04-2002

XP P94-429	P94-429	Rocks – Resistance under point load – Franklin test.	01-12-2002
NF P94-430-1	P94-430-1	Rocks – Determination of the abrasive power of a rock – Part 1: Scratch test with a point.	01-10-2000
NF P94-430-2	P94-430-2	Rocks – Determination of the abrasive power of a rock – Part 2: Test with a rotating tool.	01-10-2000
XP P94-444	P94-444	Rocks – Static pull-out test of an anchor sealed in a rock mass, performed under an axial tensile force – Step test.	01-12-2002
FD P94-447-1	P94-447-1	Rocks – Guides for the measurement of ground-transmitted vibration in geotechnical works – Part 1: Vibrations caused by explosive blasts.	01-06-2003
FD P94-447-2	P94-447-2	Rocks – Guides for the measurement of ground-transmitted vibration in geotechnical works – Part 2: Vibration induced by mechanical equipment.	01-06-2003

S.8.2. *European and international standards (NF EN ISO)*

Official references	*Classification index*	*English title*	*Date of publication*
NF EN 12372		Test methods for natural stone – Determination of bending resistance under centered load.	01-05-2007
NF EN ISO 14689-1	P94-401-1	Geotechnical assessment and testing – Rock naming and classification – Part 1: Naming and description.	01-09-2004
NF EN ISO 22282-3	P94-523-1	Geotechnical assessment and testing – Geohydraulic testing – Part 3: Water pressure tests in rocks.	18-01-2014

S.9. Waste – Permeability testing

Official references	*Classification index*	*English title*	*Date of publication*
NF X30-418	X30-418	Waste – Determination of the vertical permeability coefficient of a soil by an open-type double ring infiltrometer test.	01-12-2012
NF X30-420	X30-420	Waste – Determination of the vertical permeability coefficient of a soil by a closed-type single ring infiltrometer test – Constant load test and variable load test.	01-09-2012

NF X30-423	X30-423	Waste – Determination of the permeability coefficient of a soil by a variable load test in an open boring.	01-09-2011
NF X30-424	X30-424	Waste – Determination of the permeability coefficient of a soil by a constant load infiltration test in a boring.	15-06-2013
NF X30-425	X30-425	Waste – Determination of the permeability coefficient of a soil by variable load test in a closed boring.	26-03-2014
BP X30-438		Waste – Good practice guide for geological, hydrogeological and geotechnical assessments of waste storage facility sites.	11-2009
NF X30-441		Waste – Laboratory determination of the saturation permeability coefficient of a material – Permeability tests with a stiff wall permeameter at a constant/variable hydraulic gradient.	11-2008
NF X30-490		Waste – Qualification of the personnel able to carry out the determination of the permeability coefficient of a geological formation in situ, of added materials, or which has been artificially reconstituted – Criteria of qualification of the companies and the personnel.	04-2012
FD X31-614		Soil quality – Methods of detection and characterization of pollution – Implementation of a drilling to control the quality of the groundwater in a site that is potentially polluted.	10-1999

S.10. Geomembranes

Official references	*Classification index*	*English title*	*Date of publication*
NF EN 13251	G38-183	Geotextiles and related products – Characteristics required for use in earthworks, foundations and retaining structures.	01-09-2001
NF EN 13251/A1	G38-183/A1	Geotextiles and related products – Characteristics required for use in earthworks, foundations and retaining structures.	01-08-2005
NF P84-500	P84-500	Geomembranes – Dictionary of terms related to geomembranes.	20-04-2013
NF P84-501	P84-501	Geomembranes – Geomembrane waterproofing device (DEG) – Determination of tensile properties.	01-09-1992

NF P84-502-1	P84-502-1	Geomembranes – Joint tests – Part 1: Determination of tensile-shear characteristics	01-02-1993
NF P84-502-2	P84-502-2	Geomembranes – Joint tests – Part 2: Determination of tensile-towing resistance.	01-08-1993
NF P84-504	P84-504	Geomembranes – Sampling.	01-09-1993
NF P84-506	P84-506	Geomembranes – Geomembrane waterproofing device (DEG) – Determination of the dynamic puncture resistance – Case of a stiff support – Pendulum method.	01-09-1992
NF P84-507	P84-507	Geomembrane testing – Determination of static puncture resistance of geomembranes and geomembrane waterproofing devices – Case of the cylindrical punch without any support.	01-09-1996
NF P84-509	P84-509	Geomembranes – Behavior in water – Accelerated test and long-term test – Gravimetric examination.	01-12-1994
NF P84-510	P84-510	Geomembranes – Geomembrane waterproofing device (DEG) – Determination of the resistance to piercing by aggregates on a stiff support.	01-04-2002
NF P84-511-2	P84-511-2	Geomembranes – Determination of flexibility characteristics – Part 2: Three-dimensional approach.	01-12-1995
XP P84-512-1	P84-512-1	Geomembranes – Geomembrane waterproofing device (DEG) – Part 1: Determination of thickness – Case of smooth geomembranes.	01-09-1996
XP P84-512-2	P84-512-2	Geomembranes – Geomembrane waterproofing device (DEG).	01-07-1998
NF P84-514	P84-514	Geomembranes – Determination of the surface mass.	01-06-1994
NF P84-520	P84-520	Geomembranes – On-site identification.	01-06-1994
XP P84-703	P84-703	Bentonite geosynthetics – Determination of the swelling capacity of clay in bentonite geosynthetics.	01-08-2002
NF EN ISO 10769	P84-704	Bentonite geosynthetics – Determination of water absorption by bentonite.	01-07-2011
NF P84-705	P84-705	Bentonite geosynthetics – Oedo-permeameter determination of swelling, flow and permeability characteristics of bentonite geosynthetics (GSB) – Characterization test and performance test.	01-11-2008

XP P84-706	P84-706	Bentonite geosynthetics – Determination of liquid flow per unit length of cover.	01-08-2002
XP P84-708	P84-708	Bentonite geosynthetics – Quantification of self-correction capacity.	01-08-2002
NF EN 13331-1	P93-323-1	Trench shoring devices – Part 1: Product specification.	01-12-2002
NF EN 13331-2	P93-323-2	Trench shoring devices – Part 2: Assessment by calculation or testing.	01-12-2002
NF EN 14653-1	P93-503-1	Trench shoring components – Part 1: Product specifications.	01-08-2005
NF EN 14653-2	P93-503-2	Trench shoring components – Part 2: Testing and assessment.	01-08-2005

S.11. Geosynthetics

S.11.1. *International standards (NF EN ISO)*

Official references	*Classification index*	*English title*	*Date of publication*
NF EN ISO 9862		Geosynthetics – Sampling and specimen preparation.	08-2005
NF EN ISO 9863-1		Geosynthetics – Determination of thickness at specified pressures – Part 1: Individual layers.	02-2006
NF EN ISO 9863-2		Geotextiles and geotextile-related products – Determination of thickness at prescribed pressures – Part 2: Method for determining the thickness of individual layers of multi-layer products.	10-1996
NF EN ISO 9864		Geosynthetics – Test method for the determination of the surface mass of geotextiles and related products.	10-2005
NF EN ISO 10318		Geosynthetics – Terms and definitions.	10-2005
NF EN ISO 10319		Geotextiles – Wide width tensile test.	08-2008
NF EN ISO 10320		Geotextiles and related products – On-site identification.	03-2000
NF EN ISO 10321		Geotextiles – Joint/seam tensile test by the wide width method.	08-2008

NF EN ISO 10722		Geosynthetics – Test procedure to assess mechanical damage under repeated loading – Damage caused by granular materials.	08-2007
NF EN ISO 10769	Replaces XP P84704	Bentonite geosynthetics – Determination of water absorption by bentonite.	07-2011
NF EN ISO 13438		Geotextiles and related products – Method for determining oxidation resistance.	07-2005
ISO/TR 20432		Guidelines for determining the long-term resistance of geosynthetics for soil reinforcement. Amended by: ISO/TR 20432 Technical correction 1 to Standard ISO/TR 20432 of December 2007.	12-2007
NF EN ISO 25619-1	Replaces: NF EN 1897	Geosynthetics – Determination of compressive behavior – Part 1: Compressive creep properties.	02-2009
NF EN ISO 25619-2		Geosynthetics – Determination of compression behavior – Part 2: Determination of short-term compression behavior.	02-2009

S.11.2. *European standards (NF EN)*

Official references	*Classification index*	*English title*	*Date of publication*
NF EN 12224		Geotextiles and related products – Determination of aging resistance to weathering.	03-2002
NF EN 12225		Geotextiles and related products – Method for determining microbiological resistance by a burial test.	03-2002
NF EN 12226		Geotextiles and related products – General tests for post-test durability.	03-2002
NF EN 12447		Geotextiles and related products – Selective test method for determining resistance to hydrolysis in water.	06-2003
NF EN 13249/A1		Geotextiles and related products – Characteristics required for use in road construction and other traffic areas.	06-2003
NF EN 13250/A1		Geotextiles and related products – Characteristics required for use in railroad construction.	08-2005
NF EN 13251/A1		Geotextiles and related products – Characteristics required for use in earthworks, foundations and retaining structures.	08-2005

NF EN 13252/A1		Geotextiles and related products – Characteristics required for use in drainage systems.	08-2005
NF EN 13253/A1		Geotextiles and related products – Characteristics required for use structures against erosion (coastal protection and bank lining).	08-2005
NF EN 13254/A1		Geotextiles and related products – Characteristics required for use in the construction of reservoirs and dams.	08-2005
NF EN 13255/A1		Geotextiles and related products – Characteristics required for use in the construction of channels.	08-2005
NF EN 13256/A1		Geotextiles and related products – Characteristics required for use in the construction of tunnels and underground structures.	08-2005
NF EN 13257/A1		Geotextiles and related products – Characteristics required for use in solid waste landfills.	08-2005
NF EN 13265/A1		Geotextiles and related products – Characteristics required for use in liquid waste containment projects.	08-2005
NF EN 13361/A1		Geomembranes, bentonite geosynthetics – Characteristics required for use in the construction of reservoirs and dams.	12-2006
NF EN 13362**		Geomembranes, bentonite geosynthetics – Characteristics required for use in the construction of canals.	02-2006
NF EN 13491/A1		Geomembranes, bentonite geosynthetics – Characteristics required for use as a liquid barrier in the construction of tunnels and underground structures.	12-2006
NF EN 13492/A1		Geomembranes, bentonite geosynthetics – Characteristics required for use in the construction of liquid waste disposal sites, transfer stations or secondary containment sites.	12-2006
NF EN 13493**		Geomembranes, bentonite geosynthetics – Characteristics required for use in the construction of solid waste storage and burial structures.	02-2006
NF EN 13562		Geotextiles and related products – Determination of resistance to water penetration (hydrostatic pressure test).	10-2000

NF EN 13719		Geotextiles and related products – Determination of the long-term protective effectiveness of geotextiles in contact with bentonitic geomembranes and geosynthetics.	08-2003
NF EN 13738		Geotextiles and geotextile-related products – Determination of soil pull-out resistance.	08-2005
NF EN 14030		Geotextiles and related products – Selective test method for determining resistance to acidic and alkaline liquids. Modified by: NF EN 14030/A1 (May 2004). Geotextiles and related products – Selective test method for determining resistance to acid and alkaline liquids.	06-2003
NF EN 14150		Geomembranes – Determination of liquid permeability.	10-2006
NF EN 14151		Geosynthetics – Determination of resistance to rupture. Removes and replaces: XP P84503.	12-2010
NF EN 14196		Geosynthetics – Test methods for the determination of the surface mass of bentonite geosynthetics.	07-2004
NF EN 14414		Geosynthetics – A selective test method for determining chemical resistance for use in waste disposal facilities.	11-2004
NF EN 14415		Geomembranes, bentonite geosynthetics – Test method for the determination of leach resistance.	01-2005
NF EN 14574		Geosynthetics – Determination of the pyramidal puncture resistance of geosynthetics on a support.	09-2005
NF EN 14575		Geomembranes – Selective test method for the determination of oxidation resistance.	10-2005
NF EN 14576		Geosynthetics – Test method for determining the resistance of polymeric geomembranes to environmental stress cracking.	11-2005
NF EN 15381**		Geotextiles and related products – Characteristics required for use in asphalt roads and surface layers.	12-2008
NF EN 15381**		Geomembranes, bentonite geosynthetics – Characteristics required for use in transport infrastructure.	12-2008
XP CEN/TS 14416		Bentonite geomembranes and geosynthetics – Test method for determining resistance to roots.	11-2006
Note: **Harmonized standards (CE mark)			

S.11.3. *French standards (NF): geotextiles and related products*

Official references	*Classification index*	*English title*	*Date of publication*
NF G00-001		Textiles – Dictionary of standardized terms. Modified by: NF G00-001/A1.	09-1985
NF G38-019		Textiles – Articles for industrial use – Testing of geotextiles – Determination of puncture resistance.	12-1988
NF G38-021-2		Geotextiles – Determination of flexibility characteristics – Part 2: Three-dimensional approach.	12-1995
G38060		Textiles – Articles for industrial use – Recommendations for the use of geotextiles and related products – Implementation – Specifications – Control of geotextiles and related products.	06-1994
G38061***		Industrial products – Recommendations for the use of geotextiles and related products – Determination of the hydraulic characteristics and application of geotextiles, and related products, used in drainage and filtration systems.	02-1993
G38063***		Articles for industrial use – Recommendations for the use of geotextiles and related products – Use of geotextiles and related products under embankment on compressible soils.	02-1993
XP G38030		Geotextiles – Determining the number of constrictions.	10-2008
XP G38030		Use of geotextiles and related products – Inclined walls and slopes reinforced with geosynthetic layers – Justification of the sizing and design elements.	10-2008
XP G38067***		Use of geotextiles and related products – Inclined walls and slopes reinforced with geosynthetic layers – Justification of the sizing and design elements.	07-2010
NF F53-777		Fixed railroad installations – Geotextiles for railroad bed structures.	10-1989
Note: ***Sizing standards			

S.12. Earthworks, compacted or treated soils and roadways

S.12.1. *Earthwork and compaction*

Official references	*Classification index*	*English title*	*Date of publication*
NF P94-078	P94-078	CBR index after immersion. Immediate CBR index. Immediate Bearing Index – Measurement on compacted sample in CBR mold.	01-05-1997
NF P94-093	P94-093	Determination of the compaction references of a material – Normal Proctor test. Modified Proctor test.	01-10-1999
NF P94-100	P94-100	Materials treated with lime and/or hydraulic binders – Test to assess the suitability of a soil for treatment.	01-08-1999
NF P94-102-1	P94-102-1	Soil treated with hydraulic binder, possibly combined with lime, for use in sub-base – Part 1: Definition – Composition – Classification.	01-07-2001
NF P94-102-2	P94-102-2	Soil treated with hydraulic binder, possibly associated with lime, for use in sub-base – Part 2: Methodology of laboratory formulation studies.	01-07-2001
NF P94-103	P94-103	Materials treated with lime and/or hydraulic binders – Test to assess the ability of a dry mineral treatment product to emit dust.	01-12-2004
NF P11-300	P11-300	Earthworks – Classification of materials for use in the construction of embankments and sub-bases of road infrastructures	01-09-1992
NF P11-301	P11-301	Execution of earthworks – Terminology.	01-12-1994
NF P-98-705		Road construction and maintenance equipment – Compactors – Terminology and commercial specifications.	07-1992
NF P98-736		Road construction and maintenance equipment – Compactors – Classification.	12-1991
NF P98-760		Road construction and maintenance equipment – Pneumatic compactors – Assessment of ground contact pressure.	12-1991
NF P98-761		Road construction and maintenance equipment – Compactors – Eccentric moment assessment.	12-1991

S.12.2. *Compacted or treated materials*

Official references	*Classification index*	*English title*	*Date of publication*
FD P15-010		Hydraulic binders – Guide for the use of cements.	10-11-1997
NF P15-108	*Under review*	Hydraulic binders – Hydraulic road binders – Composition, specifications and conformity criteria.	12-2000
NF P15-301		Hydraulic binders – Common cements – Composition, specifications and conformity criteria.	06-1994
PR NF EN 13282-1		Hydraulic road binders – Part 1: Composition, specifications and conformity criteria for fast hardening hydraulic road binders.	*Project*
PR NF EN 13282-2		Hydraulic road binders – Part 2: Composition, specifications and conformity criteria for normal hardening hydraulic road binders.	*Project*
PR NF EN 13282-3		Road hydraulic binders – Part 3: Conformity assessment.	*Project*
NF EN 13286-40	P98-846-40	Treated and untreated mixtures with hydraulic binders – Part 40: Test method for determining the direct tensile resistance of mixtures treated with hydraulic binders.	07-2003
NF EN 13286-41	P98-846-41	Treated and untreated mixtures with hydraulic binders – Part 41: Test method for determining the compressive resistance of mixtures treated with hydraulic binders.	07-2003
NF EN 13286-42	P98-846-42	Treated and untreated mixtures with hydraulic binders – Part 42: Test method for determining the indirect tensile resistance of mixtures treated with hydraulic binders.	07-2003
NF EN 13286-43	P98-846-43	Treated and untreated mixtures with hydraulic binders – Part 43: Test method for determining the modulus of elasticity of mixtures treated with hydraulic binders (Replaces NF P98-232-1, NF P98-232-2, and partially replaces NF P98-232-3, of September 2001).	09-2003
NF EN 13286-45	P98-846-45	Treated and untreated mixtures with hydraulic binders – Part 45: Test methods for determining the handling time.	05-2004

NF EN 13286-47	P98-846-47	Treated and untreated mixtures with hydraulic binders – Part 47: Test methods for determining the California Bearing Ratio (CBR), Immediate Bearing Ratio (IBR) and the linear swelling.	07-2012
NF EN 13286-48	P98-846-48	Treated and untreated mixtures with hydraulic binders – Part 48: Test method for determining the degree of pulverization.	12-2005
NF EN 13286-49	P98-846-49	Treated and untreated mixtures with hydraulic binders – Part 49: Accelerated swelling test for soil treated with lime and/or a hydraulic binder.	10-2004
NF EN 13286-50	P98-846-50	Treated and untreated mixtures with hydraulic binders – Part 50: Method of preparation by compaction with a Proctor device or a vibrating table of specimens of materials, treated with hydraulic binders (replaces NF P98-230-1).	05-2005
NF EN 13286-51	P98-846-51	Gravel treated with hydraulic binders and untreated gravel. Part 51: Method of compaction, with a hammer vibrating specimens of materials treated with hydraulic binders.	05-2005
NF EN 13286-52	P98-846-52	Treated and untreated mixtures with hydraulic binders – Part 52: Method of making material specimens treated with hydraulic binders by vibro-compression.	05-2005
NF EN 13286-53	P98-846-53	Treated and untreated mixtures with hydraulic binders – Part 53: Method of making material specimens treated with hydraulic binders by axial compression (replaces NF P98-230-2).	05-2005
PR NF EN 13286-54	P98-846-54	Treated and untreated mixtures with hydraulic binders – Part 54: Test method for determining frost sensitivity – Freeze-thaw resistance of mixtures treated with hydraulic binders.	09-2011 (*Non-applicable*)

S.12.3. *European standards under review*

Official references	***Classification index***	***English title***	***Date of publication***
EN 16907-1		Earthworks – Part 1: Principles and general rules.	*Project*
NF EN 16907-2	P11-303-2	Earthworks – Part 2: Classification of materials.	12-2018
EN 16907-3		Earthworks – Part 3: Construction procedures.	*Project*
EN 16907-4		Earthworks – Part 4: Soil treatment (with binders).	*Project*

EN 16907-5		Earthworks – Part 5: Quality control.	*Project*
EN 16907-6		Earthworks – Part 6: Dredging and hydraulic backfill.	*Project*
EN 16907-7		Earthworks – Part 7: Hydraulic placement of soil and mineral waste.	*Project*
NF EN 459-1		Building lime – Part 1: Definitions, specifications and conformity criteria.	03-2012
NF EN 459-2		Building lime – Part 2: Test methods.	08-2012
NF EN 459-3		Building lime – Part 3: Conformity assessment.	04-2012
NF P98-232-3		Road testing – Determination of mechanical properties of materials treated with hydraulic binders – Part 3: Diameter compression test on materials treated with hydraulic and pozzolanic binders (partially replaced by NF EN 13286-43).	09-2001

S.12.4. *Roadways*

Official references	*Classification index*	*English title*	*Date of publication*
NF P98-086		Structural sizing of (new) roadways.	05-2019
NF P98-100		Road beds – Water for road bed. Classification.	11-1991
NF P98-101		Road beds – Calcium air lime for soils and roads – Specifications.	07-1991
NF P98-102		Road beds – Calcium air lime for soils and roads – Quicklime reactivity test with water.	07-1991
NF P98-234-1		Road beds – Freezing behavior of materials treated with hydraulic binders – Part 1: Freeze-thaw resistance test of treated gravel and sand.	04-1992
NF P98-234-2		Road testing – Behavior to frost – Part 2: Frost swelling test of treated or untreated soils and granular materials of D less than or equal to 20 mm.	02-1996
NF P98-331		Roadways and outbuildings. Trenches: Opening, backfilling, repair.	02-2005
XP P98-333		Roadways and outbuildings – Small trenches.	06-2009

S.12.5. *Granulates*

Official references	*Classification index*	*English title*	*Date of publication*
NF EN 13 043		Aggregates for hydrocarbon mixes and surface coatings.	08-2003
NF EN 13 242+A1		Aggregates for materials treated with hydraulic binders and untreated materials.	03-2008
NF EN 13 055-2		Lightweight aggregates for hydrocarbon mixtures, surface coatings, and for use in treated and untreated layers.	02-2005
NF EN 12 620+A1		Aggregates for concrete.	06-2008
NF EN13 139		Aggregates for mortar.	01-2003
NF EN 13 055-1		Lightweight aggregates for concrete and mortar.	12-2002
NF EN 13 383-1		Rockfill – Specifications.	05-2003
NF EN 13 383-2		Rockfill – Test methods.	06-2003
NF EN 13 450		Aggregates for railroad stones.	08-2003
NF EN 932 1 to 6		General properties of aggregates.	-
NF EN 933 1 to 11		Determination of the geometric characteristics of aggregates.	-
NF EN 1097 1 to 11		Determination of the mechanical and physical characteristics of aggregates.	-
NF EN 1367 1 to 8		Determination of the thermal and weathering properties of aggregates.	-
NF EN 1744 1 to 8		Determination of the chemical properties of aggregates.	-
NF EN 1926		Test methods for natural stone – Determination of uniaxial compressive resistance.	04-2007

NF P18-545		Aggregates – Elements of definition conformity and coding.	09-2011
FD P18-542		Aggregates – Natural aggregates qualification criteria for hydraulic concrete with respect to the alkali reaction.	02-2004
XP P18-566		Aggregates – Sieve analysis, flattening, elongation – Test with a shadowgraph device.	07-2002
NF P18-576		Aggregates – Determination of the friability coefficient of sands.	02-2013
NF P18-579		Aggregates – Determination of abrasiveness and grindability coefficients.	02-2013
XP P18-580		Aggregates – Determination of the resistance to accelerated polishing of gravel – Projection method.	10-1997
XP P18-581		Aggregates – Rapid determination of soluble sulfates in water – Spectrophotometry method.	10-1997
XP P18-594		Aggregates – Alkali reactivity test method.	02-2004
NF EN 13036-4		Method for measuring surface adhesion – Pendulum test.	03-2012

Index

A, B

C

D

E, F, G, H

I, L

M, O

P

R, S

T, V, W

Summary of Volume 2

Chapter 2. Fundamental Principles of Soil Mechanics

Chapter 3. Geotechnical Expertise

Summary of Volume 3

Summary of Volume 4

Chapter 2. Soil Reinforcement and Improvement

Other titles from

iSTE

in

Civil Engineering and Geomechanics

2022

MAZARS Jacky, GRANGE Stéphane
Damage and Cracking of Concrete Structures: From Theory to Practice

2020

SALENÇON Jean
Elastoplastic Modeling

2019

KOTRONIS Panagiotis
Risk Evaluation and Climate Change Adaptation of Civil Engineering Infrastructures and Buildings: Project RI-ADAPTCLIM

LAMBERT David Edward, PASILIAO Crystal L., ERZAR Benjamin, REVIL-BAUDARD Benoit, CAZACU Oana
Dynamic Damage and Fragmentation

PERROT Arnaud
3D Printing of Concrete: State of the Art and Challenges of the Digital Construction Revolution

SALENÇON Jean
Viscoelastic Modeling for Structural Analysis

ZEMBRI-MARY Geneviève
Project Risks: Actions Around Uncertainty in Urban Planning and Infrastructure Development

2018

FROSSARD Etienne
Granular Geomaterials Dissipative Mechanics: Theory and Applications in Civil Engineering

KHALFALLAH Salah
Structural Analysis 1: Statically Determinate Structures
Structural Analysis 2: Statically Indeterminate Structures

VERBRUGGE Jean-Claude, SCHROEDER Christian
Geotechnical Correlations for Soils and Rocks

2017

LAUZIN Xavier
Civil Engineering Structures According to the Eurocodes

PUECH Alain, GARNIER Jacques
Design of Piles Under Cyclic Loading: SOLCYP Recommendations

SELLIER Alain, GRIMAL Étienne, MULTON Stéphane, BOURDAROT Éric
Swelling Concrete in Dams and Hydraulic Structures: DSC 2017

2016

BARRE Francis, BISCH Philippe, CHAUVEL Danièle *et al.*
Control of Cracking in Reinforced Concrete Structures: Research Project CEOS.fr

PIJAUDIER-CABOT Gilles, LA BORDERIE Christian, REESS Thierry, CHEN Wen, MAUREL Olivier, REY-BETHBEDER Franck, DE FERRON Antoine
Electrohydraulic Fracturing of Rocks

TORRENTI Jean-Michel, LA TORRE Francesca
Materials and Infrastructures 1
Materials and Infrastructures 2

2015

AÏT-MOKHTAR Abdelkarim, MILLET Olivier
Structure Design and Degradation Mechanisms in Coastal Environments

MONNET Jacques
In Situ Tests in Geotechnical Engineering

2014

DAÏAN Jean-François
Equilibrium and Transfer in Porous Media – 3-volume series
Equilibrium States – Volume 1
Transfer Laws – Volume 2
Applications, Isothermal Transport, Coupled Transfers – Volume 3

2013

AMZIANE Sofiane, ARNAUD Laurent
Bio-aggregate-based Building Materials: Applications to Hemp Concretes

BONELLI Stéphane
Erosion in Geomechanics Applied to Dams and Levees

CASANDJIAN Charles, CHALLAMEL Noël, LANOS Christophe,
HELLESLAND Jostein
Reinforced Concrete Beams, Columns and Frames: Mechanics and Design

GUÉGUEN Philippe
Seismic Vulnerability of Structures

HELLESLAND Jostein, CHALLAMEL Noël, CASANDJIAN Charles,
LANOS Christophe
Reinforced Concrete Beams, Columns and Frames: Section and Slender Member Analysis

LALOUI Lyesse, DI DONNA Alice
Energy Geostructures: Innovation in Underground Engineering

LEGCHENKO Anatoly
Magnetic Resonance Imaging for Groundwater

2012

BONELLI Stéphane
Erosion of Geomaterials

JACOB Bernard *et al.*
ICWIM6 – Proceedings of the International Conference on Weigh-In-Motion

OLLIVIER Jean-Pierre, TORRENTI Jean-Marc, CARCASSES Myriam
Physical Properties of Concrete and Concrete Constituents

PIJAUDIER-CABOT Gilles, PEREIRA Jean-Michel
Geomechanics in CO_2 Storage Facilities

2011

BAROTH Julien, BREYSSE Denys, SCHOEFS Franck
Construction Reliability: Safety, Variability and Sustainability

CREMONA Christian
Structural Performance: Probability-based Assessment

HICHER Pierre-Yves
Multiscales Geomechanics: From Soil to Engineering Projects

IONESCU Ioan R. *et al.*
Plasticity of Crystalline Materials: from Dislocations to Continuum

LOUKILI Ahmed
Self Compacting Concrete

MOUTON Yves
Organic Materials for Sustainable Construction

NICOT François, LAMBERT Stéphane
Rockfall Engineering

PENSÉ-LHÉRITIER Anne-Marie
Formulation

PIJAUDIER-CABOT Gilles, DUFOUR Frédéric
Damage Mechanics of Cementitious Materials and Structures

RADJAI Farhang, DUBOIS Frédéric
Discrete-element Modeling of Granular Materials

RESPLENDINO Jacques, TOUTLEMONDE François
Designing and Building with UHPFRC

2010

ALSHIBLI A. Khalid
Advances in Computed Tomography for Geomechanics

BUZAUD Eric, IONESCU Ioan R., VOYIADJIS Georges
Materials under Extreme Loadings / Application to Penetration and Impact

LALOUI Lyesse
Mechanics of Unsaturated Geomechanics

NOVA Roberto
Soil Mechanics

SCHREFLER Bernard, DELAGE Pierre
Environmental Geomechanics

TORRENTI Jean-Michel, REYNOUARD Jean-Marie, PIJAUDIER-CABOT Gilles
Mechanical Behavior of Concrete

2009

AURIAULT Jean-Louis, BOUTIN Claude, GEINDREAU Christian
Homogenization of Coupled Phenomena in Heterogenous Media

CAMBOU Bernard, JEAN Michel, RADJAI Fahrang
Micromechanics of Granular Materials

MAZARS Jacky, MILLARD Alain
Dynamic Behavior of Concrete and Seismic Engineering

NICOT François, WAN Richard
Micromechanics of Failure in Granular Geomechanics

2008

BETBEDER-MATIBET Jacques
Seismic Engineering

CAZACU Oana
Multiscale Modeling of Heterogenous Materials

HICHER Pierre-Yves, SHAO Jian-Fu
Soil and Rock Elastoplasticity

JACOB Bernard *et al.*
HVTT 10

JACOB Bernard *et al.*
ICWIM 5

SHAO Jian-Fu, BURLION Nicolas
GeoProc2008

2006

BALAGEAS Daniel, FRITZEN Claus-Peter, GÜEMES Alfredo
Structural Health Monitoring

DESRUES Jacques *et al.*
Advances in X-ray Tomography for Geomaterials

FSTT
Microtunneling and Horizontal Drilling

MOUTON Yves
Organic Materials in Civil Engineering

2005

PIJAUDIER-CABOT Gilles, GÉRARD Bruno, ACKER Paul
Creep Shrinkage and Durability of Concrete and Concrete Structures CONCREEP – 7